IMAGE ENCRYPTION

A Communication Perspective

IMAGE ENCRYPTION

A Communication Perspective

Fathi E. Abd El-Samie ◆ Hossam Eldin H. Ahmed
Ibrahim F. Elashry ◆ Mai H. Shahieen
Osama S. Faragallah ◆ El-Sayed M. El-Rabaie
Saleh A. Alshebeili

CRC Press
Taylor & Francis Group
Boca Raton London New York

CRC Press is an imprint of the
Taylor & Francis Group, an **informa** business

CRC Press
Taylor & Francis Group
6000 Broken Sound Parkway NW, Suite 300
Boca Raton, FL 33487-2742

© 2014 by Taylor & Francis Group, LLC
CRC Press is an imprint of Taylor & Francis Group, an Informa business

No claim to original U.S. Government works

Printed on acid-free paper
Version Date: 20131009

International Standard Book Number-13: 978-1-4665-7698-8 (Hardback)

Library of Congress Cataloging-in-Publication Data

Abd el-Samie, Fathi E.
   Image encryption : a communication perspective / Fathi E. Abd El-Samie, Hossam Eldin H. Ahmed, Ibrahim F. Elashry, Mai H. Shahieen, Osama S. Faragallah, El-Sayed M. El-Rabaie, Saleh A. Alshebeili.
     pages cm
   Includes bibliographical references and index.
   ISBN 978-1-4665-7698-8 (hardback)
   1. Image processing--Security measures. 2. Data encryption (Computer science) 3. Wireless communication systems--Security measures. I. Title.

TA1637.A255 2013
006.6--dc23   2013023827

Visit the Taylor & Francis Web site at
http://www.taylorandfrancis.com

and the CRC Press Web site at
http://www.crcpress.com

Contents

Preface

We try in this book to look at image encryption with the eyes of communication researchers. Traditional studies of encryption concentrate on the strength of the encryption algorithm without taking into consideration what is after encryption. What is after encryption is the question we must answer to select the appropriate encryption algorithm. For real-life applications, what is after encryption is communication of encrypted images. With the advances in mobile and TV applications, we have to transmit encrypted images wirelessly. So, "Do our encryption algorithms tolerate the wireless communication impairments?" is the question we are trying to answer in this book.

We can summarize the main contributions in this book as follows:

This book is devoted to the issue of image encryption for the purpose of wireless communications.

Diffusion as well as permutation ciphers are considered in this book, with a comparison between them using different evaluation metrics.

Modifications are presented to existing block ciphers either to speed up or to enhance their performance.

The wireless communication environment in which the encrypted images needs to be communicated is studied.

Simulation experiments are presented for the validation of the discussed algorithms and their modifications and for investigating the performance of algorithms over wireless channels.

MATLAB® codes for most of the simulation experiments in this book are included in two appendices at the end of the book.

Finally, we hope that this book will be helpful for the image-processing and wireless communication communities.

MATLAB® and Simulink® are registered rademarks of The Math Works, Inc. For product information, please contact:

The Math Works, Inc.
3 Apple Hill Drive
Natick, MA 01760-2098 USA
Tel: 508 647 7000
Fax: 508-647-7001
E-mail: info@mathworks.com
Web: http://www.mathworks.com/

About the Authors

Fathi E. Abd El-Samie received the BSc(Honors), MSc, and PhD degrees from Menoufia University, Menouf, Egypt in 1998, 2001, and 2005, respectively. Since 2005, he has been a teaching staff member with the Department of Electronics and Electrical Communications, Faculty of Electronic Engineering, Menoufia University. He is currently a researcher at KACST-TIC in radio frequency and  photonics for the e-Society (RFTONICs). He is a co-author of about 200 papers in international conference proceedings and journals and 4 textbooks. His current research interests include image enhancement, image restoration, image interpolation, superresolution reconstruction of images, data hiding, multimedia communications, medical image processing, optical signal processing, and digital communications.

Dr. Abd El-Samie was a recipient of the Most Cited Paper Award from the *Digital Signal Processing* journal in 2008.

Hossam Eldin H. Ahmed received a BSC(Honors) in nuclear engineering in June 1969 (Faculty of Engineering, Alexandria

University, Egypt); an MSc in microelectronic electron diffraction in April 1977 (Nuclear Department, Faculty of Engineering, Alexandria University); and a PhD in June 1983 (High Institute of Electronic and Optics, Paul Sabatier University, Toulouse, France). From 1970 to 1977, he was in the Egyptian marine forces. He was a demonstrator in 1977, giving lectures to staff members in the Department of Electronic and Electrical Communications, Faculty of Engineering and Technology, Menoufia University. In 1993, he became a professor of microelectronics, VLSI design technology, communication systems, and computer networks. From 1993 until 1999, he was vice dean for education and student affairs at the Faculty of Electronic Engineering. In 2001, he became a chairman of the Electronics and Electrical Communications Department. He is a member of the Menoufia periodic electronic faculty journal and since 1995 has been the director, designer, and constructor of the Menoufia University wide-area network (WAN) (21 LANs). He is the developer of the Menoufia University libraries and FRCU universities libraries. His current research interests are electron microscopy; transmission and backscattering of electrons and ion beams into amorphous or polycrystalline targets; optical fibers; VLSI design; nanotechnology; lithography; digital, optical, and multimedia communications; digital images; multimedia and database communications; security applications; telemetry microcomputer applications in satellites; and OBC and satellite communications.

Ibrahim F. Elashry graduated from the Faculty of Engineering, Kafrelshiekh University, Egypt in 2007. He is now a teaching assistant and PhD student at the University of Wollongong (UOW), Australia. His research interests are security over wired and wireless networks and image processing.

Mai H. Shahieen graduated in May 2005 from the Faculty of Electronic Engineering, Menoufia University. Her 2011 MSc degree is in encrypted image transmission over wireless channels. She is now a PhD student at the Faculty of Electronic Engineering. Her research interests are broadband wireless distribution systems, image and video compression, multimedia systems, and wireless networks.

Osama S. Faragallah received BSc, MSc, and PhD degrees in computer science and engineering from Menoufia University, Menouf, Egypt in 1997, 2002, and 2007, respectively. He is currently associate professor with the Department of Computer Science and Engineering, Faculty of Electronic Engineering, Menoufia University, where he was a demonstrator from 1997 to 2002 and assistant lecturer from 2002 to 2007. Since 2007, he has been a teaching staff member with the same department. His research interests are network security, cryptography, Internet security, multimedia security, image encryption, watermarking, steganography, data hiding, medical image processing, and chaos theory.

El-Sayed M. El-Rabaie (senior member, IEEE 1992, MIEE chartered electrical engineer) was born in Sires Elian (Menoufia), Egypt in 1953. He received the PhD degree in microwave device engineering from the Queen's University of Belfast in 1986. He was a postdoctoral fellow at Queen's (Department of Electronic Engineering) until February 1989. In his doctoral research, he constructed a CAD (computer-aided design) package in nonlinear circuit simulations based

on harmonic balance techniques. Since then, he has been involved in different research areas, including CAD of nonlinear microwave circuits, nanotechnology, digital communication systems, and digital image processing. He was invited in 1992 as a research fellow at North Arizona University (College of Engineering and Technology) and in 1994 as a visiting professor at Ecole Polytechnique of Montreal, Quebec, Canada. Professor El-Rabaie has authored and co-authored more than 130 papers and technical reports and 15 books. In 1993, he was awarded the Egyptian Academic Scientific Research Award (Salah Amer Award of Electronics), and, in 1995, he received the award of the Best Researcher on CAD from Menoufia University. He acts as a reviewer and member of the editorial board for several scientific journals. He participated in translating the first part of the Arabic encyclopedia. Professor El-Rabaie was the head of the Electronics and Electrical Communications Engineering Department, Faculty of Electronic Engineering, Menoufia University, and then the vice dean of postgraduate studies and research in the same faculty. He currently is the vice dean of the Scientific Committee for Professors and Assistant Professors promotion in Egypt.

Saleh A. Alshebeili is professor and chairman (2001–2005) of the Electrical Engineering Department, King Saud University, Riyadh, Saudi Arabia. He has more than 20 years of teaching and research experience in the area of communications and signal processing. Dr. Alshebeili is a member of the board of directors of Prince Sultan Advanced Technologies Research Institute (PSATRI) and has been the vice president of PSATRI (2008–2011), the director of Saudi-Telecom Research Chair (2008–2012), and the director (2011–present) of the Technology Innovation Center, RF and Photonics in the e-Society (RFTONICS), funded by King Abdulaziz City for Science and Technology (KACST). Dr. Alshebeili has been on the editorial board of the *Journal of Engineering Sciences* of King Saud University (2009–2012). He also has active involvement in the review process of a number of research journals, KACST general directorate grants programs, and national and international symposiums and conferences.

1
INTRODUCTION

The presence of communication networks has prompted new problems with security and privacy. Having a secure and reliable means for communicating with images and video is becoming a necessity, and its related issues must be carefully considered. Hence, network security and data encryption have become important. Images can now be considered one of the most usable forms of information. Image and video encryption have applications in various fields, including wireless communications, multimedia systems, medical imaging, telemedicine, and military communications [1–5].

In this book, image encryption algorithms are studied for the purpose of wireless communication of images in a secure form. The main objective is to come to a conclusion whether a certain image encryption algorithm is suitable for wireless communication applications. In this regard, we explore in this book some of the number-theory-based encryption algorithms such as the Data Encryption Standard (DES), the Advanced Encryption Standard (AES), and the RC6 algorithms [6–12]. These algorithms are known in the literature as strong encryption algorithms. Our main concern is not only the strength of the encryption algorithm, but also its ability to work with the limitations of wireless communication systems. Unfortunately, some of the ciphers used currently were not designed for image encryption. For example, the AES implemented in the Electronic Code-Book (ECB) mode is not recommended for encrypting images because of the repetitive data patterns existing in images. So, we either modify these ciphers to work for image encryption or search for other approaches suitable for this task. This book tries to investigate both directions.

For this purpose, we have also explored another branch of cryptography that depends on chaos theory. The usage of chaotic cryptography for image encryption is promising. The idea of using chaos for encryption can be traced to Shannon's classical article [13]. Chaotic maps have been applied to cryptography in several different ways. Chaotic sequences have several good properties, including ease of generation, sensitive dependence on the initial conditions, and noise-like behavior. Applying the chaos to cryptography was a great contribution to improve the security of data due to the adequate properties of chaotic sequences [14,15].

To make the study in this book comprehensive, we switch in Chapters 7 and 8 to communication concepts concentrating on the orthogonal frequency division multiplexing (OFDM) system. In these chapters, we present a simplified model for the OFDM communication system with its different implementations, with concentration on the limitations of this system, and how these limitations affect the quality of received images after decryption. An extensive comparison study is presented in Chapter 9, considering different possible scenarios and different encryption algorithms to determine which encryption algorithm is suitable for which scenario. Finally, we include all MATLAB® codes utilized in the experiments in this book to open the door for researchers who wish to complete this study using different ideas.

2

FUNDAMENTALS OF IMAGE ENCRYPTION

2.1 Introduction

We are living in the information age; we need to keep information about every aspect of our lives. In other words, information is an asset that has a value like any other asset. As an asset, information needs to be secured from attacks. Since 1990, communication networks created a revolution in the use of information. Authorized people can send and receive information from a distance using communication networks. To be secure, information needs to be hidden from unauthorized access (confidentiality), protected from unauthorized change (integrity), and available to an authorized entity when it is needed (availability). Although these three requirements have not changed, they now have some new dimensions. Not only should information be confidential when it is stored in a computer, but also there should be a way to maintain its confidentiality when it is transmitted over a communication network [16].

Information transmitted over computer networks currently is not only text but also audio, images, and other multimedia types. The field of multimedia security has matured in the past decade to provide a class of tool sets and design insights for the protection and enhancement of digital media under a number of diverse attack scenarios. Research in multimedia security was first motivated partly by the increasing use of digital means to communicate, store, and represent entertainment information such as music and video. The digital form allowed the perfect duplication of information and almost seamless manipulation and tampering of the data. This created new types of security attacks not addressed in the past by the entertainment industry. The paradigm shift from analog to digital multimedia for entertainment has had an enormous impact on artists, publishers, copyright holders, and consumers [17,18].

This chapter gives the principles underlying the design of cryptographic algorithms. A review of the different issues that arise when selecting, designing, and evaluating a cryptographic algorithm is presented. Diffusion-based cipher algorithms along with their modes of operation are covered. Chaotic encryption with a Baker map is also discussed.

2.2 Basic Concepts of Cryptography

Image security is based on cryptography. In fact, some basic concepts from cryptography are used as building blocks (primitives) for applications in image security. For a better understanding of issues concerning image security, an overview of cryptography is presented first.

2.2.1 Goals of Cryptography

Cryptography is a study of techniques (called *cryptosystems*) that are used to accomplish the following four goals [1,16–18]:

- Confidentiality
- Data integrity
- Authentication
- Nonrepudiation

The study of the techniques used to break existing cryptosystems is called *cryptanalysis*. Since cryptography and cryptanalysis are greatly dependent on each other, people refer to *cryptology* as a joint study of cryptography and cryptanalysis. Let us try to understand all four goals of cryptography.

Confidentiality refers to the protection of information from unauthorized access. An undesired communicating party, called an adversary, must not be able to access the communication material. This goal of cryptography is a basic one that has always been addressed and enforced throughout the history of cryptographic practice.

Data integrity ensures that information has not been manipulated in an unauthorized way. If the information is altered, all communicating parties can detect this alteration.

Authentication methods are classified into two categories: entity authentication and message authentication. *Entity authentication* is the process by which one party is assured of the identity of a second party involved in a protocol, and that the second has actually participated immediately prior to the time the evidence is acquired. *Message authentication* is a term used analogously with data origin authentication. It provides data origin authentication with respect to the original message source and data integrity but with no uniqueness and timeliness guarantees.

Nonrepudiation means that the receiver can prove to everyone that the sender did indeed send the message. That is, the sender cannot claim that he or she did not encrypt or sign certain digital information.

Fortunately, modern cryptography has developed techniques to handle all four goals of cryptography.

2.2.2 Principles of Encryption

The basic idea of *encryption* is to modify the message in such a way that only a legal recipient can reconstruct its content [16–18]. A discrete-value cryptosystem can be characterized by

> a set of possible plaintexts P
> a set of possible ciphertexts C
> a set of possible cipher keys K
> a set of possible encryption and decryption transformations E and D

An *encryption system* is also called a *cipher,* or a cryptosystem. The message for encryption is called *plaintext,* and the encrypted message is called *ciphertext.* Denote the plaintext and the ciphertext by P and C, respectively [16–19]. The encryption procedure of a cipher can be described as

$$C = E_{K_e}(P) \tag{2.1}$$

where K_e is the encryption key, and E is the encryption function. Similarly, the decryption procedure is defined as

$$P = D_{K_d}(C) \tag{2.2}$$

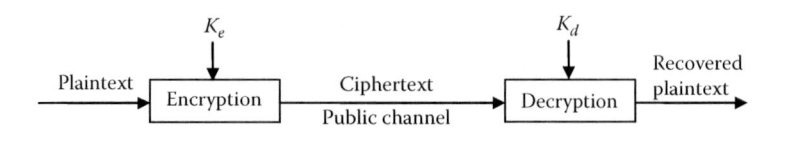

Figure 2.1 Encryption/decryption of a cipher.

where K_d is the decryption key, and D is the decryption function. The security of a cipher should only rely on the decryption key K_d as an adversary can recover the plaintext from the observed ciphertext once the adversary gets K_d. Figure 2.1 shows a block diagram for encryption/decryption of a cipher.

2.3 Classification of Encryption Algorithms

Encryption algorithms can be classified in different ways: according to structures of the algorithms, according to keys, or according to the percentage of the data encrypted [19,20].

2.3.1 Classification According to Encryption Structure

Encryption algorithms can be classified according to encryption structure into block ciphers and stream ciphers.

A *block cipher* is a type of symmetric-key encryption algorithm that transforms a fixed-length block of plaintext data into a block of ciphertext data of the same length. The fixed length is called the *block size*. For several block ciphers, the block size is 64 or 128 bits. The larger the block size, the more secure is the cipher, but the more complex are the encryption and decryption algorithms and devices. Modern block ciphers have the following features [21]:

1. Variable key size
2. Mixed arithmetic operations, which can provide nonlinearity
3. Data-dependent rotations and key-dependent rotations
4. Lengthy key schedule algorithms
5. Variable plaintext/ciphertext block sizes and variable number of rounds

Block ciphers can be characterized by

1. Block size: Larger block sizes mean greater security.
2. Key size: Larger key sizes mean greater security.

3. Number of rounds: Multiple rounds increase security.
4. Encryption modes: They define how messages larger than the block size are encrypted.

Unlike block ciphers that operate on large blocks of data, *stream ciphers* typically operate on smaller units of plaintext, usually bits. So, stream ciphers can be designed to be exceptionally fast, much faster than a typical block cipher. Generally, a stream cipher generates a sequence of bits as a key (called key stream) using a pseudorandom number generator (PRNG) that expands a short secret key (e.g., 128 bits) into a long string (key stream) (e.g., 10^6 bits), and the encryption is accomplished by combining the key stream with the plaintext. Usually, the bitwise XOR operation is chosen to perform ciphering, basically for its simplicity [6–9]. Stream ciphers have the following properties [22, 23]:

1. They do not have perfect security.
2. Security depends on the properties of the PRNG.
3. The PRNG must be unpredictable; given a consecutive sequence of output bits, the next bit must be hard to predict.
4. Typical stream ciphers are very fast.

2.3.2 Classification According to Keys

According to keys, there are two kinds of ciphers following the relationship of K_e and K_d. When $K_e = K_d$, the cipher is called a private-key cipher or a symmetric cipher. For private-key ciphers, the encryption/decryption key must be transmitted from the sender to the receiver via a separate secret channel. When $K_e \neq K_d$, the cipher is called a public-key cipher or an asymmetric cipher. For public-key ciphers, the encryption key K_e is published, and the decryption key K_d is kept private, for which no additional secret channel is needed for key transfer. In conventional encryption as shown in Figure 2.2, the sender encrypts the data (plaintext) using the encryption key, and the receiver decrypts the encrypted data (ciphertext) into the original data (plaintext) using the decryption key. In symmetric encryption, both encryption and decryption keys are identical. Figure 2.3 shows the public-key encryption (asymmetric encryption), in which the encryption and decryption keys are different. Public-key cryptography solves the problem of conventional cryptosystems by distributing the key

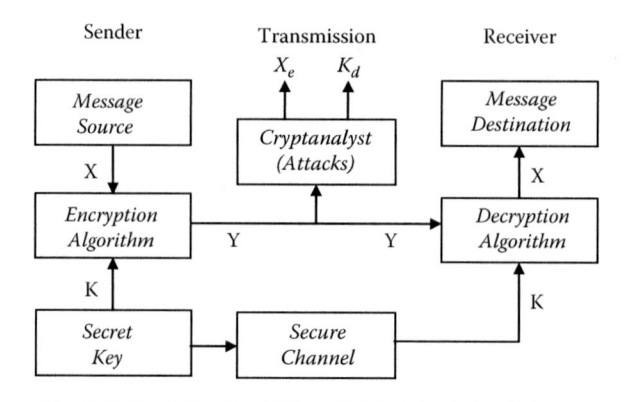

Figure 2.2   Model of symmetric encryption.

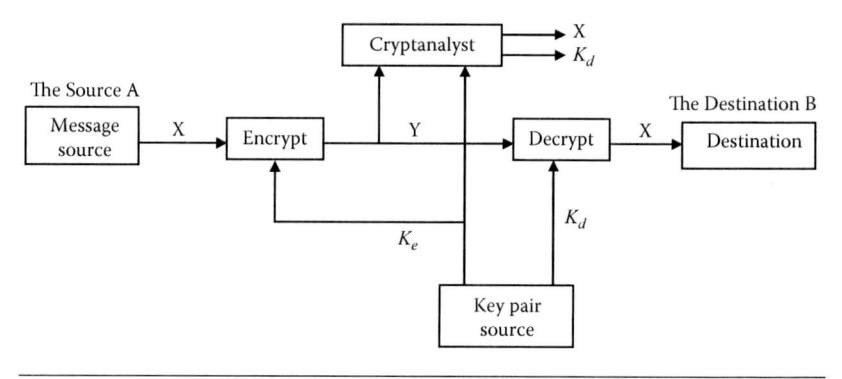

Figure 2.3   Asymmetric-key encryption.

[24–27]. Table 2.1 shows a comparison between symmetric encryption and asymmetric encryption.

In general, there are two types of cryptosystems: (1) symmetric- (private-) key cryptosystems and (2) asymmetric- (public-) key cryptosystems. Most people have chosen to call the first group simply symmetric-key cryptosystems, and the popular name for the second group is just public-key cryptosystems.

2.3.3 Classification According to Percentage of Encrypted Data

With respect to the amount of encrypted data, the encryption can be divided into full encryption and partial encryption (also called selective encryption) according to the percentage of the data encrypted.

Table 2.1   Comparison between Symmetric Encryption and Asymmetric Encryption

CONVENTIONAL ENCRYPTION (SYMMETRIC ENCRYPTION)	PUBLIC-KEY ENCRYPTION (ASYMMETRIC ENCRYPTION)
Requirements to work: 1. The same algorithm with the same key can be used for encryption and decryption. 2. The sender and receiver must share the algorithm and the key.	Requirements to work: 1. One algorithm is used for encryption and decryption with a pair of keys, one for encryption and one for decryption. 2. The sender and receiver must each have one of the matched pair of keys.
Requirements for security: 1. The key must be kept secret. 2. It must be impossible or at least impractical to decipher a message if no other information is available. 3. Knowledge of the algorithm plus samples of the ciphertext must be insufficient to determine the key.	Requirements for security: 1. The decryption key must be kept secret. 2. It must be impossible or at least impractical to decipher a message if no other information is available. 3. Knowledge of the algorithm, the encryption key, and samples of the ciphertext must be insufficient to determine the decryption key.

2.4 Cryptanalysis

Cryptanalysis is the art of deciphering an encrypted message as a whole or in part when the decryption key is not known. Depending on the amount of known information and the amount of control over the system by the adversary (cryptanalyst), there are several basic types of cryptanalytic attacks [28–30]. Some of the most important ones, for a system implementer, are described next, and they are summarized in Table 2.2.

Ciphertext-only attack. The adversary only has access to one or more encrypted messages. The most important goal of a proposed cryptosystem is to withstand this type of attack.

Brute-force attack. This is a type of ciphertext-only attack. It is based on an exhaustive key search, and for well-designed cryptosystems, it should be computationally infeasible.

Known-plaintext attack. In this type of attack, an adversary has some knowledge about the plaintext corresponding to the given ciphertext. This may help the adversary determine the key or a part of the key.

Chosen-plaintext attack. Essentially, an adversary can feed the chosen plaintext into the black box that contains the encryption algorithm and the encryption key. The black box gives the corresponding ciphertext, and the adversary may use the

Table 2.2 Types of Attacks on Encrypted Images

TYPE OF ATTACKS	PREREQUISITES FOR THE CRYPTANALYST
Ciphertext only	1. Encryption algorithm
	2. Ciphertext to be decoded
Known plaintext	1. Encryption algorithm
	2. Ciphertext to be decoded
	3. Plaintext message together with its corresponding ciphertext generated with the secret key
Chosen ciphertext	1. Encryption algorithm
	2. Ciphertext to be decoded
	3. Reported ciphertext chosen by cryptanalyst together with its corresponding plaintext generated with the decryption algorithm and the decryption key
Chosen plaintext	1. Encryption algorithm
	2. Ciphertext to be decoded
	3. Reported plaintext chosen by cryptanalyst together with its corresponding ciphertext generated with the encryption algorithm and the encryption key

accumulated knowledge about the plaintext–ciphertext pairs to obtain the secret key or at least a part of it.

Chosen-ciphertext attack. Here, an adversary can feed the chosen ciphertext into the black box that contains the decryption algorithm and the decryption key. The black box produces the corresponding plaintext, and the adversary tries to obtain the secret key or a part of the key by analyzing the accumulated ciphertext–plaintext pairs.

2.5 Features of Image Encryption Schemes

Unlike text messages, image data have their special features, such as high redundancy and high correlation among pixels. Also, they are usually huge in size. Together, these make traditional encryption methods difficult to apply and slow to process. Sometimes, image applications have their own requirements, like real-time processing, fidelity reservation, image format consistency, data compression for transmission, and so on. Simultaneous fulfillment of these requirements along with high-security and high-quality demands has presented great challenges to real-time imaging practice. For studying image encryption, we must first analyze the differences between implementations for image data and text data encryption. Basically, there are some differences between image and text data encryption [31–34]:

1. When the ciphertext is produced, it must be decrypted to the original plaintext in a full lossless manner. However, the cipherimage can be decrypted to the original plainimage in some lossy manner.
2. Text data are sequences of words. They can be encrypted directly by using block or stream ciphers. However, digital images are usually represented as two-dimensional (2D) arrays. For protecting the stored 2D arrays of data with text-processing algorithms, they must be converted to 1D arrays before using various traditional encryption techniques.
3. Because the storage space of a picture is very large, it is sometimes inefficient to encrypt or decrypt images directly. One of the best methods is to encrypt/decrypt information that is used by image compression only for reducing both its storage space and transmission time.

2.6 Conventional Symmetric Block Ciphers

This section gives a brief overview of the construction of some popular conventional encryption algorithms. Each of the following encryption algorithms is a symmetric block cipher algorithm. *Symmetric* means that the key used for encryption and decryption is the same; *block* means that the data (information) to be encrypted is divided into blocks of equal length [35–45].

2.6.1 Data Encryption Standard

The Data Encryption Standard (DES) is the most well-known symmetric-key block cipher. It was selected by the National Bureau of Standards as an official Federal Information Processing Standard (FIPS) for the United States in 1976, and it subsequently enjoyed widespread use internationally [46].

The DES is a block cipher, which encrypts data in 64-bit blocks. A 64-bit block of plaintext goes at one end of the algorithm, and a 64-bit block of ciphertext comes out at the other end. The same

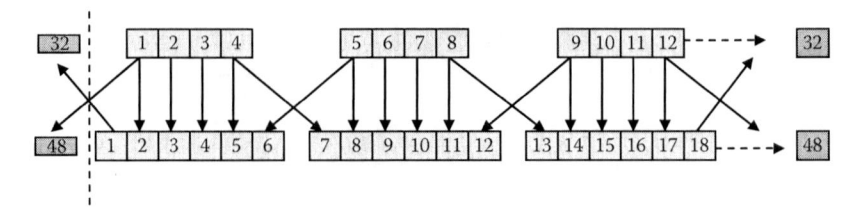

Figure 2.4 Expansion process.

algorithm and key of size 56 bits are used for both encryption and decryption except for minor differences in the key schedule. The key is usually expressed as a 64-bit number, but every eighth bit, one bit is used for parity checking and is ignored. These parity bits are the least-significant bits of the key bytes. The key can be any 56-bit number and can be changed at any time, although some selections can be considered weak keys.

The DES is based on four basic operations: expansion, permutation, XOR, and substitution. The data to be encrypted are first divided into 64-bit blocks and fed into an Initial Permutation (IP) stage, in which each block is divided into two subblocks, each with a 32-bit length. The right subblock is fed into a Feistel function (f-function), which is depicted in Figure 2.4. It operates on half a block (32 bits) at a time and consists of four stages as shown in Figure 2.5.

1. Expansion. The 32-bit half block is expanded to 48 bits using the expansion permutation, denoted as E in the diagram, by duplicating half of the bits. The output consists of eight 6-bit ($8 \times 6 = 48$ bits) pieces, each containing a copy of 4 corresponding input bits plus a copy of the immediately adjacent bit from each of the input pieces to either side.

2. Key mixing. The result is combined with a subkey using an XOR operation. Sixteen 48-bit subkeys, one for each round, are derived from the main key using a key-schedule mechanism.

3. Substitution. After mixing with the subkey, the block is divided into eight 6-bit pieces before processing by the Substitution boxes (S-boxes). Each of the eight S-boxes replaces its six input bits with four output bits according to a nonlinear transformation, provided in the form of a lookup table. The S-boxes provide the core of security of the DES.

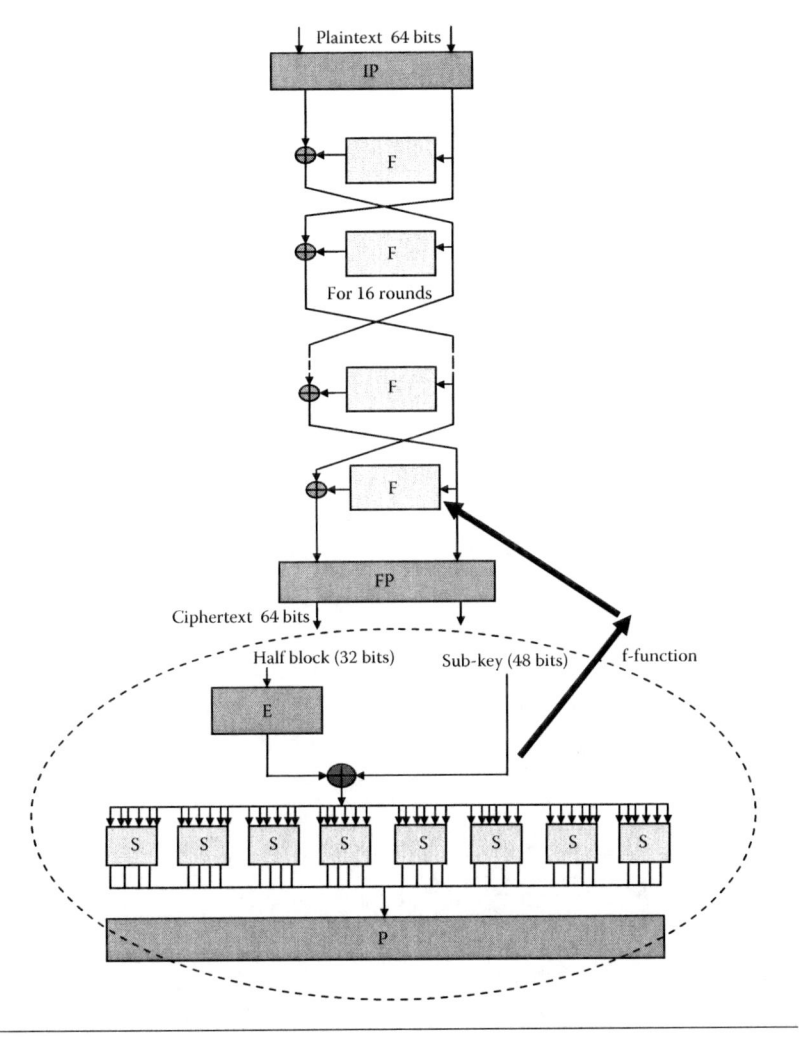

Figure 2.5 The DES algorithm and f-function.

Without them, the cipher would be linear and trivially breakable.

4. Permutation. Finally, the 32 outputs from the S-boxes are rearranged according to a fixed permutation, the P-box, which is designed so that, after expansion, each group of S-box output bits is spread across six different S-boxes in the next round.

At its simplest level, the DES algorithm is nothing more than a combination of the two basic techniques of encryption: confusion and diffusion. The fundamental building block of the DES is a single combination of these techniques: a substitution followed by a permutation on the data based on the key. After an initial permutation, each block is broken into a right half and a left half, each with 32 bits. Then, there are 16 rounds of identical operations, called f functions, in which the data are combined with the key. After these rounds, the right and left halves are joined, and a final permutation, the inverse of the initial permutation, finishes the algorithm.

2.6.2 Double DES

A naive way of improving the security of a block cipher algorithm is to encrypt each block twice with two different keys. First, encrypt a block with the first key and then encrypt the resulting ciphertext with the second key. Decryption is the reverse process. In the Double DES encryption algorithm, each 64-bit block of data is encrypted twice with the DES algorithm, first with a key K_1 and then with another key K_2. The scheme involves a key of 112 bits.

The resultant doubly encrypted ciphertext block should be much harder to decrypt using an exhaustive search. Instead of 2^{56} attempts, it requires 2^{128} attempts to find the key and 2^{112} attempts to break the encryption. In 1981, Merkle and Hellman declared their "meet-in-the-middle" attack, which proved the weakness of the Double DES algorithm [47]. The "meet-in-the-middle" attack is a known plaintext attack that requires that an attacker have both a known piece of plaintext and the corresponding ciphertext. The attack requires storing 2^{56} intermediate results when trying to crack a message that has been encrypted with the double DES. Merkle and Hellman developed a time-memory trade-off that could break this double-DES encryption scheme in 2^{56+1} trials, not in 2^{112} trials.

2.6.3 Triple DES

The dangers of the Merkle-Hellman "meet-in-the-middle" attack can be circumvented by performing three block encryption operations.

This method is called Triple DES. Triple DES is performed by executing the DES three times, producing an effective key size of 168 bits. In the Triple DES, each 64-bit block of data is encrypted three times with the DES algorithm. In practice, the most common way to perform the Triple DES is

1. Encrypt with key 1.
2. Decrypt with key 2.
3. Encrypt with key 3.

To decrypt, reverse the steps:

1. Decrypt with key 3.
2. Encrypt with key 2.
3. Decrypt with key 1.

For several applications, we can use the same key for both key 1 and key 3 without creating a significant vulnerability. The choice between Single, Double, and Triple DES is a trade-off between performance and security requirements [45].

2.6.4 International Data Encryption Algorithm

The International Data Encryption Algorithm (IDEA) cipher was first presented by Lai and Massey in 1990 under the name Proposed Encryption Standard (PES). After Biham and Shamir presented differential cryptanalysis, the authors strengthened their cipher against the attack and named it the Improved Proposed Encryption Standard (IPES). The IPES name was changed to the International Data Encryption Algorithm (IDEA) in 1992. This algorithm was intended as a replacement for the DES. IDEA is a block cipher that operates on 64-bit plaintext blocks. The key is 128 bits long. There are eight identical rounds, and the same algorithm is used for both encryption and decryption. It uses both confusion and diffusion. The design philosophy behind the algorithm is based on mixing operations from different algebraic groups. Three algebraic groups (XOR, addition modulo, and multiplication modulo) are mixed in this algorithm [16,18], and they are all easily implemented in both hardware and software. All of these operations operate on 16-bit subblocks.

2.6.5 Blowfish

Blowfish is a 64-bit block cipher with a variable-length key. It was designed in 1993 by Schaneier [16]. This algorithm consists of two parts: key expansion and data encryption. Key expansion converts a key of up to 448 bits into several subkey arrays totaling 4168 bytes. Data encryption consists of a simple function iterated 16 times. Each round consists of a key-dependent permutation and a key- and data-dependent substitution. All operations are additions and XORs on 32-bit words. The only additional operations are four indexed array data lookups per round. The keys must be precomputed before any data encryption or decryption, with decryption exactly the same as encryption except that the subkeys are used in reverse order.

2.6.6 RC5 Algorithm

The iterated block RC5 was introduced by Rivest, Shamir, and Adleman in 1994 [48]. The main feature of the RC5 is the heavy use of data-dependent rotations. RC5 has a variable word size w, a variable number of rounds r, and a variable secret key with b bytes. It is represented as RC5 $w/r/b$. The nominal value of w is 32 bits, and RC5 encrypts blocks of two words. The RC5 is composed of encryption, decryption, and key expansion. The expanded key contains $t = 2 \times (r + 1)$ words. The primitive operations of the RC5 are illustrated in Table 2.3. Generally, RC5 is a fast symmetric block cipher that is suitable for hardware and software implementations with low memory requirements. It provides high security when good parameters are chosen.

Table 2.3 Primitive Operations of RC5

$a + b$	Integer addition modulo $2w$
$a - b$	Integer subtraction modulo $2w$
$a \oplus b$	Bitwise XOR of w-bit words
$a * b$	Integer multiplication modulo $2w$
$a <<< b$	Rotate the w-bit word a to the left by the amount given by the least-significant $lg\ w$ bits of b
$a >>> b$	Rotate the w-bit word a to the right by the amount given by the least-significant $lg\ w$ bits of b

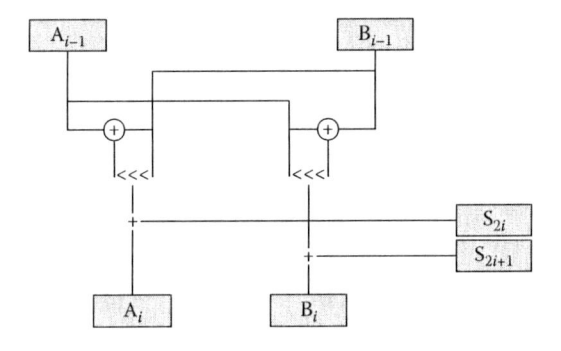

Figure 2.6   RC5$^{w/t/b}$ symmetric block cipher diagram.

2.6.6.1 RC5 Encryption Algorithm   We assume that the input block is given in two w-bit registers A and B, and we also assume that key expansion has already been performed so that $S[0]$, $S[1]$, ..., $S[t - 1]$ have been computed. The steps of the encryption algorithm can be summarized as follows:

```
A = A+S [0];
B = B+S [1];
For i = 1 to r do
A = ((A ⊕ B) <<< B) +S [2i];
B = ((B ⊕ A) <<< A) +S [2i+1];
End
```

In each round of RC5, both registers A and B are updated as shown in Figure 2.6.

2.6.6.2 RC5 Decryption Algorithm   The decryption step can be summarized as follows:

```
For i = r downto 1 do
B = ((B - S [2i+1]) >>> A) ⊕ A;
A = ((A - S [2]) >>> B) ⊕ B;
End
B = B-S [1];
A = A-S [0];
```

2.6.6.3 RC5 Key Expansion   Key expansion expands the user's secret key K to fill the expanded key array S, which makes S similar to an array of $t = 2(r + 1)$ random binary words. Two magic constants, P_w and Q_w, are used in this process. These constants are defined as

$$P_w = Odd\left((e-2)2^w\right) \tag{2.3}$$

$$Q_w = Odd\left((\phi-1)2^w\right) \tag{2.4}$$

where
    e = 2.718281828459....(base of natural logarithms)
    ϕ = 1.618033988749....(golden ratio)
and *Odd(x)* is the odd integer nearest to x. For $w = 16$ and 32, these constants are given in hexadecimal:

$$P_{16} = b7e1; Q_{16} = 9e37$$

$$P_{32} = b7e15163; Q_{32} = 9e3779b9,$$

The expansion begins by copying the secret key *K[0....b-1]* into an array *L[0....c-1]* that has $c = \lceil b / u \rceil$ words, where $u = w/8$ is the number of bytes per word. u consecutive key bytes of K are used to fill each successive word in L in a low-order to high-order byte manner. All unfilled byte positions of L are zeroed.

To initialize the array S, we use the following steps:

```
S [0] = P_w;
For i = 1 to t-1 do
S [i] = S [i-1] + Q_w;
End
```

The last step is to mix the user secret key in three passes over the arrays S and L as follows:

```
i = j = 0;
A = B = 0;
Do 3*max (t, c) times:
A = S [i] = (S [i] +A+B) <<<3;
B = L[j] = (L[j] + A+ B) <<< (A+B);
i = (i+1) mod (t);
j = (j+1) mod (c);
```

2.6.7 RC6 Algorithm

The RC6 block cipher is a modified version of RC5 that uses four working registers instead of two and integer multiplication as an additional primitive operation. The integer multiplication process greatly enhances the diffusion achieved per round, which leads to greater security, fewer rounds, and increased throughput. The key schedule of RC6-*w/r/b* is similar to the key schedule of RC5-*w/r/b*. The only difference is that for RC6-*w/r/b*, more words are derived from the user-supplied key for use during encryption and decryption. The user supplies a key of b bytes, where $0 \le b \le 255$. From this key, $2r + 4$ words (w bits each) are derived and stored in the array $S[0, \dots, 2r + 3]$. This array is used in both encryption and decryption [12]. Generally, RC6 consists of two Feistel networks whose data are mixed via data-dependent rotations. The operations in a single round of RC6 contain two applications of the squaring function $f(x) = x(2x + 1)$ mod 2^{32}, two fixed 32-bit rotations, two data-dependent 32-bit rotations, two XORs, and two additions modulo 2^{32}. The steps of RC6 encryption and decryption are summarized next, and the block diagrams of RC6 encryption and decryption are shown in Figures 2.7 and 2.8, respectively.

2.6.7.1 RC6 Encryption Algorithm

```
Input: Four w-bit plaintext values stored in registers
A, B, C, and D
              Number r of rounds
              w-bit round keys S[0, ..., 2r+3]
Output:       Four w-bit ciphertext values stored in
registers A, B, C, and D.
Procedure:    B = B + S [0];
              D = D + S [1];
              For i = 1 to r do
              {t = (B × (2B + 1)) <<< lg w;
              u = (D × (2D + 1)) <<< lg w;
              A = ((A ⊕ t) <<< u) + S [2i];
              C = ((C ⊕ u) <<< t) + S [2i + 1];
              (A, B, C, D) = (B, C, D, A);}
              End
        A = A + S [2r + 2];
        C = C + S [2r + 3];
```

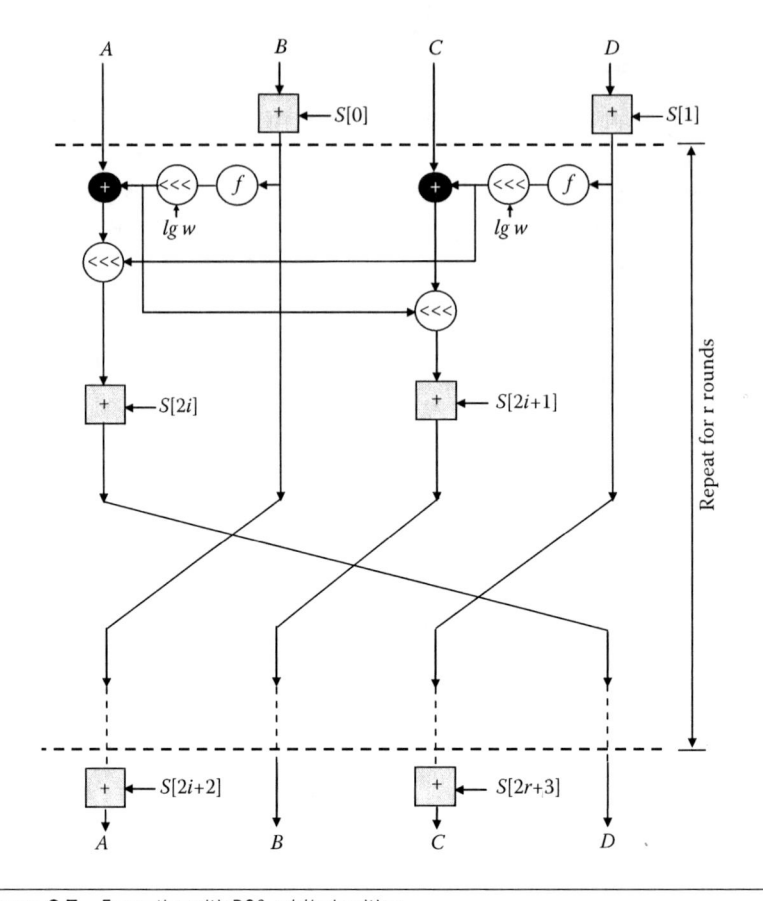

Figure 2.7   Encryption with RC6-*w*/*r*/*b* algorithm.

2.6.7.2 RC6 Decryption Algorithm

```
Input : Four w-bit ciphertext values stored in registers
A, B, C, and D
            Number of rounds r
       w-bit round keys S[0, …, 2r+3]
Output :       Four w-bit plaintext values stored in
registers A, B, C, and D.
Procedure:    C = C - S [2r + 3];
       A = A - S [2r + 2];
       for i = r downto 1 do
                   {(A, B, C, D) = (D, A, B, C);
       u = (D × (2D + 1)) <<< log(w);
       t = (B × (2B + 1)) <<< log(w);
       C = ((C - S[2i + 1]) >>> t) ⊕ u;
```

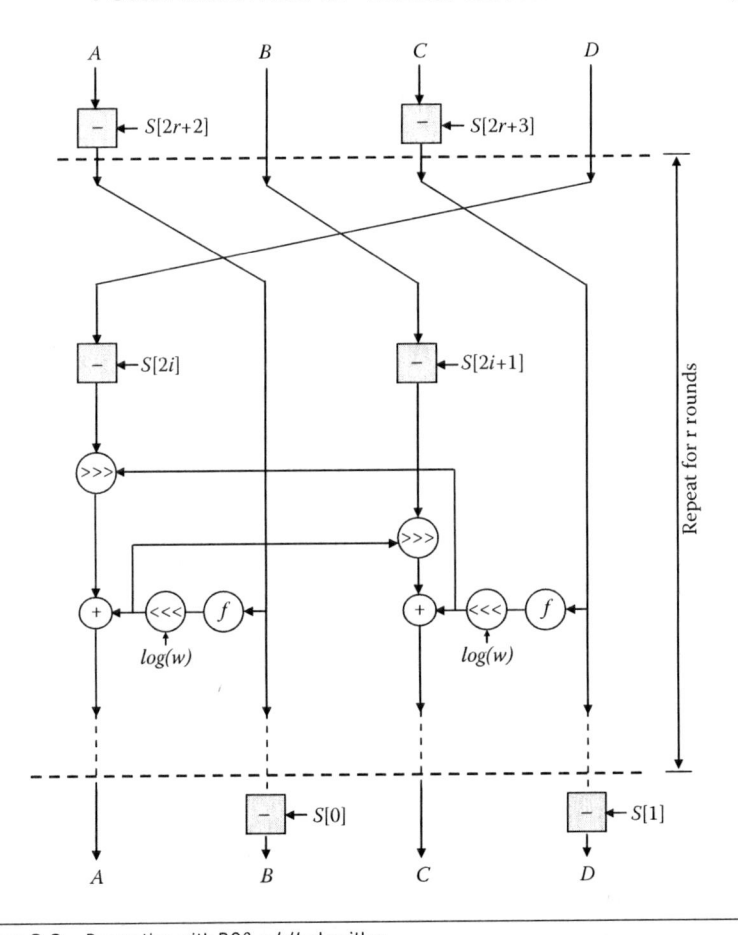

Figure 2.8   Decryption with RC6-*w*/*r*/*b* algorithm.

$$A = ((A - S[2i]) >>> u) \oplus t;\}$$
$$\text{End}$$
$$D = D - S[1];$$
$$B = B - S[0];$$

2.6.8 *The Advanced Encryption Standard*

The Advanced Encryption Standard (AES) is based on the Rijndael algorithm, which is an iterated block cipher algorithm with a variable block size and a variable key size. The block size and the key size can be independently 128, 192, or 256 bits. The intermediate resulting ciphertext is called a *state,* and it is in the form of a rectangular array of four rows and a number of columns equal to the block size divided by 32. The cipher key is similarly a rectangular array with four rows and a number of columns equal to the key size divided by 32. The number of rounds performed on the intermediate state is related to the key size. For key sizes of 128, 192, and 256 bits, the number of rounds is 10, 12, and 14, respectively. Each round consists of a fixed sequence of transformations, except the first and the last round [16–18].

The AES consists of rounds. Any round, except the final one, consists of SubBytes, ShiftRows, MixColumns, and AddRoundKey operations. In the final round, no MixColumns operation is performed. In the SubBytes step, a linear substitution for each byte is performed according to Figure 2.9. Each byte in the array is updated using an 8-bit S-box, which provides the nonlinearity in the cipher system [16–18].

The S-box is derived from the multiplicative inverse over the finite Galois field GF(2^8), known to have good nonlinearity properties. To avoid attacks based on simple algebraic properties, the S-box is chosen to avoid any fixed points and any opposite fixed points [16–18].

The ShiftRows step operates on the rows of the state. It cyclically shifts the bytes in each row. For the AES, the first row is left unchanged. Each byte of the second row is shifted a single byte to the left. Similarly, the third and fourth rows are shifted by offsets of 2 and 3 bytes, respectively. For the block of size 128 bits and 192 bits, the shifting pattern is the same [16–18].

In this way, each column of the output state of the ShiftRows step is composed of bytes from each column of the input state. In the case of the 256-bit blocks, the first row is unchanged, and the shifting for

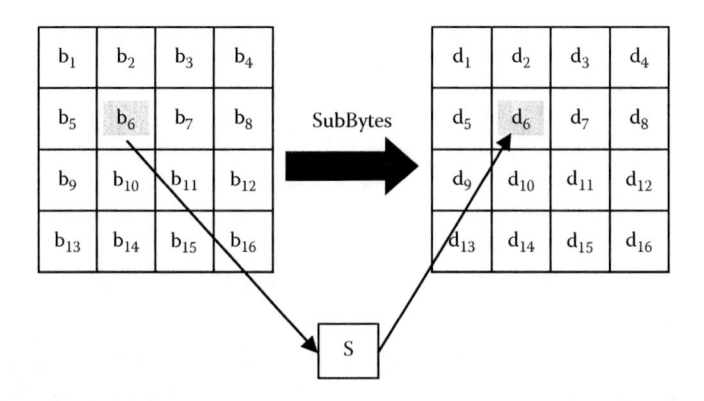

Figure 2.9 SubBytes step.

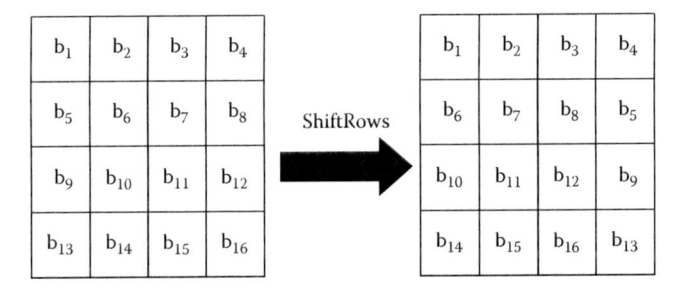

Figure 2.10   ShiftRows step.

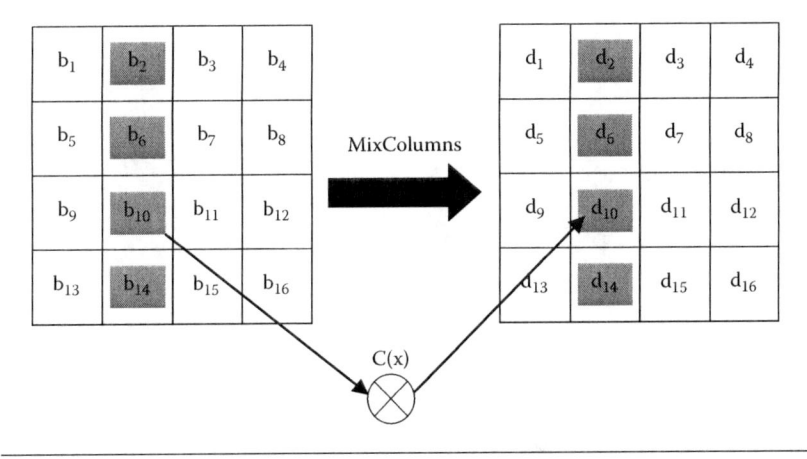

Figure 2.11   MixColumns step.

second, third, and fourth rows is 1 byte, 3 bytes, and 4 bytes, respectively, as shown in Figure 2.10.

In the MixColumns step, the 4 bytes of each column of the state are combined using an invertible linear transformation. The MixColumns function takes 4 bytes as input and outputs 4 bytes, where each input byte affects all 4 output bytes. With ShiftRows, MixColumns provides diffusion in the cipher system. Each column is treated as a polynomial over $GF(2^8)$ and is then multiplied with a fixed polynomial $C(x) = 3x^3 + x^2 + x + 2$. The MixColumns step can also be viewed as multiplication by a particular matrix as shown in Figure 2.11 [16–18].

In the AddRoundKey step, the subkey is combined with the state. For each round, a subkey is derived from the main key using the algorithm key schedule. Each subkey has the same size as the state.

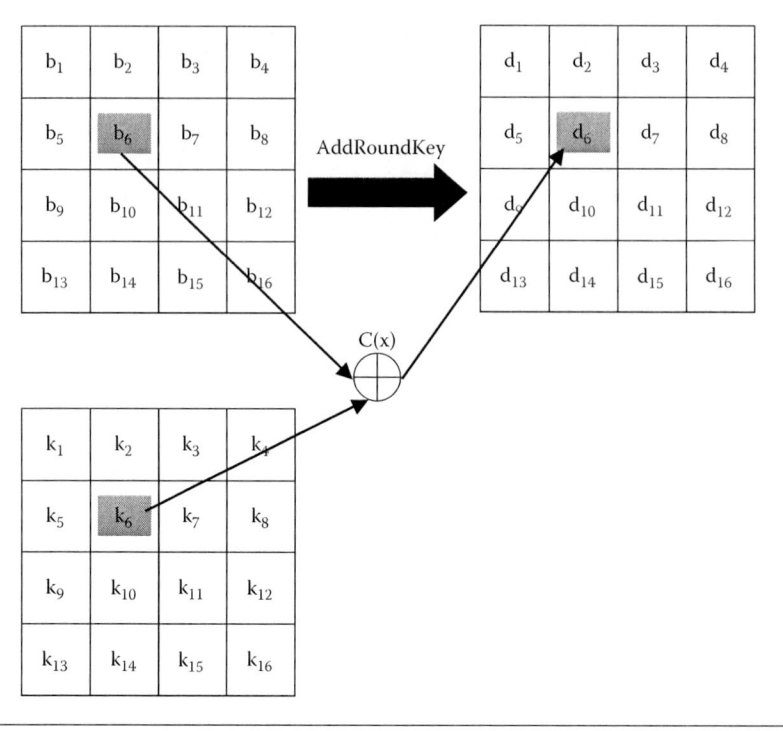

Figure 2.12   AddRoundKey step.

The subkey is added by combining each byte of the state with the corresponding byte of the subkey using a bitwise XOR [16–18]. The AddRoundKey step is shown in Figure 2.12. We apply the AES with a fixed block size of 128 bits and a key size of 128 bits.

2.7 Modes of Operation

Block ciphers can be run in different modes of operation, allowing users to choose appropriate modes to meet the requirements of their applications. Using a certain mode in the encryption process restricts the decryption process to use the same mode. In this section, we discuss different possible ways in which block codes can be utilized to implement a cryptosystem. The possible block cipher modes of operation that we treat are identified by the acronyms electronic codebook (ECB), cipher-block chaining (CBC), cipher feedback (CFB), and output feedback (OFB). In each case, we assume that we have a block cipher of block length n with enciphering map E_K and deciphering map D_K for each key K.

2.7.1 The ECB Mode

ECB is the simplest mode of operation for encryption algorithms; the data sequence is divided into blocks of equal size, and each block is encrypted, separately, with the same encryption key. As illustrated in Figure 2.13, the plaintext is divided into blocks $(P_1, P_2, P_3, ...)$ of size n bits that are encrypted to ciphertext blocks $(C_1, C_2, C_3, ...)$. The encryption algorithm is

$$C_j = E_K (P_j) \qquad (2.5)$$

and the decryption algorithm is

$$P_j = D_K (C_j) \qquad (2.6)$$

where $j = 1, 2, 3, ...$; E_K is the encryption map with the key K; and D_K is the decryption map with the same key K.

The ECB mode has several advantages. There is no need to encrypt a file progressively; the middle blocks can be encrypted first, then the blocks at the end, and finally the blocks at the beginning. This is important for encrypted files that are accessed randomly, like a database. If a database is encrypted in the ECB mode, then any record can be added, deleted, encrypted, or decrypted independently, assuming that a record consists of independent encryption blocks.

The disadvantage of this mode is that identical plaintext blocks are encrypted to identical ciphertext blocks; it does not hide data patterns. The advantage is that error propagation is limited to a single block. The disadvantage of the ECB mode appears in image encryption if there is an image with large areas of the same color or repeated patterns so that there are many blocks of the same plaintext.

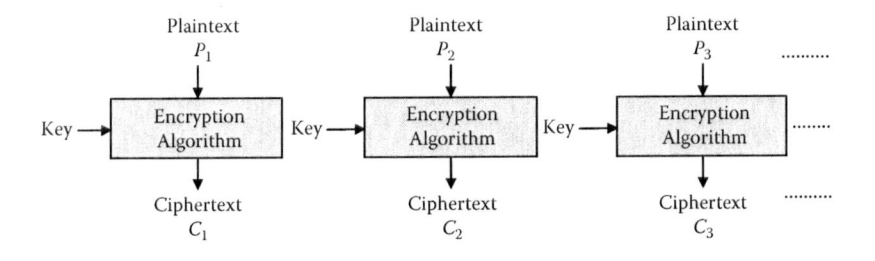

Figure 2.13 Using a block cipher in the ECB mode.

This may reveal much information about the original image from the encrypted image. This disadvantage is treated in CBC, CFB, and OFB modes.

2.7.2 The CBC Mode

The CBC mode uses an initialization vector (IV) of size equal to the size of each block of pixels. In this mode, each block of plaintext is XORed with the previous ciphertext block before being encrypted. By this way, each ciphertext block is dependent on all plaintext blocks up to that point. In decryption, the same XOR operation is repeated so that its effect is cancelled. This mechanism is shown in Figure 2.14.

The main disadvantage of the CBC mode is that an error in (or attack on) one ciphertext block impacts two plaintext blocks on decryption. On the other hand, if there is an image that has blocks of the same input data, these blocks are encrypted to totally different ciphertext data. So, the CBC mode is a better approach in encrypting images in the spatial domain, especially when these images contain large areas of the same activity. In the CBC mode, the encryption algorithm is

$$C_j = E_K(C_{j-1} \oplus P_j) \tag{2.7}$$

and the decryption algorithm is

$$P_j = D_K(C_j) \oplus C_{j-1}, j = 1, 2, 3, \dots \tag{2.8}$$

$$C_0 = IV \tag{2.9}$$

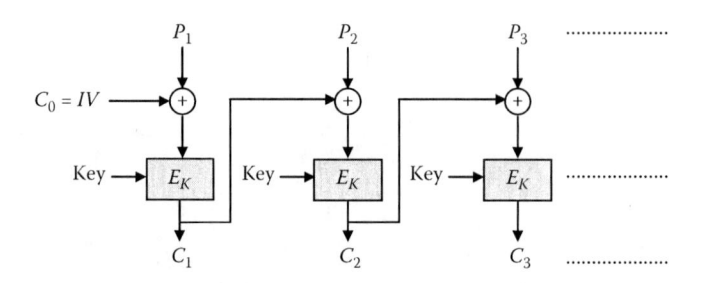

Figure 2.14 Using a block cipher in the CBC mode.

2.7.3 The CFB Mode

In contrast to the CBC mode, the CFB mode begins by encrypting the IV, and then an XOR operation is performed between the bits of the encrypted IV and the corresponding bits of the first block of the image. The result is the encrypted version of the first block. For the encryption of each of the next plaintext blocks, the previous ciphertext block is encrypted, and the output is XORed with the current plaintext block to create the current ciphertext block. The XOR operation conceals plaintext patterns.

Common to the CBC mode, changing the IV to the same plaintext block results in different outputs. Although the IV need not be secret, some applications would see this as desirable [16–18,49,50]. Figure 2.15 shows the CFB mode. The encryption algorithm is

$$C_j = P_j \oplus I_j \qquad (2.10)$$

and the decryption algorithm is

$$P_j = C_j \oplus I_j \qquad (2.11)$$

$$I_j = E_K(C_{j-1}), j = 1, 2, 3, \ldots \qquad (2.12)$$

$$C_0 = IV \qquad (2.13)$$

2.7.4 The OFB Mode

The OFB mode is similar to the CFB mode. It begins by encrypting the IV. The bits of the encrypted IV are XORed with the corresponding bits of the first plaintext block to obtain the corresponding ciphertext block. Also, the output of the encryption algorithm is used as an input to the next encryption step instead of the IV. This process

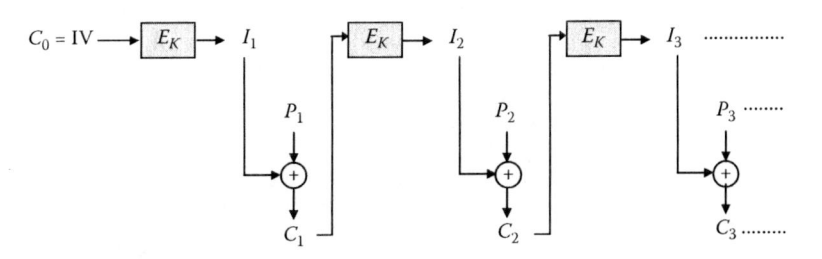

Figure 2.15   Using a block cipher in the CFB mode.

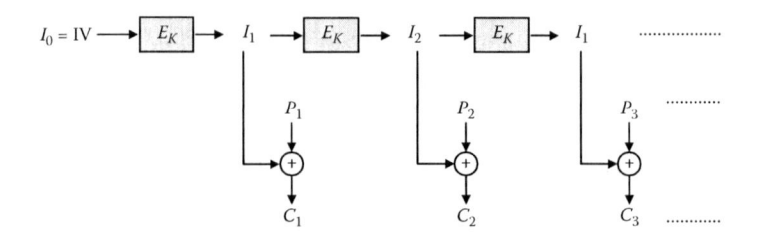

Figure 2.16   Using a block cipher in the OFB mode.

continues until the last block. Changing the IV for the same plaintext block results in different ciphertext blocks. Figure 2.16 shows the OFB mode [16–18,49,50]. The encryption algorithm is

$$C_j = P_j \oplus I_j \tag{2.14}$$

and the decryption algorithm is

$$P_j = C_j \oplus I_j \tag{2.15}$$

$$I_j = E_K(I_{j-1}), j = 1,2,3,..... \tag{2.16}$$

$$I_0 = IV \tag{2.17}$$

2.8 Chaos and Cryptography

Chaos theory has been established since the 1970s in many different research areas, such as physics, mathematics, engineering, biology, and others. The most well-known characteristics of chaos are the so-called butterfly effect (sensitivity to initial conditions) and the pseudo-randomness generated by deterministic equations. Many fundamental properties of chaotic systems have their corresponding counterparts in traditional cryptosystems. Chaotic systems have several significant features favorable to secure communications, such as ergodicity, sensitivity to initial conditions and control parameters, and random-like behavior [38–40]. With all these advantages, scientists expected to introduce new and powerful tools of chaotic cryptography. So, chaos has become a new rich source of new ciphers [31–35].

Chaotic dynamic systems are dimensional nonlinear dynamic systems that are capable of complex and unpredictable behavior. Chaos describes a system that is sensitive to initial conditions to generate an apparently random behavior but at the same time it is completely deterministic. These properties of chaos have much potential for applications in cryptography as it is hard to make long-term predictions on chaotic systems. First, being completely deterministic means that we can always obtain the same set of values provided we have exactly the same mapping function and initial conditions. Because chaotic functions are sensitive to initial conditions, any slight difference in the initial values used means that the ciphertext produced using chaos will be completely different. This means that the system will be strong against brute-force attacks as the number of possible keys is large.

The basic properties of chaotic systems are the deterministicity, the sensitivity to initial conditions and parameters, the topological transitivity, and the ergodicity [51–55].

Deterministicity means that chaotic systems have some determining mathematical equations ruling their behavior.

Sensitivity to initial conditions means that when a chaotic map is iteratively applied to two initially close points, the iterations quickly diverge and become uncorrelated in the long term. Sensitivity to parameters causes the properties of the map to change quickly when slightly perturbing the parameters on which the map depends. Hence, a chaotic system can be used as a pseudorandom number generator.

Topological transitivity is the tendency of the system to quickly scramble up small portions of the state space into an intricate network of filaments. Local, correlated information becomes scattered over the state space.

The ergodicity of a chaotic map means that if the state space is partitioned into a finite number of regions, no matter how many, any orbit of the map will pass through all these regions. The ergodicity of a chaotic map is ensured by the topological transitivity property.

Many properties of chaotic systems, such as mixing and sensitivity to initial conditions, have their corresponding counterparts in traditional cryptosystems. Table 2.4 contains a partial list of

Table 2.4 Similarities and Differences between Chaotic Systems and Cryptographic Algorithms

CHAOTIC SYSTEMS	CRYPTOGRAPHIC ALGORITHMS
Set of real numbers	Finite set of integers
Iterations	Rounds
Parameters	Key
Sensitivity to a change in initial conditions	Diffusion
—	Security

these properties. The main difference between chaos theory and cryptography is that cryptosystems work on a finite field, while chaos works on a continuum.

Several chaos-based image encryption schemes been developed in recent years. In 1992, Bourbakis and Alexopoulos proposed an image encryption scheme that utilizes the SCAN language to encrypt and compress an image simultaneously [56]. Fridrich demonstrated the construction of a symmetric block cipher algorithm based on a 2D standard Baker map [52]. There are three basic steps in Fridrich's method [52]: Choosing a chaotic map and generalizing it by introducing some parameters, discretizing the chaotic map to a finite square lattice of points that represent pixels, and extending the discredited map to three dimensions and further composing it with a simple diffusion mechanism.

Scharinger designed a chaotic Kolmogorov-flow-based image encryption technique in which the whole image is taken as a single block and permuted through a key-controlled chaotic system [57]. In addition, a shift register pseudorandom generator is adopted to introduce the confusion in the data. Yen and Guo proposed an encryption method called bit recirculation image encryption (BRIE) based on a chaotic logistic map [58]. The basic principle of this BRIE method is bit recirculation of pixels, which is controlled by a chaotic pseudorandom binary sequence. The secret key of the BRIE method consists of two integers and an initial condition of the logistic map. Yen and Guo [59] also proposed an encryption method called CKBA (chaotic key-based algorithm), in which the key binary sequence is generated using a chaotic system. The image pixels are rearranged according to the generated binary sequence and then XORed and XNORed with the selected key. In 2002, Li and Zheng [29] pointed out some defects in the encryption schemes presented in the references [58] and discussed some possible improvements for them. Chen et al. [60] proposed a symmetric algorithm in which a 2D chaotic map is generalized to three dimensions for designing a real-time secure image encryption scheme. This approach employs the 3D cat map to shuffle the positions of the image pixels and uses another chaotic map to confuse the relationship between the encrypted image and its original image.

In chaotic encryption with 1D maps, the encryption key is generated with a chaotic map based on selected initial conditions. Each map produces various random keys from various orbits of the map, which guarantees the security. Based on the key, a binary sequence is generated to control the encryption algorithm. The input image of two dimensions is transformed into a 1D array and then divided into various subblocks. Then, the position permutation and value permutation are applied to each binary matrix representing a subblock. Finally, the receiver uses the same subkeys to decrypt the encrypted images. On the other hand, chaotic encryption with 2D maps like the Baker map is different, as shown in the next section.

2.9 The 2D Chaotic Baker Map

The Baker map stretches the image horizontally and then folds it vertically. Repeating this process, the positions of all pixels of the plainimage are changed [51,52,61]. Let $B(n_1, \ldots, n_k)$ denote the discretized map, where the vector $[n_1, \ldots, n_k]$ represents the secret key S_{key}. Defining N as the number of data items in one row, the secret key is chosen such that each integer n_i divides N, and $n_1 + \ldots + n_k = N$.

Let $N_i = n_1 + \ldots + n_i$. The data item at the indices (q, z) is moved to the indices:

$$B_{(n_1,\ldots,n_k)}(q,z) = \left(\frac{N}{n_i}(q - N_i) + z \bmod\left(\frac{N}{n_i} \right), \right.$$

$$\left. \frac{n_i}{N}\left(z - z \bmod\left(\frac{N}{n_i} \right) \right) + N_i \right) \qquad (2.18)$$

where $N_i \leq q < N_i + n_i$, and $0 \leq z < N$.

In steps, the chaotic permutation is performed as follows:

1. An $N \times N$ square image is divided into k rectangles of width n_i and number of elements N.
2. The elements in each rectangle are rearranged to a row in the permuted rectangle. Rectangles are taken from left to right beginning with upper rectangles, then lower ones.
3. Inside each rectangle, the scan begins from the bottom left corner toward upper elements.

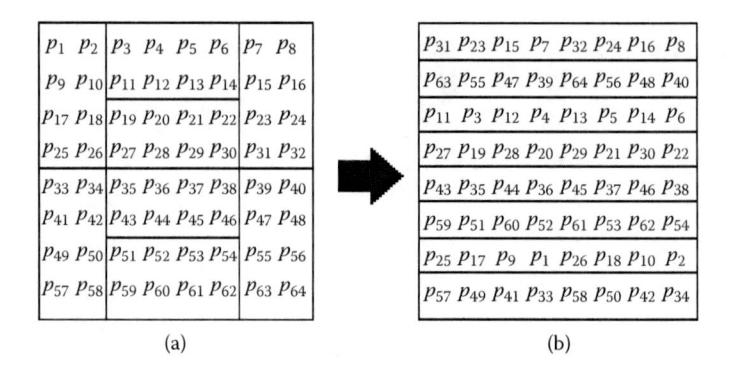

Figure 2.17   (a) The 8 × 8 matrix divided into rectangles. (b) The matrix after applying the 2D Baker map.

Figure 2.17 shows an example for the chaotic map of an (8 × 8) square image (i.e., $N = 8$). The secret key, $S_{key} = [n_1, n_2, n_3] = [2, 4, 2]$.

The cipher resulting from the chaotic Baker map encryption is a permutation cipher, which does not change the histogram of the original image. Although this cryptosystem is simple and fast to be used for video encryption, it lacks the high degree of security, and its computation time grows as the image size increases [51,52,61].

3
ENCRYPTION EVALUATION METRICS

3.1 Introduction

With the application of an encryption algorithm to an image, its pixel values change when compared with the original image. A good encryption algorithm must make these changes in an irregular manner and maximize the difference in pixel values between the original and the encrypted images. Also, to obtain a good encrypted image, it must be composed of totally random patterns that do not reveal any of the features of the original image. The encrypted image has to be independent of the original image. It should have a low correlation with the original image [62–66].

One of the important metrics in examining an encrypted image is the visual inspection: The more hidden the features of the image are, the better the encryption algorithm is. Unfortunately, visual inspection only is not enough to judge the complete hiding of the contents of the image, so other metrics are considered to evaluate the degree of encryption quantitatively [67,68].

Diffusion is an important parameter that must be measured to judge the encryption algorithm randomization. If an algorithm has a good diffusion characteristic, the relation between the encrypted image and the original image is too complex, and it cannot be predicted easily. To measure the diffusion of any algorithm, a bit is changed in the plainimage, and the difference between the encrypted image obtained from the original plainimage and the encrypted image obtained from the modified one is obtained [69].

Other tests can be used to determine some specific characteristics of the encryption algorithms, such as the noise immunity and the processing time.

3.2 Encryption Evaluation Metrics

This section discusses, in detail, two families of encryption metrics; the first family evaluates the ability of the encryption algorithm to substitute the original image with an uncorrelated encrypted image. In this family, five metrics are studied: the histogram deviation D_H, the correlation coefficient r_{xy}, the irregular deviation D_I, the histogram uniformity, and the deviation from ideality. The second family evaluates the diffusion characteristics of the encryption algorithm. In this family, three metrics are studied: the Avalanche effect, number of pixel change rate (NPCR), and unified average changing intensity (UACI).

3.2.1 Histogram Deviation

The histogram deviation measures the quality of encryption in terms of how it maximizes the deviation between the original and the encrypted images [70]. The steps for calculating this metric are as follows:

1. Estimate the histogram of both the original and the encrypted images.
2. Estimate the absolute difference between both histograms.
3. Estimate the area under the absolute difference curve divided by the total area of the image as follows:

$$D_H = \frac{\left(\dfrac{d_0 + d_{255}}{2} + \sum_{i=1}^{254} d_i \right)}{M \times N} \tag{3.1}$$

where d_i is the amplitude of the absolute difference at the gray level i. M and N are the dimensions of the image to be encrypted. The higher the value of D_H is, the better the quality of the encrypted image will be [70].

Although this metric gives good results about how the encrypted image deviates from the original image, it cannot be used alone to measure the quality of encryption as it has some limitations, explained further in this chapter.

3.2.2 Correlation Coefficient

A useful metric to assess the encryption quality of any image cryptosystem is the correlation coefficient between pixels at the same indices in the plain- and the cipherimages [70]. This metric can be calculated as follows:

$$r_{xy} = \frac{\text{cov}(x, y)}{\sqrt{D(x)}\sqrt{D(y)}} \tag{3.2}$$

where x and y are the plain- and cipherimages. In numerical computations, the following discrete formulas can be used:

$$E(x) = \frac{1}{L}\sum_{l=1}^{L} x_l \tag{3.3}$$

$$D(x) = \frac{1}{L}\sum_{l=1}^{L} (x_l - E(x))^2 \tag{3.4}$$

$$\text{cov}(x, y) = \frac{1}{L}\sum_{l=1}^{L} (x_l - E(x))(y_l - E(y)) \tag{3.5}$$

where L is the number of pixels involved in the calculations. The closer the value of r_{xy} to zero, the better the quality of the encryption algorithm will be.

3.2.3 Irregular Deviation

The irregular deviation measures the quality of encryption in terms of how much the deviation caused by encryption (on the encrypted image) is irregular [71]. The steps for calculating this metric are as follows:

1. Calculate the absolute difference between the encrypted image and the original image.
2. Estimate the histogram H of this absolute difference matrix.
3. Estimate the mean value M_H of this histogram.

4. Estimate the absolute of the histogram deviations from this mean value as follows:

$$H_D(i) = |H(i) - M_H| \qquad (3.6)$$

The irregular deviation D_I is calculated as follows:

$$D_I = \frac{\sum_{i=0}^{255} H_D(i)}{M \times N} \qquad (3.7)$$

The lower the value of D_I, the better the encryption quality will be.

3.2.4 Histogram Uniformity

A histogram uses a bar graph to profile the occurrence of each gray level of the image. The horizontal axis represents the gray-level value. It begins at zero and goes to the number of gray levels. Each vertical bar represents the number of occurrences of the corresponding gray level in the image [72].

For image encryption algorithms, the histogram of the encrypted image should have two properties:

1. It must be totally different from the histogram of the original image.
2. It must have a uniform distribution, which means that the probability of occurrence of any grayscale value is the same.

3.2.5 Deviation from Ideality

The deviation from ideality measures the quality of encryption in terms of how the encryption algorithm minimizes the deviation of the encrypted image from an assumed ideal encryption case. An ideally encrypted image C_I must have a completely uniform histogram distribution, which means that the probability of existence of any gray level is constant. From this definition of the ideal encrypted image histogram, it can be formulated as

$$H(C_I) = \begin{cases} \dfrac{M \times N}{256} & 0 \le C_I \le 255 \\ 0 & elsewhere \end{cases} \qquad (3.8)$$

The deviation from ideality can be represented as follows:

$$D = \frac{\sum_{C_I=0}^{255} |H(C_I) - H(C)|}{M \times N} \tag{3.9}$$

where $H(C)$ is the histogram of encrypted image. Of course, the lower the value of D, the better the encryption quality will be.

3.2.6 Avalanche Effect

We can use the Avalanche effect metric [73,74] to test the efficiency of the diffusion mechanism. A single bit change can be made in the image P to give a modified image P'. Both P and P' are encrypted to give C and C'. The Avalanche effect metric is the percentage of different bits between C and C'. If C and C' differ from each other in half of their bits, we can say that the encryption algorithm possesses good diffusion characteristics.

3.2.7 NPCR and UACI

To test the influence of a one-pixel change on the whole image encrypted by any encryption algorithm, two common metrics may be used: NPCR and UACI [75]. Let the two ciphered images, whose corresponding plainimages have only one pixel difference, be denoted by C_1 and C_2. Label the grayscale values of the pixels at grid (i,j) in C_1 and C_2 by $C_1(i,j)$ and $C_2(i,j)$, respectively. Define a binary matrix D with the same size as the images C_1 and C_2. Then, $D(i,j)$ is determined from $C_1(i,j)$ and $C_2(i,j)$. If $C_1(i,j) = C_2(i,j)$, then $D(i,j) = 0$; otherwise, $D(i,j) = 1$.

The NPCR is defined as

$$\text{NPCR} = \frac{\sum_{i,j} D(i,j)}{M \times N} \times 100\% \tag{3.10}$$

The *NPCR* measures the percentage of different pixels in the two images.

The UACI is defined as

$$\text{UACI} = \frac{1}{M \times N}\left[\sum_{i,j}\frac{C_1(i,j)-C_2(i,j)}{255}\right]\times 100\% \qquad (3.11)$$

It measures the average intensity of differences between the two images. The higher the values of *NPCR* and *UACI* are, the better the encryption will be.

3.3 Other Tests

3.3.1 Noise Immunity

The noise immunity reflects the ability of the image cryptosystem to tolerate noise. To test the noise immunity, noise with different signal-to-noise ratios (SNRs) is added to the encrypted image, and then the decryption algorithm is performed. If the decrypted image is close to the original image, we can say that the cryptosystem at hand is immune to noise. This closeness can be verified visually or numerically with the value of r_{xyd}, which represents the correlation coefficient between the original image and the decrypted image, and the peak signal-to-noise ratio (PSNR) of the decrypted image, which is defined as follows [60,66]:

$$PSNR = 10\times\log_{10}\left(\frac{M\times N\times 255^2}{\sum_{m=1}^{M}\sum_{n=1}^{N}\left|\left(f(m,n)-f_d(m,n)\right)\right|^2}\right) \qquad (3.12)$$

where $f(m,n)$ is the original image, and $f_d(m,n)$ is the decrypted image.

3.3.2 The Processing Time

The processing time is the time required to encrypt and decrypt an image. The smaller value the processing time has, the better the encryption efficiency will be.

3.4 Testing the Evaluation Metrics

We have tested these metrics by evaluating three encrypted images with well-known encryption quality. The first one is obtained by flipping the Cameraman image using the Caesar cipher, which is a straightforward ciphering technique. The image size is 256 × 256 pixels. If we apply this cipher on an image I to obtain the image J, then we will add a constant offset to the pixel values such that the values exceeding 255 are rotated back starting from the 0 level again. This is shown by the equation

$$J(m, n) = \{I(m, n)+K(m, n)\}\bmod 256, \quad K(m, n) = 254 \times I(m, n) + 255 \tag{3.13}$$

where K is the cipher offset, and it is considered the key as well. We choose the offset K of the Caesar cipher such that the location containing a peak in the histogram of the original image contains a very low value in the encrypted image, and the location containing a low value in the original image contains a peak in the encrypted image as shown in Figure 3.1.

Note here that the Caesar cipher is useless in image encryption, because only a change in the image gray scale occurs keeping the entire image features visible to the attacker, who does not know the decryption key. It is noticed that the image was not actually encrypted. Besides, this cipher does not possess any diffusion at all.

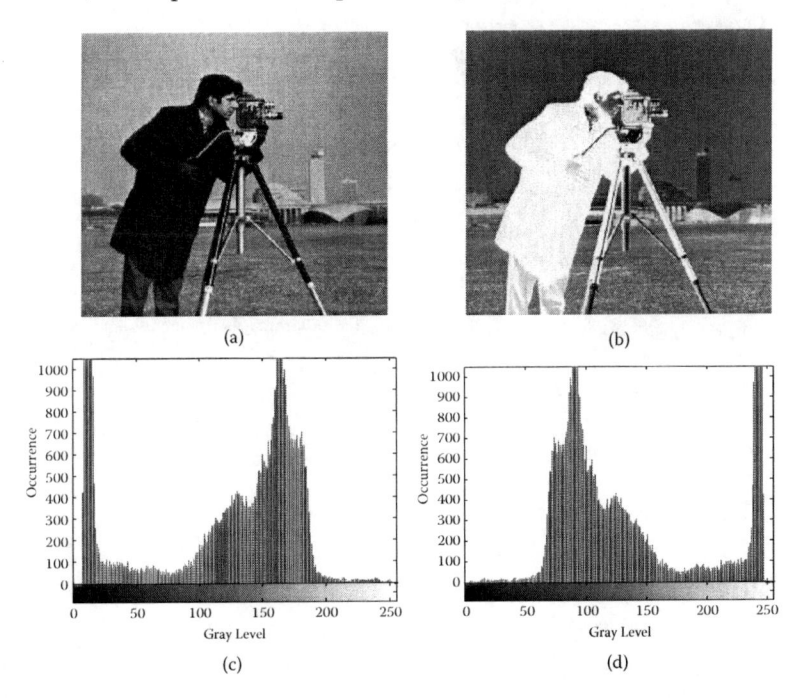

Figure 3.1 The Cameraman image: (a) original version; (b) flipped version; (c) original image histogram; (d) flipped image histogram.

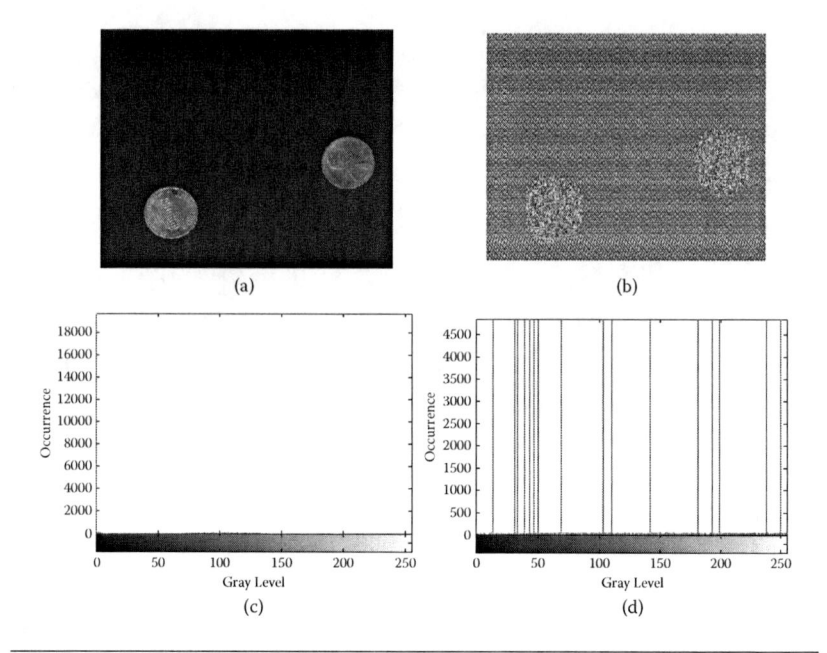

Figure 3.2 The Coin image: (a) original version; (b) encrypted version; (c) original image histogram; (d) encrypted image histogram.

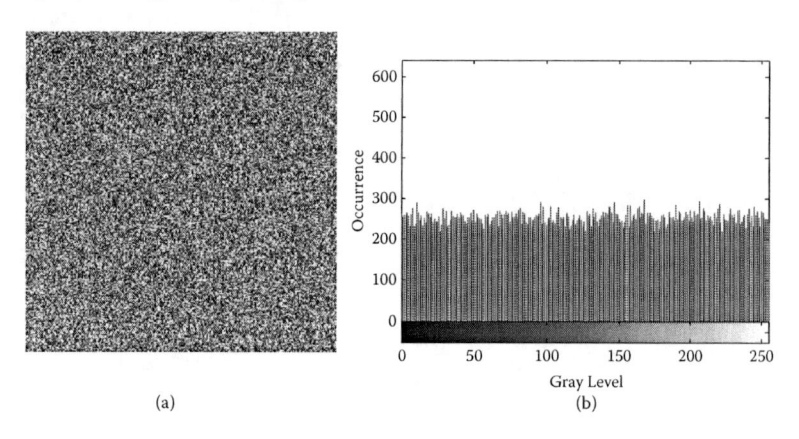

Figure 3.3 (a) The encrypted image with the RC6 algorithm; (b) the histogram.

The second image is obtained by encrypting the Coin image, which is a black-and-white image using the RC6 algorithm in the ECB mode. The image size is 340 × 400 pixels. Algorithms implemented in the ECB mode cannot encrypt a black-and-white image because of its localized histogram. This is shown in Figure 3.2. The third image is obtained by encrypting the Cameraman image using the RC6 algorithm in the CBC mode as shown in Figure 3.3.

Table 3.1 Encryption Quality Results: Substitution

	D_H	C.C	D_I	D
Case 1	1.29	−1	1.26	0.98
Case 2	1.45	0.0554	1.7	1.7
Case 3	0.98	−0.0076	0.6	0.05

Table 3.2 Encryption Quality Results: Diffusion

	AVALANCHE	NPCR	UACI
Case 1	≅0%	≅0%	≅0%
Case 2	0.0059%	0.0118%	0.0013%
Case 3	50.07%	99.62%	16.7%

It is obvious that the first two images were encrypted badly without any diffusion characteristics, while the last image was encrypted well, with good diffusion characteristics. The results of each evaluation metric for every case are tabulated in Table 3.1, and the diffusion test results are tabulated in Table 3.2.

From these results, we can see the following for the first case:

1. The histogram deviation gives a high result of 1.29, indicating the good performance of the Caesar cipher, which is completely a wrong decision as the entire image features are still visible. The histogram deviation depends on the difference between the histograms of the plainimage and the cipherimage, which does not necessarily mean a good encryption.
2. The correlation coefficient of −1 means that the cipherimage is the reverse of the plainimage, which is totally true.
3. The rest of the results, including the histogram uniformity, have judged the bad performance of the Caesar cipher correctly.
4. For the diffusion tests, the results are almost zero because the Caesar cipher does not possess any diffusion characteristics.

For the second case,

1. The histogram deviation, again, could not judge the encryption performance correctly because the plainimage has a localized histogram.
2. For the correlation coefficient, the result indicates that the plainimage is uncorrelated with the cipherimage, which is

true, but this does not mean that the plainimage is encrypted correctly.
3. The rest of the tests have judged the encryption quality correctly.
4. For the ECB mode, the diffusion characteristics are bad, and this is indicated correctly with the diffusion tests.

For the third case, all test results correctly judged the good encryption and diffusion characteristics of the CBC mode.

3.5 Summary

After all these tests, we can see that

1. Visual inspection is the first test to be used. If the specifications of the plainimage were not completely hidden, we can say that this encryption algorithm is not confident and cannot be used to encrypt images regardless of all other tests.
2. The irregular deviation, the deviation from ideality, and the histogram uniformity perform well for judging encryption quality.
3. The three tests of diffusion can be used effectively to judge the diffusion characteristics of the encryption algorithm.

4

HOMOMORPHIC IMAGE ENCRYPTION

4.1 Overview

This chapter presents a new image cryptosystem. This system is based on homomorphic image processing, which has evolved primarily as a tool of image enhancement for images captured in bad lighting conditions. The main idea of homomorphic image processing is based on modeling the image as a product of constant illumination and varying reflectance. The product is dealt with as a summation using the logarithmic operation. The reflectance component can be separated using a high-pass filter, while the illumination component is separated using a low-pass filter. Most of the image details lie in the reflectance component, while the illumination component is approximately constant [76,77].

We can carry out the encryption process in the homomorphic domain on the reflectance component, which is the most significant component of the image. Rather than encrypting the illumination component, which causes redundancy in image information, it is appended as a least-significant bit (LSB) watermark in the encrypted reflectance component. Two algorithms are used for the encryption of the reflectance component: the RC6 block cipher algorithm and the chaotic Baker map scrambling algorithm. A comparison is made between them.

4.2 Homomorphic Cryptosystem

The idea of the homomorphic cryptosystem is based on homomorphic image processing. It is known that the image intensity can be represented as follows [78–80]:

$$I(m,n) = i(m,n)r(m,n) \qquad (4.1)$$

where $i(m,n)$ is the light illumination, and $r(m_1,n_2)$ is the reflectance of the object to be imaged. Taking the natural logarithm of both sides leads to

$$\ln[I(m,n)] = \ln[i(m,n)] + \ln[r(m,n)] \tag{4.2}$$

The illumination is approximately constant, while the reflectance is variable from object to object. Thus, the term $\ln[i(m,n)]$ is approximately constant. We can perform the encryption process on the $\ln[r(m,n)]$ term. To avoid the redundancy resulting from the existence of two components of the image in the homomorphic domain, we can embed the illumination component as an LSB watermark to the encrypted reflectance component. The homomorphic cryptosystem is illustrated in Figure 4.1. The 3 × 3 averaging filter shown in Figure 4.2 is used as the low-pass filter, and the reflectance component is obtained by subtracting the log illumination component from the log image intensity.

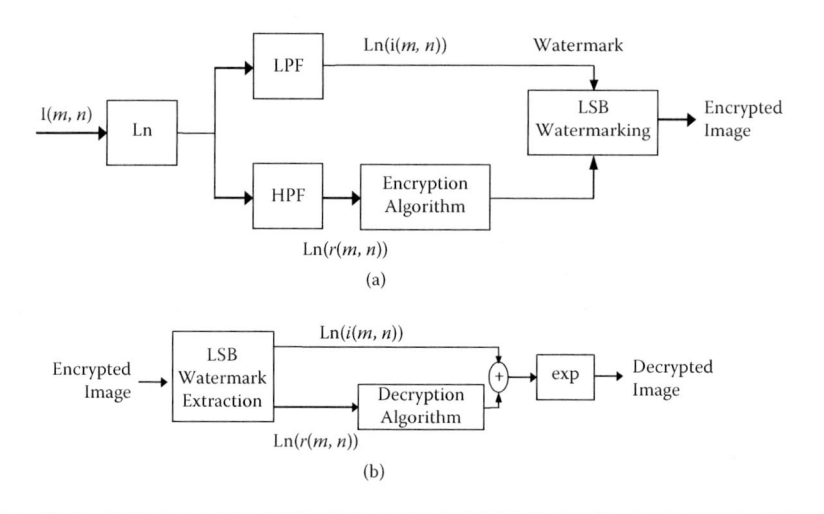

Figure 4.1   Homomorphic image cryptosystem: (a) encryption subsystem and (b) decryption subsystem.

$$\frac{1}{9} \times \begin{bmatrix} 1 & 1 & 1 \\ 1 & 1 & 1 \\ 1 & 1 & 1 \end{bmatrix}$$

Figure 4.2   The 3 × 3 averaging filter.

The RC6 block cipher algorithm and the chaotic Baker map scrambling algorithm have been chosen as they are representatives of different encryption families. The first one belongs to the family of diffusion algorithms, and the second one belongs to the family of permutation algorithms. The objective is to decide which family is more appropriate for the homomorphic cryptosystem.

4.3 Security Analysis and Test Results

A good encryption scheme should resist all kinds of known attacks, such as the known-plaintext attack, the ciphertext-only attack, the statistical attack, the differential attack, and the various brute-force attacks. The security of the homomorphic image cryptosystem was investigated for digital images under the brute-force attack, the statistical attacks, and the differential attacks [81–85].

Some security analysis results, including the key space analysis, the statistical analysis, and the differential analysis have been presented [60,86,87]. Tests are made on the image of Lena shown in Figure 4.3.

4.3.1 Statistical Analysis

In [13], Shannon mentioned that, "It is possible to solve many kinds of ciphers by statistical analysis." Statistical analysis has been performed on the homomorphic image cryptosystem, demonstrating its superior confusion and diffusion properties, which strongly resist statistical attacks.

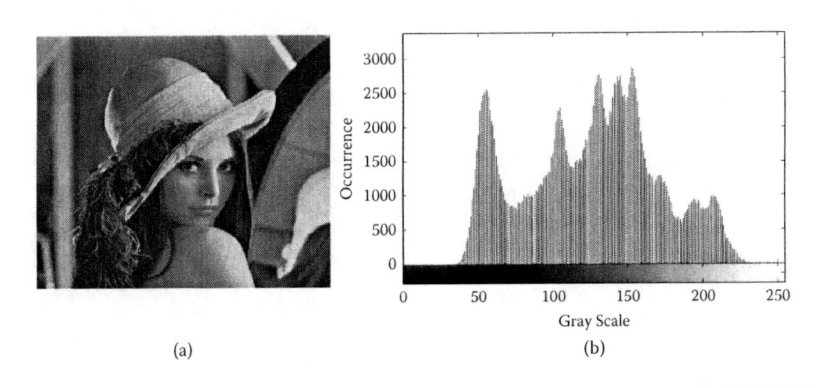

(a)   (b)

Figure 4.3 Lena: (a) image and (b) histogram.

4.3.1.1 Histograms of Encrypted Images   A typical example of the histogram test is shown in Figures 4.4 to 4.7. From these figures, one can see that the histogram of the encrypted image (cipherimage) by the homomorphic cryptosystem using the RC6 algorithm is fairly uniform and is significantly different from that of the original image (plainimage) as shown in Figure 4.4. This result implies to the RC6 block cipher as shown in Figure 4.5. For the homomorphic cryptosystem using chaotic Baker map scrambling, the result is shown in Figure 4.6. It is clear that the histogram is different from that of the individual chaotic Baker map encryption shown in Figure 4.7. It is known that chaotic Baker map encryption does not change the histogram of the encrypted image from that of the original image. On the other hand, the homomorphic cryptosystem using the chaotic Baker map scrambling significantly changes the histogram of the encrypted image.

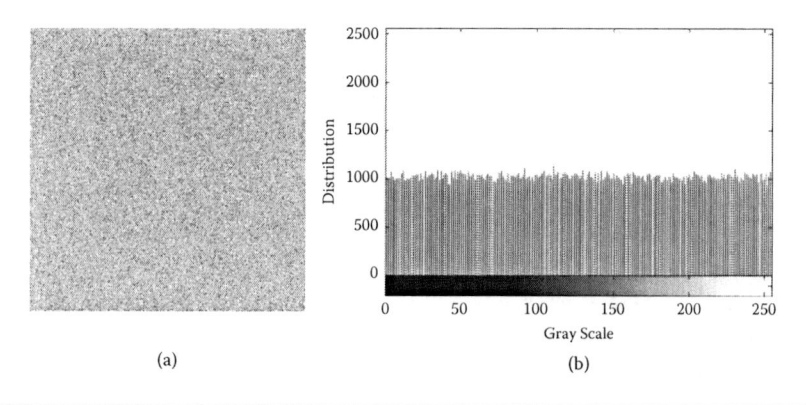

(a)   (b)

Figure 4.4   Encrypted image using the homomorphic cryptosystem with the RC6 algorithm: (a) encrypted image and (b) histogram.

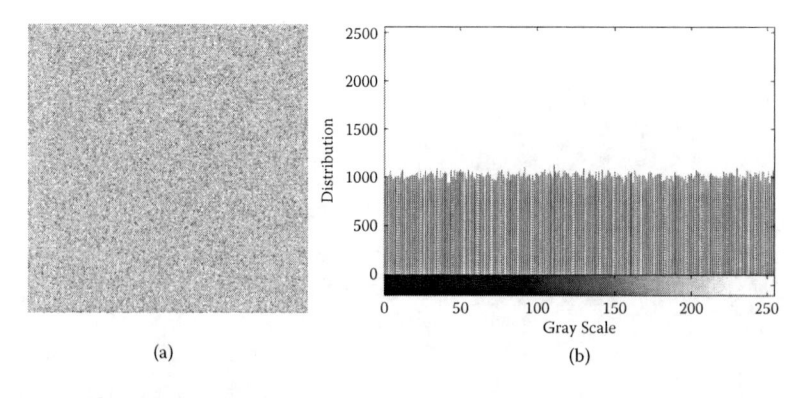

(a)   (b)

Figure 4.5   Encrypted image using the RC6 algorithm: (a) encrypted image and (b) histogram.

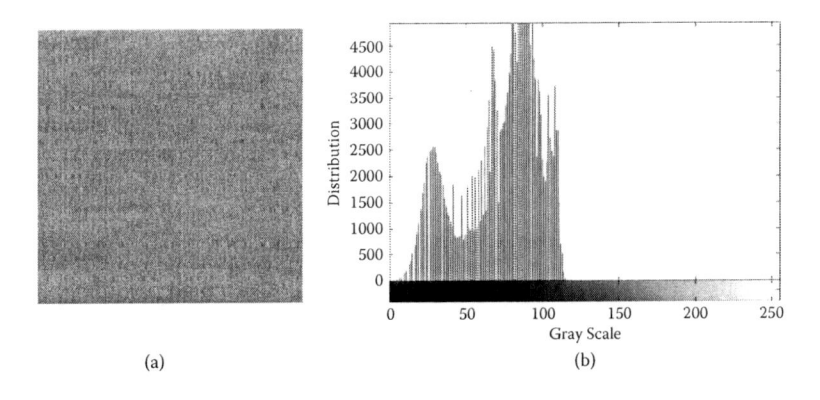

(a) (b)

Figure 4.6   Encrypted image using the homomorphic cryptosystem with the chaotic Baker map scrambling algorithm: (a) encrypted image and (b) histogram.

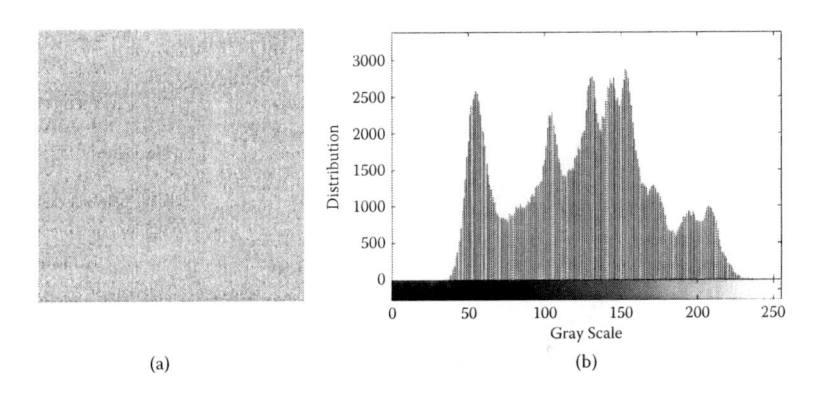

(a) (b)

Figure 4.7   Encrypted image using the chaotic Baker map algorithm: (a) encrypted image and (b) histogram.

4.3.1.2 Encryption Quality Measurements   The correlation coefficient between the plain- and cipherimages, the irregular deviation, and the deviation from ideality metrics have been used to test the quality of each cipher algorithm. The results of these tests are shown in Table 4.1.

From the correlation test, we can see that, in all cases, the plainimage was uncorrelated with the cipherimage, but as discussed in Chapter 3, we cannot completely depend on this test to judge the encryption quality correctly, as we can see that the chaotic Baker map scrambling algorithm achieved a good result of 0.0032 although it is only a permutation cipher.

For the irregular deviation and the deviation from ideality, we can see that the quality of the homomorphic encryption depends mainly

Table 4.1   Encryption Quality Test Results for Each Cipher Algorithm

ENCRYPTION ALGORITHM	R_{xye}	D_I	D
Homomorphic cryptosystem with the RC6 algorithm	0.0033	0.707	0.0254
RC6 algorithm	0.0013	0.705	0.0233
Homomorphic cryptosystem with the chaotic Baker map scrambling algorithm	0.0043	0.846	1.2618
Chaotic Baker map scrambling algorithm	0.0032	0.979	0.7155

on the core algorithm used. If the core encryption algorithm is powerful, like the RC6, the homomorphic version will be powerful, and vice versa.

4.3.2 Key Space Analysis

A good image cryptosystem algorithm should be sensitive to the cipher keys. For the homomorphic image cryptosystem, the key space analysis and test are summarized in the following sections.

4.3.2.1 Exhaustive Key Search   For a secure image cryptosystem, the key space should be large enough to make the brute-force attack infeasible [88]. The RC6 algorithm is a 128-bit encryption scheme. An exhaustive key search will take 2^k operations to succeed, where k is the key size in bits. An attacker simply tries all keys, one by one, and checks whether the given plainimage encrypts to the given cipherimage.

For a practical use of the homomorphic cryptosystem, assume that the secret key length is 128 bits. Therefore, an opponent may try to bypass guessing the key and directly guess all the possible combinations. The opponent will need about 2^{128} operations to successfully determine the key. If an opponent employs a 1000-MIPS (million instructions per second) computer to guess the key by the brute-force attack, the computational load is then

$$\frac{2^{128}}{1000 \times 10^6 \times 60 \times 60 \times 24 \times 365} > 10.7902831 \times 10^{21} \text{ years} \quad (4.3)$$

This is practically infeasible.

For chaotic Baker map encryption, the key is dependent on the width (or height) of an image. This is due to the scrambling phenomena of the chaotic Baker map. For the 512 × 512 Lena image, the number of possible keys is 10^{126} [52]. So, in this case, the computational load is then

$$\frac{10^{126}}{1000 \times 10^6 \times 60 \times 60 \times 24 \times 365} > 3.1710 \times 10^{109} \text{ years} \quad (4.4)$$

4.3.2.2 Key Sensitivity Test High key sensitivity is required for secure image cryptosystems. This means that the cipherimage cannot be decrypted correctly if there is only a slight difference between encryption or decryption keys [89]. This guarantees the security of the proposed cryptosystem against brute-force attacks. Assume that a 16-character ciphering key is used. This means that the key consists of 128 bits.

For testing the key sensitivity of the homomorphic cryptosystem using the RC6 algorithm, we have performed the following steps:

(a) An image is encrypted using the secret key of 32 zeroes (in hexadecimal), and the resultant image is referred to as encrypted image A as shown in Figure 4.8a.

(b) The same image is encrypted by making a slight modification in the secret key (i.e., 8 and 31 zeroes [in hexadecimal]). The change is made in the most significant digit in the secret key. The resultant image is referred to as encrypted image B as shown in Figure 4.8b.

(c) Again, the same image is encrypted by making another slight modification in the secret key (i.e., 31 zeroes and 1 [in hexadecimal]). The change is made in the least-significant digit in

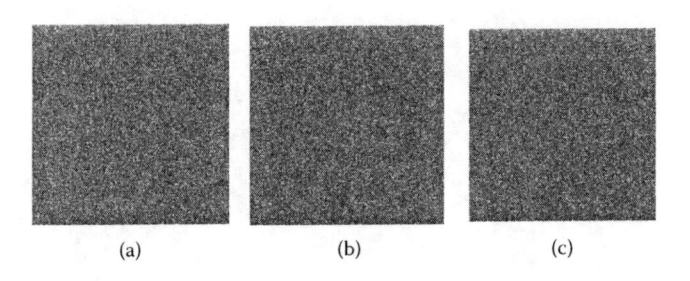

(a)   (b)   (c)

Figure 4.8 Key sensitivity test of the homomorphic cryptosystem with the RC6 algorithm: (a) encrypted image A with a key of 32 zeroes hexadecimal; (b) encrypted image B with a key of 8 and 31 zeroes hexadecimal; and (c) encrypted image C with a key of 31 zeroes and 1 hexadecimal.

the secret key. The resultant image is referred to as encrypted image C as shown in Figure 4.8c.

(d) Finally, the three encrypted images A, B, and C are compared.

It is not easy to compare the encrypted images by simply observing them. So, for comparison, we calculate the correlation coefficients between each two of the three encrypted images. Table 4.2 gives the correlation coefficient results. It is clear from the table that no correlation exists among the encrypted images even though they have been produced using slightly different secret keys. The same results are obtained using the RC6 algorithm, as shown in Figures 4.9a–4.9c. The results of the correlation coefficients are tabulated in Table 4.3.

For the homomorphic cryptosystem using the chaotic Baker map scrambling algorithm:

(a) The original image is encrypted using the secret key:

$$n = [10,5,12,5,10,8,14,10,5,12,5,10,8,14,10,5,12,$$
$$5,10,8,14,10,5,12,5,10,8,14,10,5,12,5,10,8,14,10,5,$$
$$12,5,10,8,14,10,5,12,5,10,8,14,10,5,12,5,10,8,14]$$

and the resultant image is referred to as encrypted image A as shown in Figure 4.10a.

Table 4.2   Results of the Key Sensitivity Test for the Homomorphic Cryptosystem with the RC6 Algorithm

IMAGE 1	IMAGE 2	R_{xy}
Encrypted image A	Encrypted image B	0.0034
Encrypted image B	Encrypted image C	0.00007
Encrypted image C	Encrypted image A	0.0008

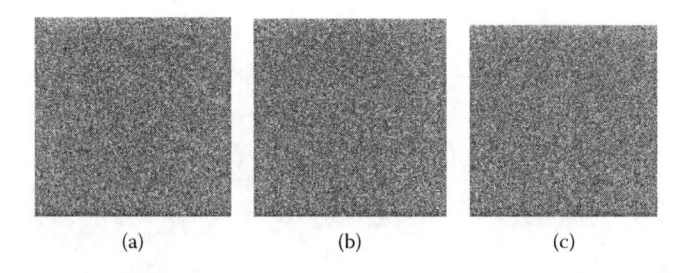

(a)   (b)   (c)

Figure 4.9   Key sensitivity test of RC6 algorithm: (a) encrypted image A with a key of 32 zeroes hexadecimal; (b) encrypted image B with a key of 8 and 31 zeroes hexadecimal; and (c) encrypted image C with a key of 31 zeroes and 1 hexadecimal.

Table 4.3   Results of the Key Sensitivity Test for the RC6 Algorithm

IMAGE 1	IMAGE 2	R_{xy}
Encrypted image A	Encrypted image B	0.0013
Encrypted image B	Encrypted image C	0.0028
Encrypted image C	Encrypted image A	0.0004

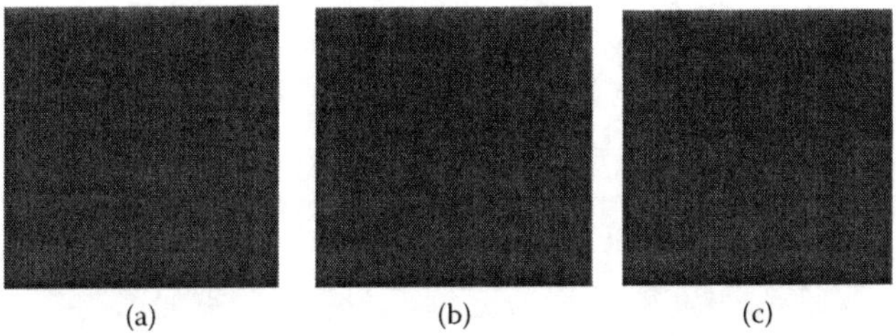

(a)   (b)   (c)

Figure 4.10   Key sensitivity test of the homomorphic cryptosystem with the chaotic Baker map scrambling algorithm: (a) encrypted image A with key n; (b) encrypted image B with key n_1; and (c) encrypted image C with key n_2.

(b) The same image is encrypted by making a slight modification in the secret key:

n_1 = [5,5,5,12,5,10,8,14,10,5,12,5,10,8,14,10,5,12, 5,10,8,14,10,5,12,5,10,8,14,10,5,12,5,10,8,14,10,5,1 2,5,10,8,14,10,5,12,5,10,8,14,10,5,12,5,10,8,14].

The change is made in (10) to (5, 5) at the beginning of the secret key. The resultant image is referred to as encrypted image B as shown in Figure 4.10b.

(c) Again, the same image is encrypted by making another slight modification in the secret key:

n_2 = [10,5,12,5,10,8,14,10,5,12,5,10,8,14,10,5,1 2,5,10,8,14,10,5,12,5,10,8,14,10,5,12,5,10,8,14, 10,5,12,5,10,8,14,10,5,12,5,10,8,14,10,5,12,5,1 0,8,7,7]

The change is made in (14) to (7, 7) at the end of the secret key. The resultant image is referred to as encrypted image C as shown in Figure 4.10c.

(d) Finally, the three encrypted images A, B, and C are compared.

The correlation coefficients between each two of the three encrypted images A, B, and C are tabulated in Table 4.4, from which it can be

Table 4.4   Results of the Key Sensitivity Test for the Homomorphic Cryptosystem with the Chaotic Baker Map Encryption Algorithm

IMAGE 1	IMAGE 2	R_{xy}
Encrypted image A	Encrypted image B	0.9533
Encrypted image B	Encrypted image C	0.8761
Encrypted image C	Encrypted image A	0.9212

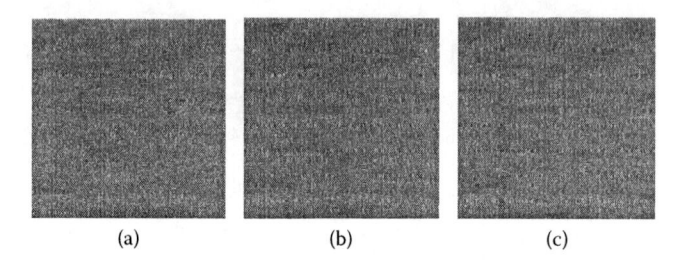

(a)   (b)   (c)

Figure 4.11   Key sensitivity test of the chaotic Baker map algorithm: (a) encrypted image A with key n; (b) encrypted image B with key n_1; and (c) encrypted image C with key n_2.

said that the correlation coefficients are worse than those obtained using the homomorphic cryptosystem with the RC6 algorithm. Similar results are obtained using the chaotic Baker map scrambling encryption algorithm only, as shown in Figures 4.11a–4.11c and Table 4.5.

Another test for the key sensitivity of the homomorphic image cryptosystem using the RC6 encryption algorithm is performed through the following steps:

1. A 512×512 image is encrypted using the secret test key of 32 zeroes.
2. The encryption key is changed by changing its LSB to be 31 zeroes and 1.
3. The two ciphered images are compared.

The result is that the image encrypted with the key of 31 zeroes and 1 is totally different from the image encrypted with the all-zeros key, although there is only 1 bit difference in the two keys. Figure 4.12a shows the difference image between the two ciphered images. A similar test was also applied to the RC6 encryption algorithm, and the result is shown in Figure 4.12b.

Table 4.5 Results of the Key Sensitivity Test for the Chaotic Baker Map Encryption Algorithm

IMAGE 1	IMAGE 2	R_{xy}
Encrypted image A	Encrypted image B	0.3247
Encrypted image B	Encrypted image C	0.8762
Encrypted image C	Encrypted image A	0.2877

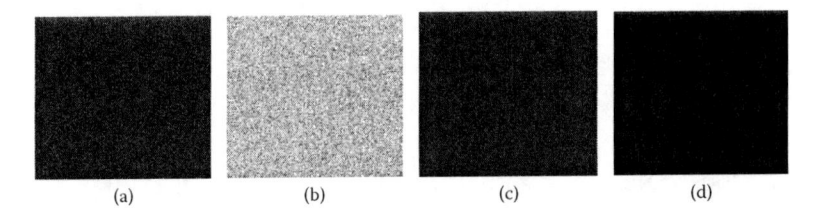

(a)　　　(b)　　　(c)　　　(d)

Figure 4.12 Difference image between the two ciphered images using (a) the homomorphic cryptosystem with the RC6 algorithm, (b) the RC6 algorithm, (c) the homomorphic cryptosystem with the chaotic Baker map algorithm, and (d) the chaotic Baker map scrambling algorithm.

For the homomorphic image cryptosystem using the chaotic Baker map scrambling algorithm, the test is done using the following steps:

1. A 512×512 image is encrypted using the secret test key n.
2. The encryption key is changed to n_1.
3. The two ciphered images are compared.

The result of this test is shown in Figure 4.12c. This test was also applied for the chaotic Baker map scrambling only, and the result is shown in Figure 4.12d.

4.3.3 Differential Analysis

A desirable property for the homomorphic cryptosystem is the high sensitivity to small changes in the plainimage (single-bit change in the plainimage) (i.e., the diffusion).

A test was performed on the 1-pixel change influence on the 256 gray-level Lena image of size 512×512, and the results are shown in Table 4.6.

With respect to the NPCR and UACI estimation results in Table 4.6, the RC6 and chaotic Baker map scrambling encryption schemes had no sensitivity to small changes in the plainimage, but the homomorphic cryptosystem using both schemes was highly sensitive

Table 4.6   NPCR and UACI Results

ALGORITHMS	NPCR	UACI
Homomorphic with RC6	99.605%	16.775%
RC6	0.0061%	0.0005%
Homomorphic with chaotic Baker map scrambling	100%	7.2314%
Chaotic Baker map scrambling	≅0%	≅0%

to small changes in the plainimage. Generally, these obtained results showed that the homomorphic cryptosystem had a very powerful diffusion mechanism.

4.4 Effect of Noise

The test for effect of noise has been performed by adding additive white Gaussian noise (AWGN) to the encrypted image prior to decryption. Test results showed that the RC6 and the homomorphic algorithm with RC6 were significantly affected by the noise in the decryption process. In other words, these algorithms can only be used in error-free scenarios. The experiment results are shown in Figures 4.13a and 4.13b.

For the chaotic Baker map encryption and the homomorphic encryption with the chaotic Baker map, the results showed that these algorithms were more robust to noise and could work in noisy environments. The experimental results are shown in Figures 4.13c and 4.13d. It is clear that the chaotic Baker map decryption process was more robust to noise than the RC6 decryption process, and this appears in Figure 4.14, which shows the variation of the peak signal-to-noise ratio (PSNR) of the decrypted image with the signal-to-noise ratio (SNR) of the encrypted image for all algorithms. This is attributed to the fact that the RC6 algorithm has a diffusion mechanism in its equation $f(x) = x(2x + 1)(\bmod 2^w)$, which leads to less noise immunity.

4.5 Summary

From these results, we can see that

1. In general, the encryption in the homomorphic domain can be considered a very powerful diffusion mechanism, but this

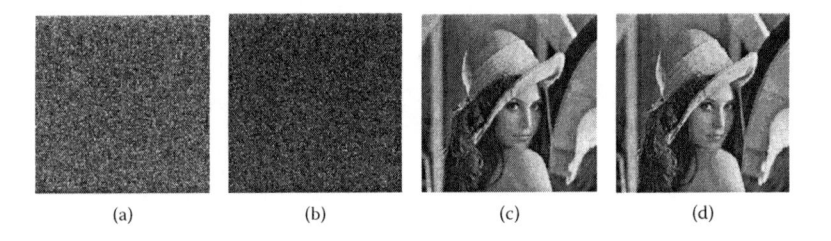

(a) (b) (c) (d)

Figure 4.13 Decrypted images for all encryption algorithms in the presence of noise with an SNR of 50 dB: (a) the RC6 algorithm; (b) the homomorphic cryptosystem with the RC6 algorithm; (c) the chaotic Baker map algorithm; and (d) the homomorphic cryptosystem with the chaotic Baker map scrambling algorithm.

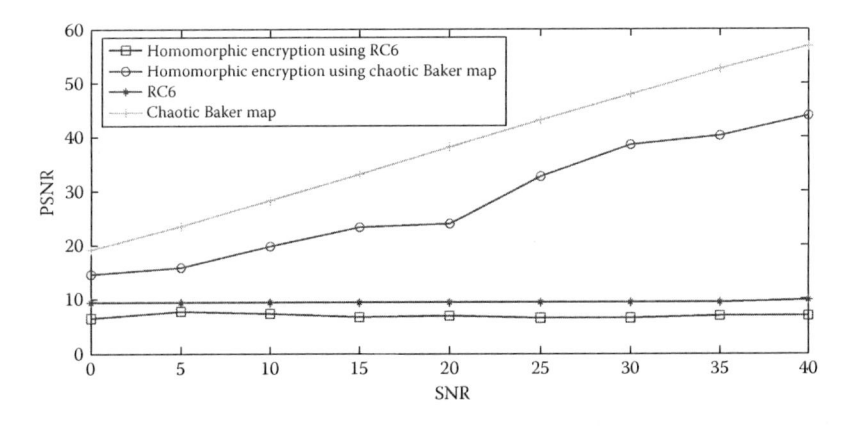

Figure 4.14 The variation of the PSNR of the decrypted image with the SNR of the encrypted image for all encryption algorithms.

comes at the expense of losses in the decrypted image due to the LSB watermark.

2. All other parameters, like encryption quality, noise immunity, and key sensitivity, depend mainly on the encryption algorithm. For the diffusion ciphers, like RC6, the encryption quality and the key sensitivity are better than those of the permutation ciphers, like chaotic Baker map. But, if the transmission medium is noisy, the permutation ciphers are preferred.

5

CHAOTIC IMAGE ENCRYPTION WITH DIFFERENT MODES OF OPERATION

5.1 Overview

Chaos theory consistently plays an active role in modern cryptography. The attractiveness of using chaos as the basis for developing a cryptosystem is mainly its random behavior and sensitivity to initial conditions and parameter settings that fulfill the classic Shannon requirements of confusion and diffusion [89]. Chaos-based algorithms have shown some exceptionally good properties in many concerned aspects regarding security, complexity, speed, computational overhead, and so on. Some chaotic cryptosystems based on ergodicity have been proposed [90,91]. A number of chaos-based image [91,92] and random number generation algorithms [67,93–95] based on discrete chaos have been proposed, but security is generally not high enough [96,97].

This chapter discusses implementing chaotic Baker map scrambling of image pixels using three different modes of operation: CBC, CFB, and OFB. This implementation depends on the block size S, where $S = N \times N$ pixels, and an IV that works as the main key. The bits of the IV must be random and uncorrelated as much as possible to yield a powerful encryption mechanism. In the chapter experiments, the image encrypted is Lena, which is a 512 × 512 grayscale image, and parts of an encrypted version of the Cameraman image are used in the IV (Figure 5.1).

5.2 Chaotic Encryption with Modes of Operation

The main objective of chaotic encryption with different modes of operation is to increase the security of chaotic encryption with moderate computation time and adjust it to encrypt images with arbitrary

Figure 5.1   Encrypted version of Cameraman image.

dimensions. We investigated chaotic encryption with three modes of operation: CBC, CFB, and OFB [16–18]. The three modes were tested to decide which one would increase the data-hiding ability of the cryptosystem. Chaotic encryption with modes of operation can be simply summarized in the following three steps:

1. Scan the image row by row.
2. Convert the scanned rows to w blocks, each with $N \times N$ pixels.
3. Encrypt these blocks using the 2D chaotic Baker map in the CBC, CFB, or OFB mode as explained in the following sections.

The block diagram of chaotic encryption with modes of operation is shown in Figure 5.2.

5.3 Implementation Issues

As stated in the previous section, chaotic encryption can be implemented in three different modes of operation. It is based on the 2D chaotic Baker map as the main encryption algorithm. It is known that the permutations induced by the Baker map behave as typical random permutations. An IV is used as the main key. This IV must be random to resist the brute-force attack. The XOR operations between the bits of the IV and the bits of the data blocks change the values of the pixels, which makes the encryption algorithm behave like a 3D Baker map. The algorithm also uses a secondary key, which is the key used by the Baker map to scramble the pixels.

Chaotic encryption with modes of operation is based on the segmentation of the image to be encrypted into blocks. The block size is

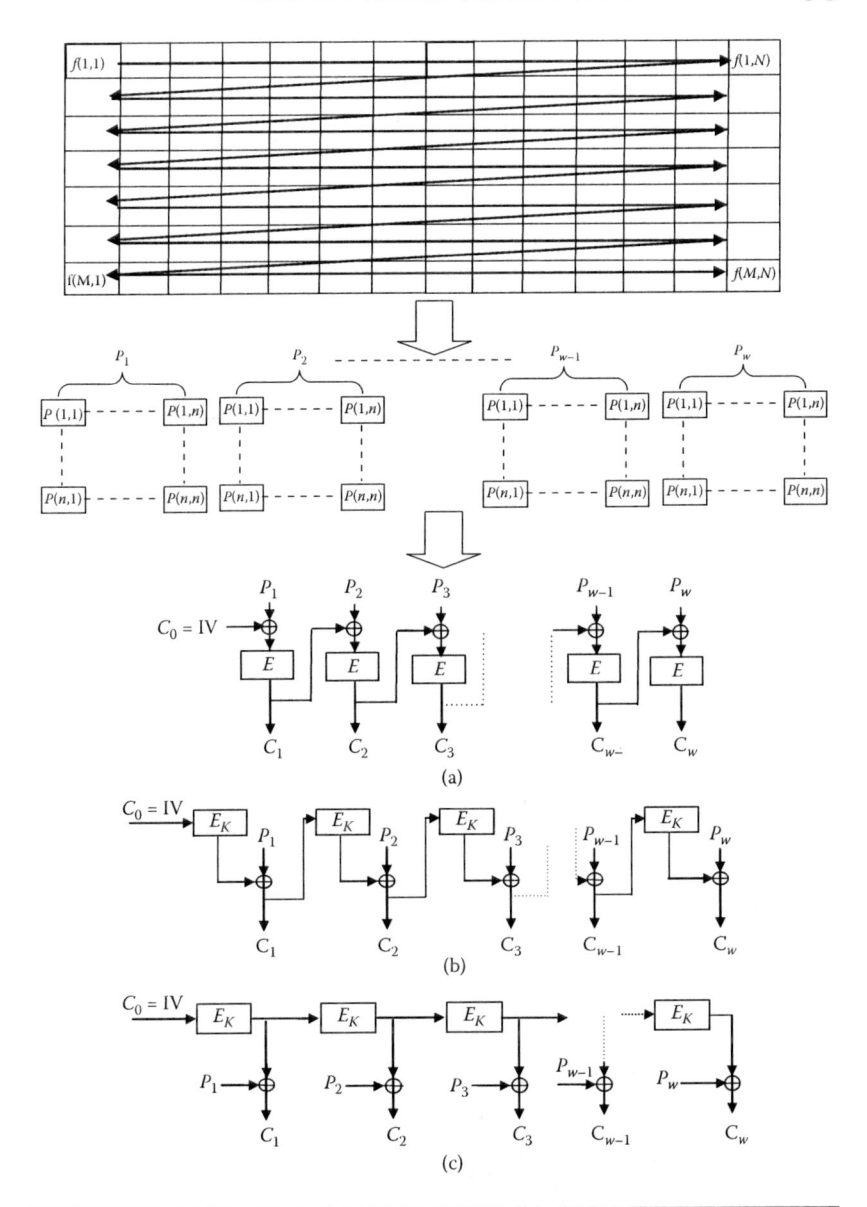

Figure 5.2 Chaotic encryption implemented in (a) CBC mode, (b) CFB mode, (c) OFB mode.

an important factor that affects the performance of the cryptosystem. The effect of the block size on the encryption quality is studied in detail in the simulation examples section. Because the IV has similar size to the plaintext blocks, the increase in the block size increases the security. In addition, the chaotic encryption algorithm with modes of operation will be able to encrypt images with arbitrary dimensions after segmentation into small blocks.

5.4 Simulation Examples and Discussion

Simulation experiments have been carried out to encrypt the Lena image with the chaotic encryption system using the different modes of operation and comparing the obtained results with the results of the 2D chaotic Baker map encryption algorithm and the RC6 algorithm. The IV has been taken as part of the encrypted Cameraman image shown in Figure 5.1, equal in size to the selected block size. Several block sizes have been tested:

1. S_1 = 128 × 128 pixels. The IV is a 128 × 128 pixel section of the encrypted Cameraman image.
2. S_2 = 64 × 64 pixels.
3. S_3 = 32 × 32 pixels.
4. S_4 = 16 × 16 pixels.
5. S_5 = 8 × 8 pixels.

The encrypted versions of the Lena image are shown using the chaotic Baker map cryptosystem and the RC6 cryptosystem (Figure 5.3) and the chaotic encryption with different modes of operation and different block sizes (Figure 5.4). It is clear from Figure 5.4 that the performance of the chaotic encryption with different modes of operation is good except for its implementation in the OFB mode with large block sizes.

To check the noise immunity of each of the cryptosystems mentioned in this chapter, additive white Gaussian noise (AWGN) with

Figure 5.3   Encrypted images using traditional cryptosystems: (a) encrypted image using the chaotic Baker map cryptosystem; (b) encrypted image using the RC6 cryptosystem.

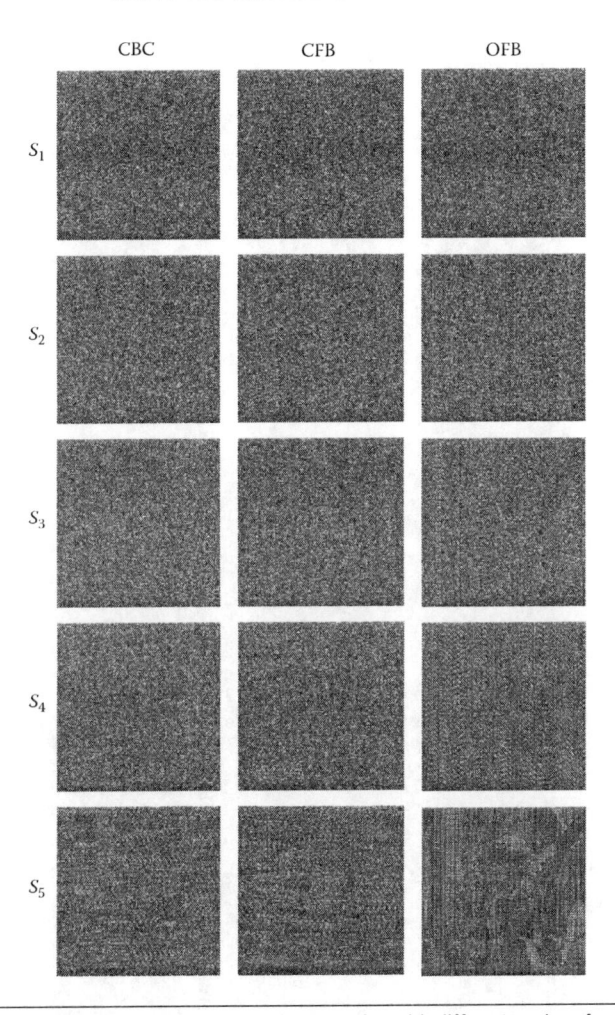

Figure 5.4 Encrypted images using chaotic encryption with different modes of operation and different block sizes.

a signal-to-noise ratio (SNR) of 5 dB was added to the encrypted images, and then the decryption was performed. The decryption results of the chaotic Baker map cryptosystem and the RC6 cryptosystem (Figure 5.5) and the chaotic encryption with different modes of operation (Figure 5.6) are shown. From visual inspection, it is clear that the chaotic Baker map cryptosystem has the highest noise immunity. It is also clear that the chaotic encryption with different modes of operation is more immune to noise than the RC6 cryptosystem.

Table 5.1 shows the values of the evaluation metrics for all cryptosystems mentioned in this chapter; r_{xye} represents the correlation

Figure 5.5 Decrypted images using the traditional cryptosystems at SNR = 5 dB: (a) decrypted image using the chaotic Baker map cryptosystem; (b) decrypted image using the RC6 cryptosystem.

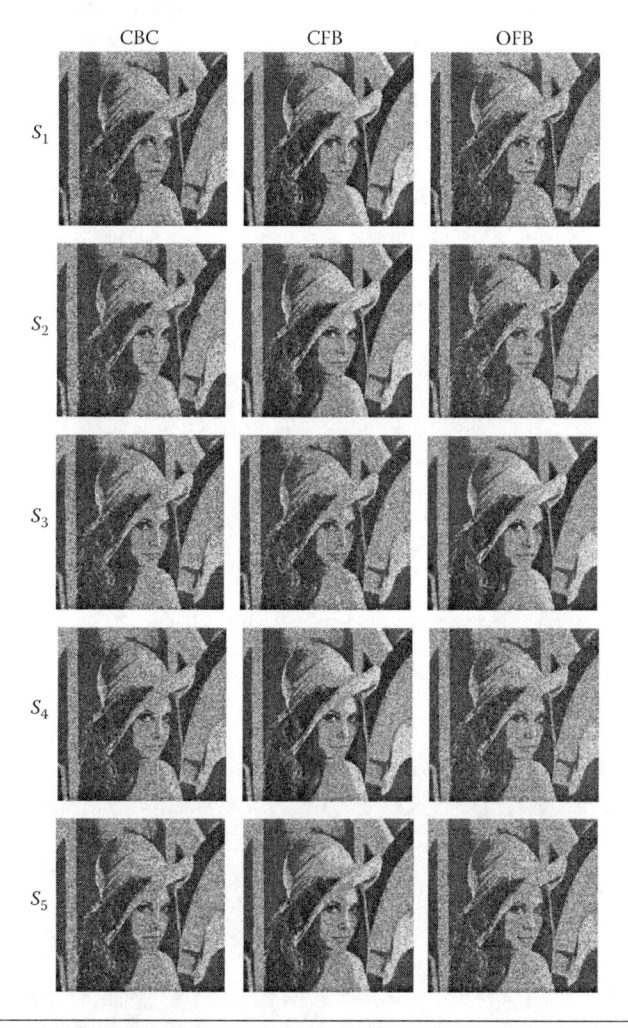

Figure 5.6 Decrypted images using the chaotic algorithm with different modes of operation and different block sizes at SNR = 5 dB.

Table 5.1 Numerical Evaluation Metrics for All Cryptosystems

| | CHAOTIC ENCRYPTION WITH MODES OF OPERATION | | | | | | | | | | | | | | | | |
| | CBC | | | | | CFB | | | | | OFB | | | | | | |
TEST	S_1	S_2	S_3	S_4	S_5	S_1	S_2	S_3	S_4	S_5	S_1	S_2	S_3	S_4	S_5	ENCRYPTION	RC6
D	0.71	0.72	0.71	0.71	0.72	0.71	0.71	0.71	0.72	0.71	0.71	0.71	0.7	0.67	0.66	0	0.71
r_{xye}	1.4×10^{-3}	4.7×10^{-6}	1.3×10^{-3}	1.7×10^{-3}	-2×10^{-3}	3.7×10^{-3}	1.5×10^{-3}	2×10^{-3}	-6.5×10^{-3}	7.4×10^{-4}	-7.1×10^{-3}	2.5×10^{-4}	2.3×10^{-3}	13.5×10^{-3}	41.8×10^{-3}	10^{-4}	1.3×10^{-3}
D_l	0.71	0.7	0.71	0.71	0.7	0.71	0.71	0.71	0.7	0.71	0.7	0.71	0.73	0.74	0.95	0.98	0.71
r_{xyd}	0.57	0.57	0.57	0.57	0.57	0.57	0.57	0.57	0.57	0.56	0.7	0.7	0.7	0.7	0.69	0.87	2.5×10^{-3}

Note: Decryption was performed in the presence of an AWGN with SNR = 5 dB.

coefficient between the original and encrypted image, and r_{xyd} represents the correlation coefficient between the original and decrypted image affected by 5-dB AWGN. From the results obtained and shown in this table, it is clear that the chaotic Baker map cryptosystem is the most immune to noise, but it is less secure than the RC6 cryptosystem and the chaotic encryption with modes of operation. It is also clear that the RC6 cryptosystem is very sensitive to the presence of noise. We can conclude that the chaotic encryption with different modes of operation achieves the trade-off between the level of security and the noise immunity.

The effect of noise on the chaotic encryption with different modes of operation and block sizes was also studied, and the results are given in Figure 5.7. This figure shows the variation of the peak signal-to-noise ratio (PSNR) of the decrypted image with the SNR of the encrypted image. It is clear from this figure that all modes of operation have approximately the same performance in the presence of noise, and the performance is improved at high SNR values. It is also clear that the block size has no effect on the noise immunity of the algorithm in all implementation modes.

The histograms of the original Lena image and the encrypted images with all cryptosystems mentioned in this chapter are shown in Figures 5.8 and 5.9, respectively. It is clear that the histogram uniformity is not achieved with the chaotic Baker map cryptosystem, which is a weakness of this cryptosystem. The RC6 cryptosystem achieves histogram uniformity. The chaotic encryption with different modes of operation also achieves histogram uniformity for all implementations except the implementations in the OFB mode with large block sizes.

It is known that the RC6 cryptosystem has long processing time as compared to the chaotic Baker map cryptosystem. As a result, our study for the processing time was restricted to the comparison between the processing time of the chaotic encryption with different modes of operation and the chaotic Baker map cryptosystem. Figure 5.10 shows the variation of the normalized processing time of the chaotic encryption algorithm implemented in the different modes of operation with the block length. Normalization was performed by dividing the processing time of the chaotic encryption algorithm with different modes of operation by the processing time of the chaotic

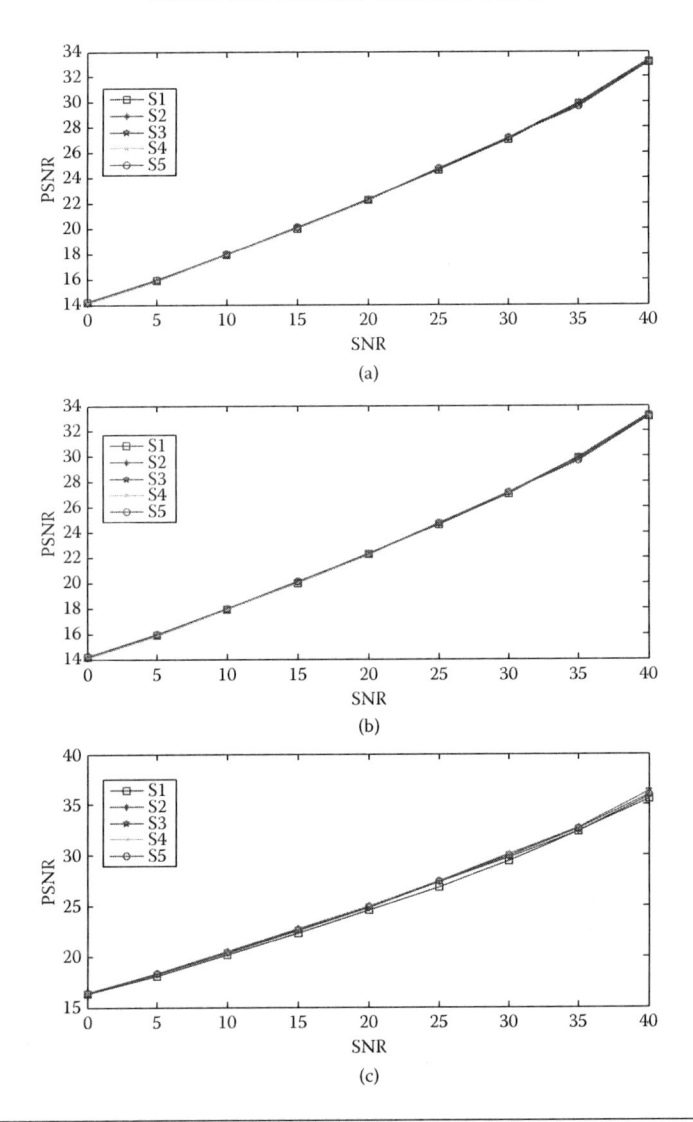

Figure 5.7 Variation of the PSNR of the decrypted image with the SNR of the encrypted image in the chaotic encryption with different modes of operation for (a) the CBC mode, (b) the CFB mode, (c) the OFB mode.

Baker map cryptosystem. The processing time includes the encryption and decryption times. It is clear from Figure 5.10 that the processing times of all implementations are approximately the same for each block length. It is also clear that large block sizes lead to shorter processing times because the encryption and decryption are performed for fewer blocks.

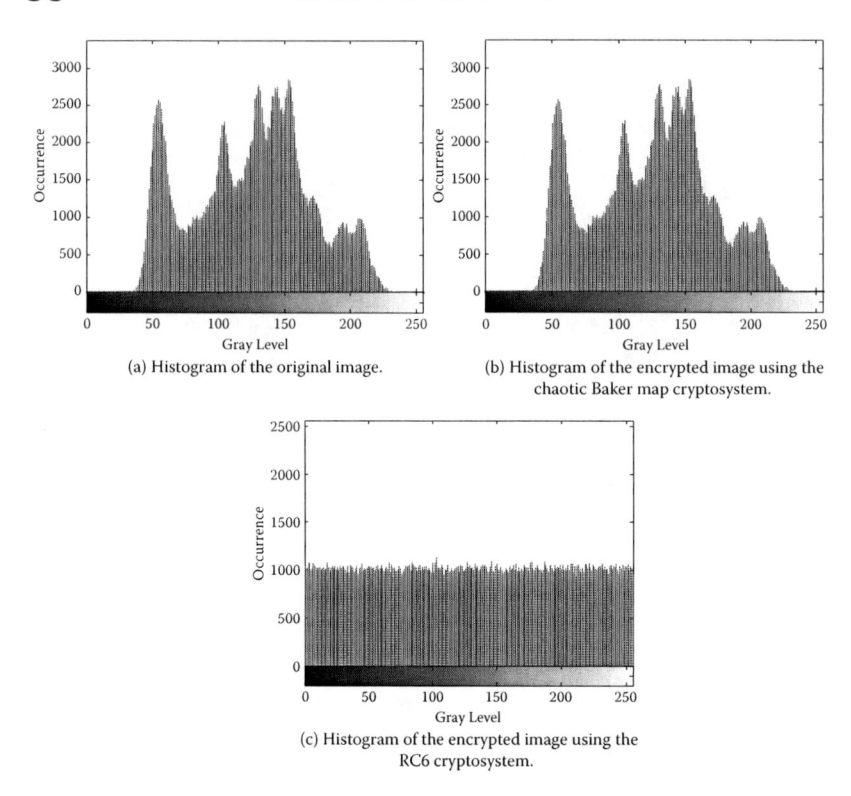

(a) Histogram of the original image.

(b) Histogram of the encrypted image using the chaotic Baker map cryptosystem.

(c) Histogram of the encrypted image using the RC6 cryptosystem.

Figure 5.8 Histograms of the original image and the encrypted images with the traditional cryptosystems.

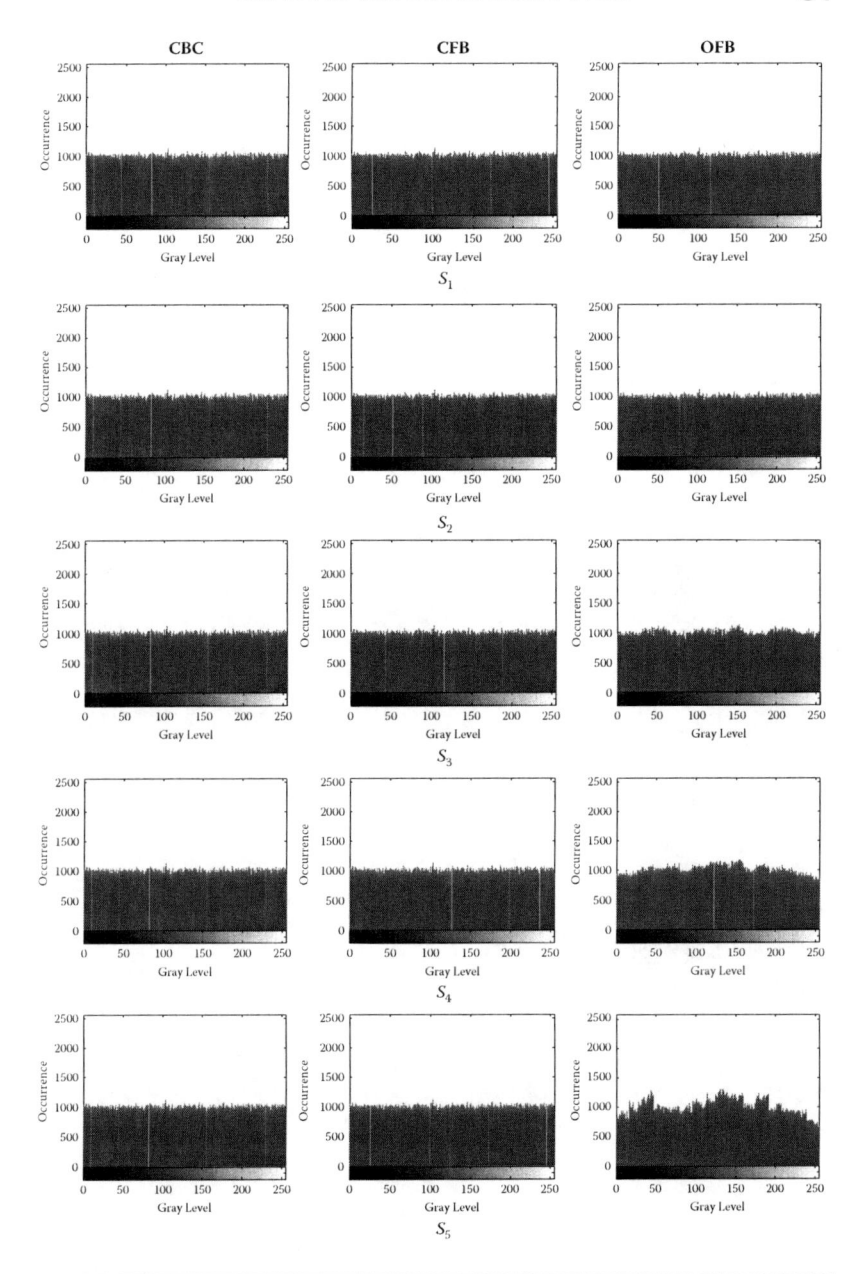

Figure 5.9 Histograms of the encrypted images using chaotic encryption with different modes of operation and different block sizes.

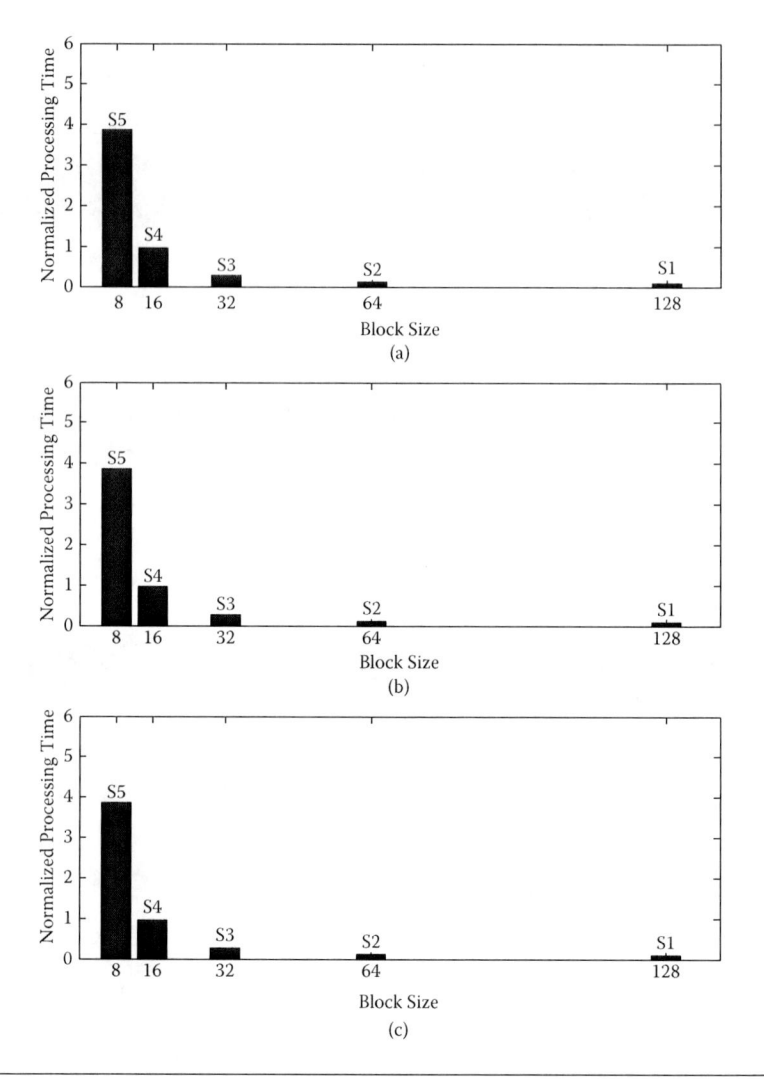

Figure 5.10 The normalized processing time for the chaotic image encryption with different modes and block sizes: (a) CBC mode; (b) CFB mode; (c) OFB mode.

5.5 Summary

1. Chaotic image encryption with different modes of operation can be used to adjust the 2D Baker map to encrypt images with arbitrary dimensions.

2. Chaotic encryption with the CBC mode is the best of all implementations tested in this chapter.

3. Generally, the encryption quality of all modes increases with the decrease of the block size, except for the OFB mode, especially with $S = S_3$, S_4, and S_5.

4. The chaotic Baker map has the worst results in all tests except for the noise immunity, and the RC6 algorithm has intermediate results.

5. The processing time increases with the decrease in the block size.

6

DIFFUSION MECHANISM FOR DATA ENCRYPTION IN THE ECB MODE

6.1 Introduction

The implementation of any block cipher algorithm depends on the mode of operation, which governs the relation between blocks during the encryption process. The ECB mode is one of the possible modes of operation. In this mode, each block is encrypted independently. Unlike different modes of operation, this mode allows parallel processing, which is a great advantage. Unfortunately, because the encryption of each block does not depend on other blocks, an adversary can replace any block with a previously intercepted block without detection; hence, the message is hacked without the need to know the key. This major security problem is called the block independency problem. In addition, identical plaintext blocks are encrypted to identical ciphertext blocks, so symmetrical large data patterns, like those in images, cannot be hidden using any encryption algorithm implemented in the ECB mode [98–102].

Two actions are discussed in this chapter to solve the block independency problem in the ECB mode [103]. The first action is to make the values of the data bytes functions of their positions in the data stream. The second action is to perform a diffusion process between these bytes. These two actions are preprocessing actions that are performed prior to the application of any block cipher algorithm implemented in the ECB mode. First, an addition step, which is used to remove any identical plaintext blocks, is performed. Then, an substitution permutation network (SPN) is used to make the diffusion between the bytes. A schematic diagram of this mechanism is shown in Figure 6.1.

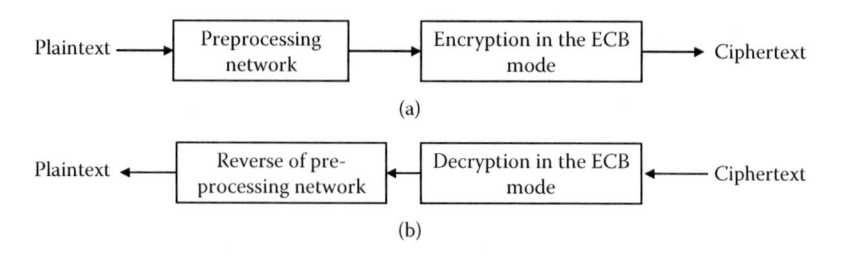

(a)

(b)

Figure 6.1   The suggested mechanism for encryption and decryption in the ECB mode: (a) encryption; (b) decryption.

6.2 The Preprocessing Network

The disadvantages of the ECB mode limit its applicability in modern cryptography despite its advantage of parallel processing [104]. If we can preprocess the data to diffuse bytes together before encryption, these disadvantages will be avoided. The preprocessing network consists of two parts: an addition part and an SPN.

6.2.1 The Addition Part

The objective of the addition process is to make the byte values functions of their positions in the data stream. This process helps in eliminating identical plaintext blocks. A function of the byte position is added to the byte value as follows:

$$B(i) = (B(i) + f(i)) \bmod 256 \qquad (6.1)$$

where i is the position of the byte in the data stream, and B is the byte value.

The function $f(i)$ is used because it satisfies the following conditions:

Randomness. The resulting values from this function for every i are random with equal probability.

Irregularity. The differences between each byte and its neighbors in the data stream after the addition of this function to the byte values are unpredictable and unrepeatable.

Nonperiodicity. This function is nonperiodic because it satisfies the following condition:

$$f(i) \neq f(i + n) \neq f(i + 2n) \neq f(i + 3n) \neq \ldots \qquad (6.2)$$

Here, we test three functions to find the one that satisfies the given conditions:

$$f_1(i) = 7 \times (i + 13) \bmod 256 \tag{6.3}$$

$$f_2(i) = (7 \times (i + 3))^5 \bmod 256 \tag{6.4}$$

$$f_3(i) = \text{fix}\left(\sqrt{7 \times (i+13)^5}\right) \bmod 256 \tag{6.5}$$

where the fix function is used to round the value of its argument toward the lower nearest integer.

The functions have been tested for values from 1 to 1024, and the results are given in Figure 6.2. The probability density functions (PDFs) of the outputs of these functions are shown in Figure 6.3.

From these figures, we can see that $f_3(i)$ gives a random, irregular, and nonperiodic output with a uniform PDF.

6.2.2 The SPN

The SPN is used to diffuse the bytes of the data together after the addition step. First, the data are divided into blocks of n bits. This network block size n is different from the encryption algorithm block size w. After that, a chain of XOR operations is performed as shown in Figure 6.4. The subkey K_1 works as an initial key for the first XOR, and then the result is XORed with the next block up to the end of the plaintext. Each block resulting from this chain of XORs is permuted bit-by-bit as shown in Figure 6.5. A block-based permutation is performed after that. Finally, another chain of XORs is implemented beginning with a subkey K_2.

6.3 Implementation Issues

The preprocessing network performance depends mainly on the correct determination of the SPN block size n and its relation to the encryption algorithm block size w. The effect of the normalized block size (n/w) on the diffusion and block dependency is studied.

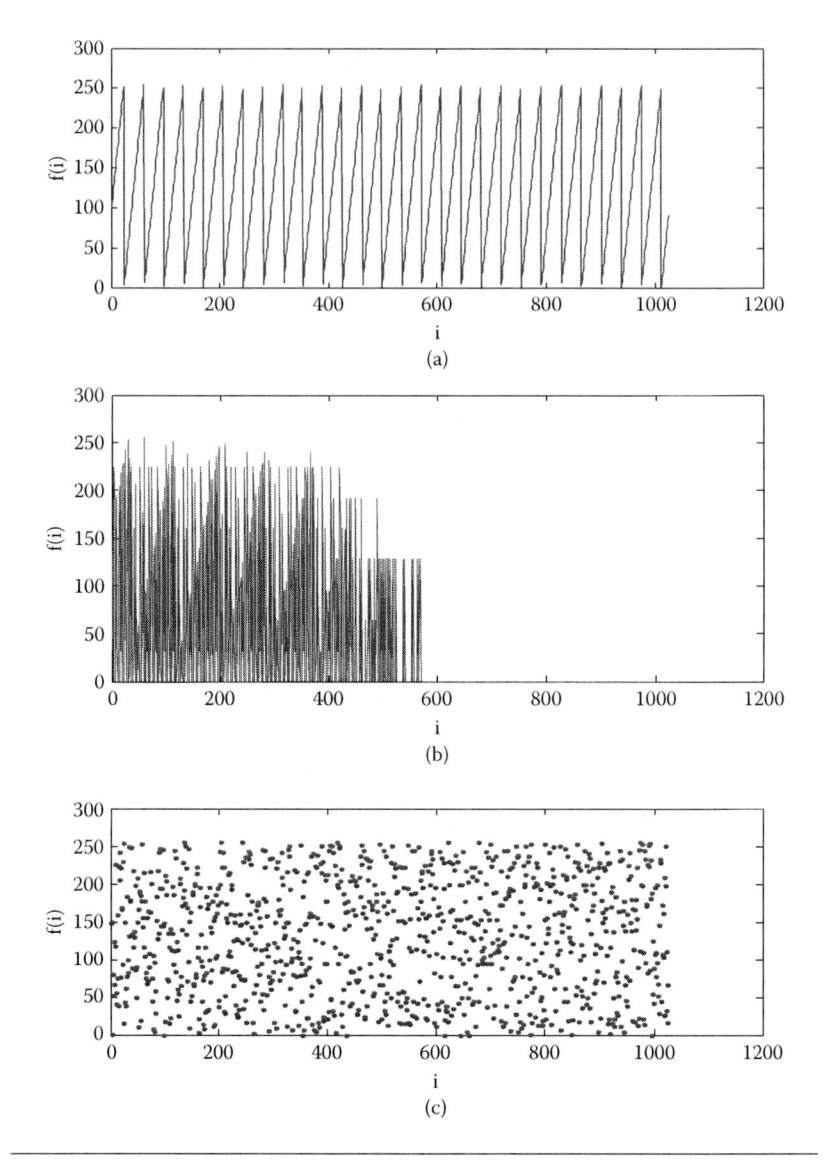

Figure 6.2 Variation of (a) $f_1(i)$, (b) $f_2(i)$, (c) $f_3(i)$ with i.

6.3.1 Effect of the Normalized Block Size on Diffusion

Diffusion is simply hiding the relation between the plaintext and the ciphertext. If a small change in the plaintext (one bit) makes a large change in the ciphertext (half of its bits), then the Avalanche effect is evident, and the algorithm has a powerful diffusion mechanism [105–110]. This is guaranteed if we can ensure that a bit change in

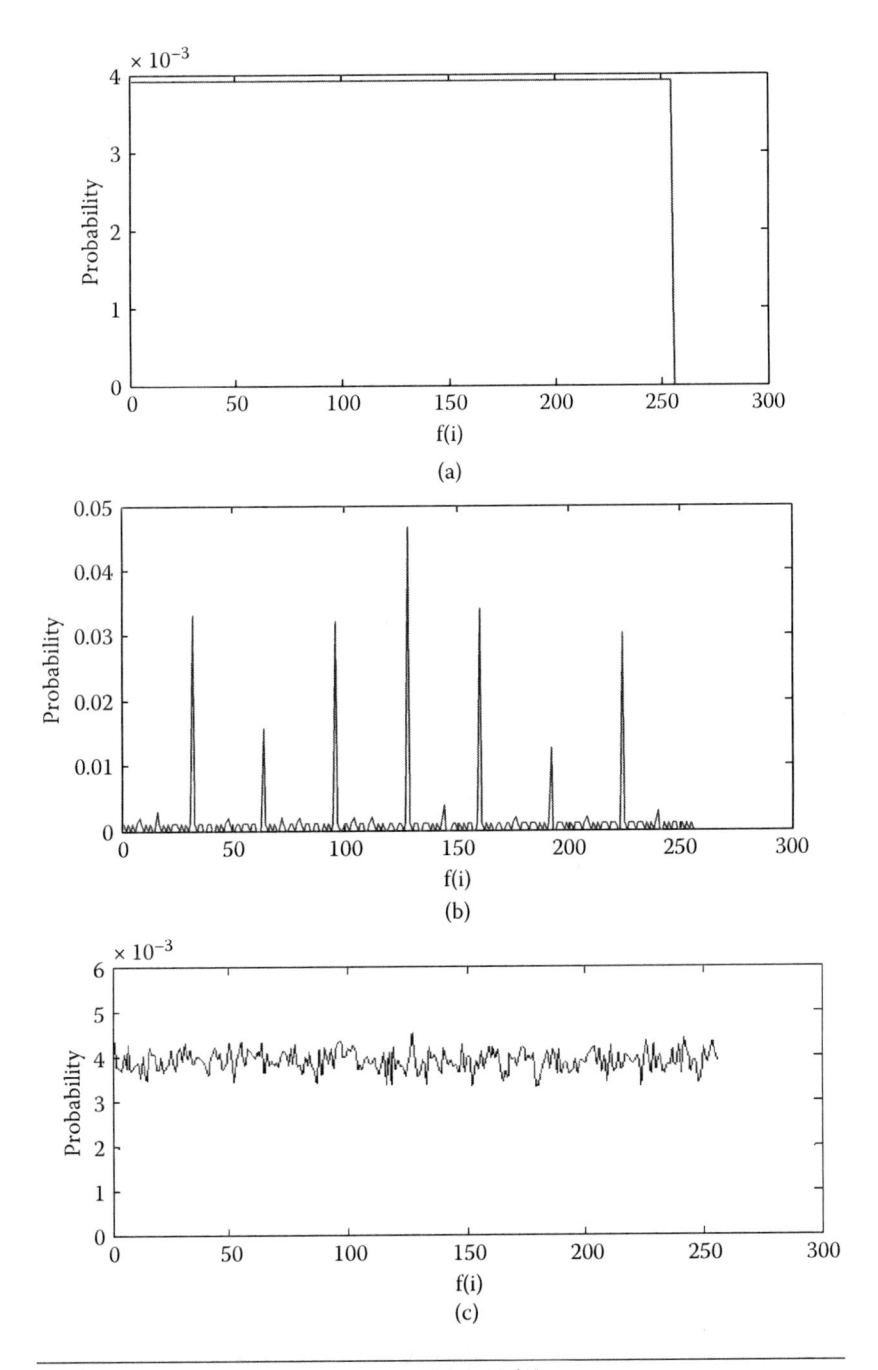

Figure 6.3   PDFs of the outputs of (a) $f_1(i)$, (b) $f_2(i)$, (c) $f_3(i)$.

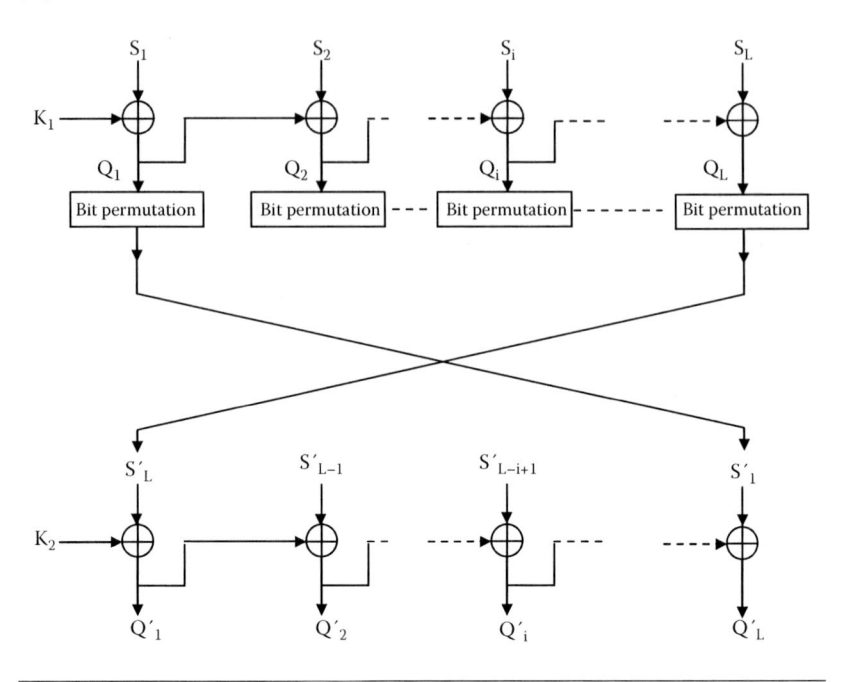

Figure 6.4 Operation of the SPN.

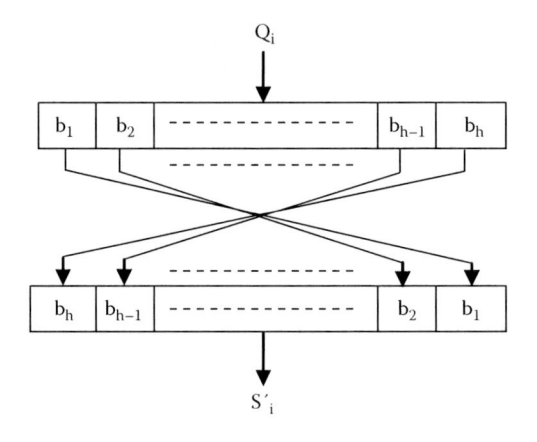

Figure 6.5 The bit permutation in the SPN.

the plaintext will affect at least one bit in each block of the input to the encryption algorithm, and the diffusion mechanism in the encryption algorithm will propagate this change through each block.

From Figure 6.6, we can see that if a change takes place in an SPN block (S_i), all the network blocks from Q'_{L-i+1} to Q'_L will be affected, where L is the number of network blocks. The output odd-number

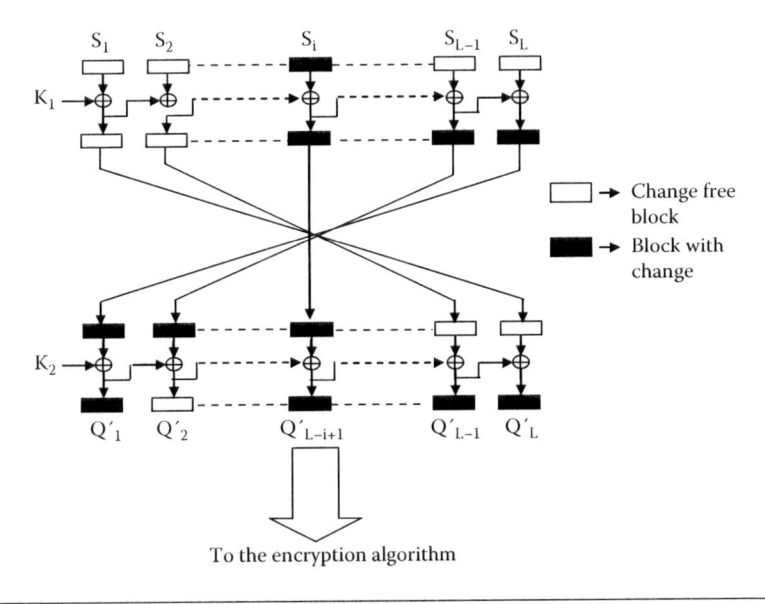

Figure 6.6   Diffusion effect of the SPN on the data blocks before encryption.

blocks from Q'_1 to Q'_i will also be affected. The even-number blocks up to Q'_i will not be affected due to performing XORs between the same patterns.

To ensure the propagation of changes in each block of the encryption algorithm, a proper choice of the SPN block size n is half the encryption algorithm block size w, $(n/w = 0.5)$. Because w in the cases of the Advanced Encryption Standard (AES) and the RC6 algorithms is 128 bits, the network block size used should be 64 bits.

This network has the ability to propagate the changes made in the addition part and diffuse the bytes of the data together.

6.3.2 Effect of the Normalized Block Size on Block Dependency

The block independency problem gives the opportunity for a hacker to exchange any cipher block without detection. The preprocessing network solves this problem by making this exchange affect other blocks besides the replaced block in the decryption process.

On the other hand, with the SPN, if a block is received in error, this error will propagate through other blocks during the decryption process, and these blocks will be decrypted incorrectly even if they

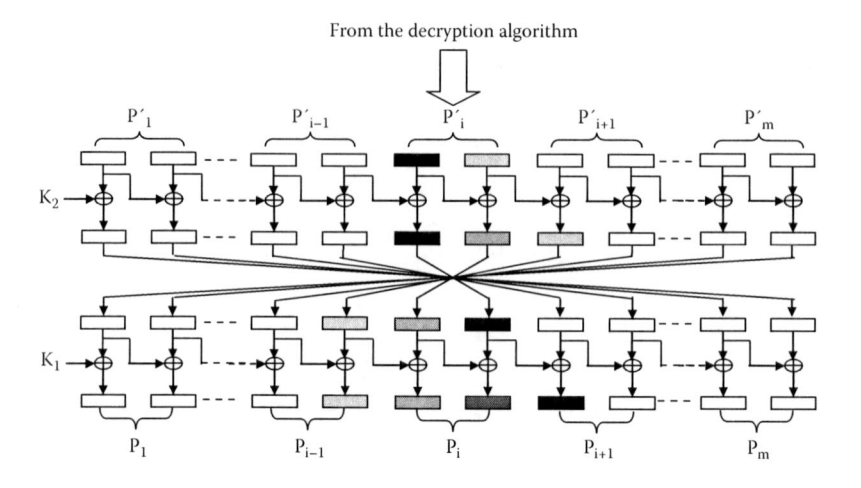

Figure 6.7   The error propagation due to the inversion of the SPN effect in the decryption process.

were received correctly. So, there is a need for a trade-off between the block dependency and the error propagation.

If an encrypted block was received in error, the application of the decryption algorithm on this block will propagate the error in two neighboring network blocks in different manners. This is attributed to the randomization of the diffusion mechanism in the decryption algorithm. If an error occurs in an intermediate encrypted block, the error spreads over three decrypted blocks. If this error occurs in the first or the last encrypted block, the error spreads over two decrypted blocks only. This is shown in Figure 6.7, where the shaded blocks refer to network blocks with different errors, and white blocks refer to the absence of errors. Every decrypted block of size w is represented by two network blocks. From this figure, we can see that the block dependency is ensured, but the error propagation slightly increases compared to the CBC mode.

6.4 Simulation Examples

Several experiments have been performed to test the effect of the suggested mechanism on image encryption in the ECB mode. Both the AES and RC6 algorithms have been used as the encryption algorithms. The Cameraman image shown in Figure 6.8 has been used in these experiments.

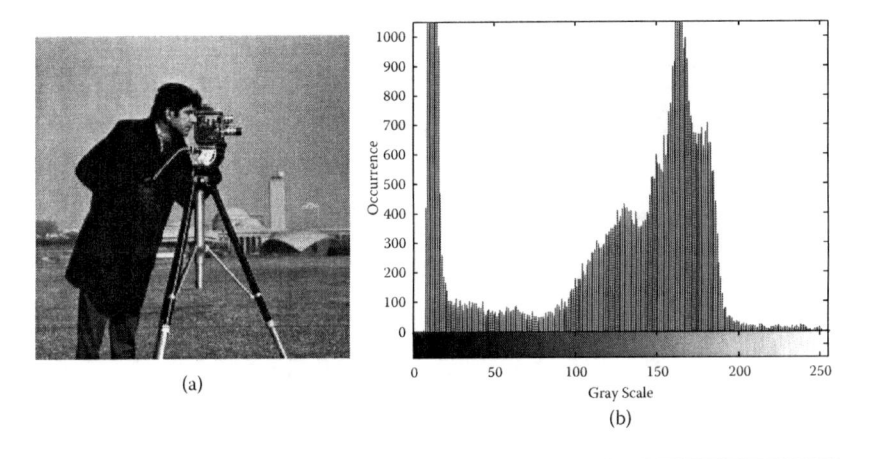

Figure 6.8   (a) The Cameraman image; (b) the histogram.

6.4.1 Encryption Quality

One of the important factors in examining the encrypted image is visual inspection. The Cameraman image has been encrypted with the AES and the RC6 algorithms in the ECB, CBC, CFB, and OFB modes and the ECB mode with preprocessing by the proposed network for comparison purposes. The results of these experiments are shown in Figures 6.9 and 6.10, respectively. From these obtained results, it is clear that the encryption in the ECB mode failed to hide the details at the upper slice of the Cameraman image because this slice is flat in intensity. The encryption in the ECB mode with the proposed network solves the problem.

The histograms of all images in Figures 6.9 and 6.10 are shown in Figures 6.11 and 6.12, respectively. The histogram uniformity in all figures ensures the success of all encryption algorithms to achieve the required randomness. The values of the encryption quality metrics D_I and D for the encrypted images in Figures 6.9 and 6.10 are tabulated in Table 6.1. These results show that all modes achieve the randomness of the data.

6.4.2 Diffusion

The diffusion metrics NPCR, UACI, and the Avalanche effect metric have been evaluated for three cases: changing a single bit in the first pixel, changing a single bit in the midpixel, and changing a single

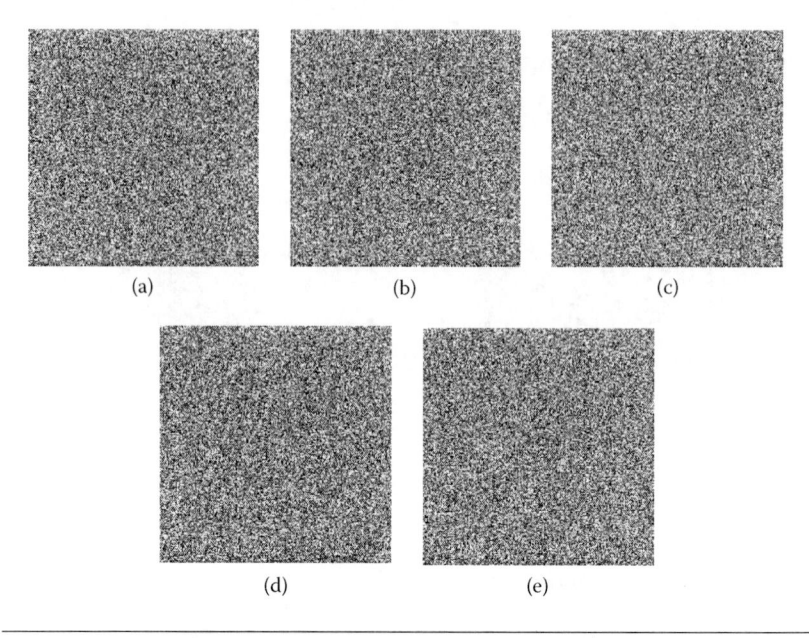

Figure 6.9 The Cameraman image encrypted with the AES in (a) ECB mode with preprocessing, (b) ECB mode, (c) CBC mode, (d) CFB mode, (e) OFB mode.

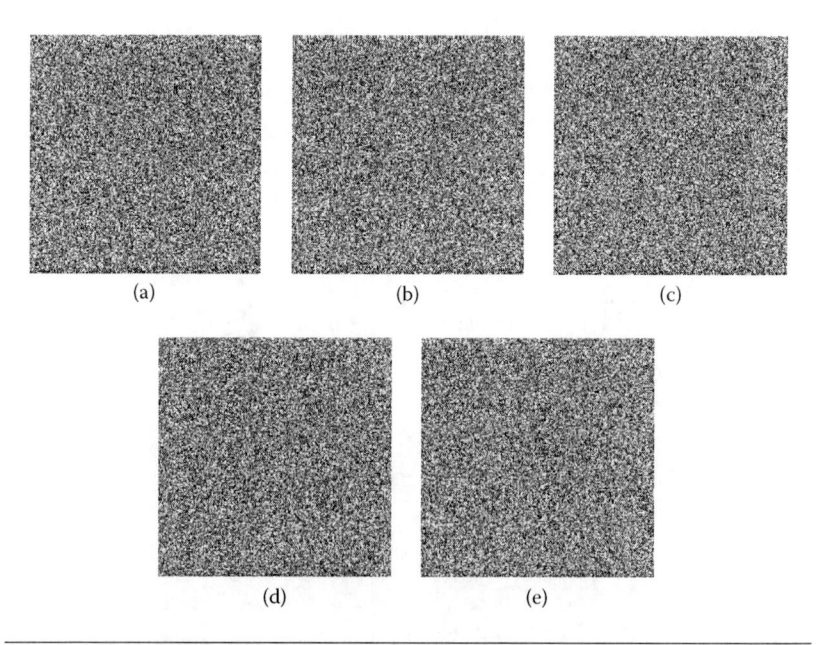

Figure 6.10 The Cameraman image encrypted with the RC6 algorithm in (a) ECB mode with preprocessing, (b) ECB mode, (c) CBC mode, (d) CFB mode, (e) OFB mode.

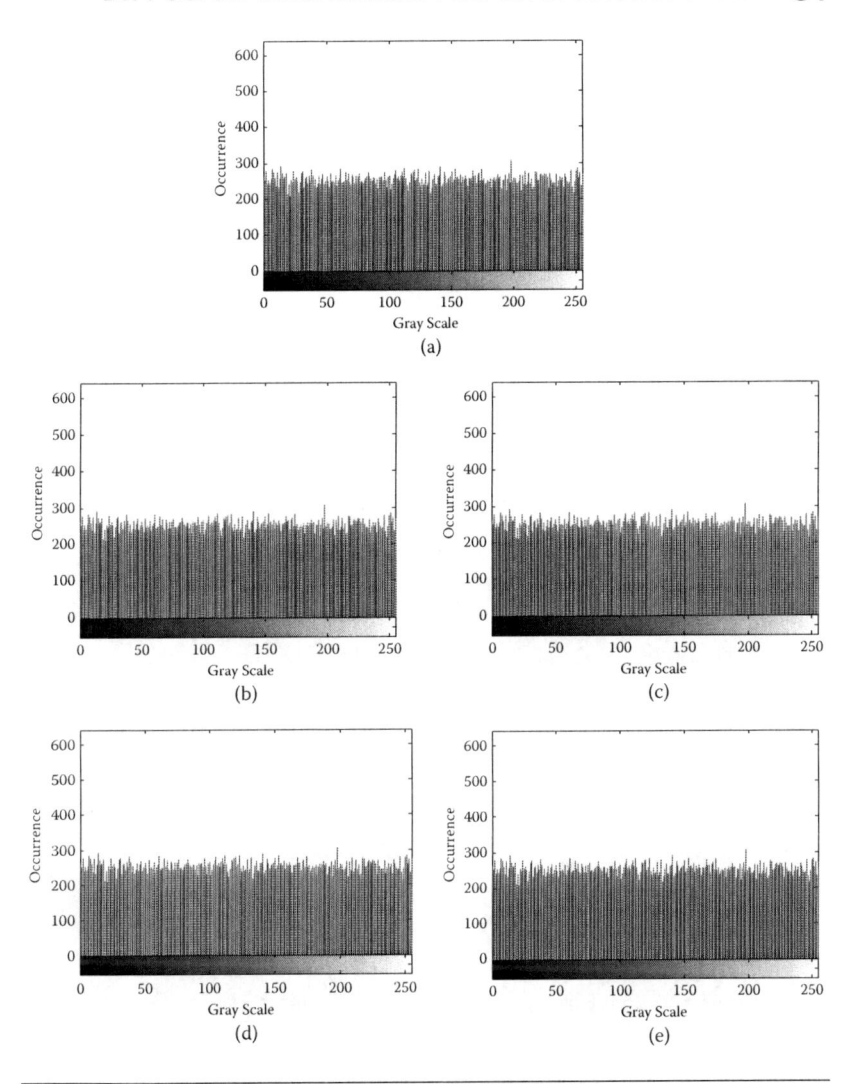

Figure 6.11   The histograms of the encrypted images in Figure 6.9.

bit in the last pixel. The results of the diffusion tests are shown in Table 6.2. From this table, we can see that the diffusion between the blocks of data depends mainly on the mode of operation, not on the encryption algorithm. The ECB and OFB modes do not possess any diffusion characteristics. The ECB mode with the preprocessing network gives the best results, unlike the CBC and CFB modes, which has low values in the Avalanche effect metric, with the changes occurring close to the end of the data.

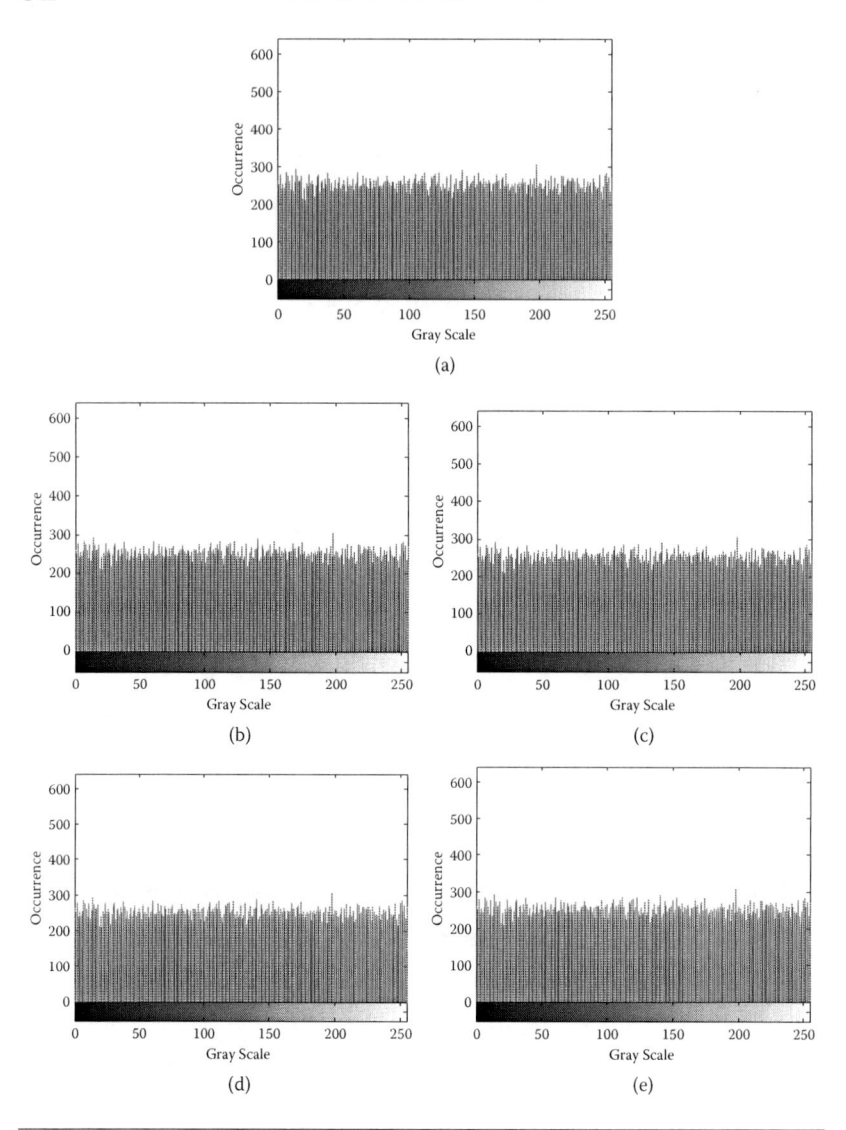

Figure 6.12   The histograms of the encrypted images in Figure 6.10.

6.4.3 Encryption of Images with Few Details

An image with few details has a large amount of adjacent pixels with similar values. Examples of these images are the medical image in Figure 6.13 and the logo image in Figure 6.14.

The encrypted versions of the medical image and the logo image with both the AES and the RC6 algorithm in the ECB mode with and without the preprocessing network are shown in Figures 6.15

Table 6.1 Encryption Quality Results

METRIC	RC6					AES				
	CBC	CFB	OFB	ECB	ECB WITH PREPROCESSING	CBC	CFB	OFB	ECB	ECB WITH PREPROCESSING
D_I	0.602	0.596	0.597	0.597	0.598	0.601	0.603	0.603	0.606	0.604
D	0.0513	0.0486	0.0503	0.0452	0.046	0.0485	0.0458	0.0472	0.052	0.048

Table 6.2 Diffusion Test Results

TEST	THE CHANGED BIT	RC6					AES				
		CBC (%)	CFB (%)	OFB (%)	ECB (%)	ECB WITH PREPROCESSING (%)	CBC (%)	CFB (%)	OFB (%)	ECB (%)	ECB WITH PREPROCESSING (%)
NPCR	Bit in the first pixel	99.61	99.58	$\cong 0$	0.02	99.6	99.65	99.6	$\cong 0$	0.02	99.6
	Bit in the middle pixel	50.03	50	$\cong 0$	0.02	99.65	50.01	50.03	$\cong 0$	0.02	99.56
	Bit in the last pixel	0.02	$\cong 0$	$\cong 0$	0.02	99.6	0.024	$\cong 0$	$\cong 0$	0.02	99.58
UACI	Bit in the first pixel	16.61	16.7	$\cong 0$	0.01	16.66	16.76	16.62	$\cong 0$	0.005	16.88
	Bit in the middle pixel	8.4	8.4	$\cong 0$	0.01	16.79	8.37	8.3	$\cong 0$	0.005	16.95
	Bit in the last pixel	0.005	$\cong 0$	$\cong 0$	0.01	16.8	0.005	$\cong 0$	$\cong 0$	0.003	16.92
Avalanche effect metric	Bit in the first pixel	50.07	49.96	$\cong 0$	0.01	50.03	50.08	49.99	$\cong 0$	0.011	50.03
	Bit in the middle pixel	25.11	25.04	$\cong 0$	0.01	50.11	25.09	25.15	$\cong 0$	0.012	50.08
	Bit in the last pixel	0.01	$\cong 0$	$\cong 0$	0.01	50.04	0.014	$\cong 0$	$\cong 0$	0.011	50.08

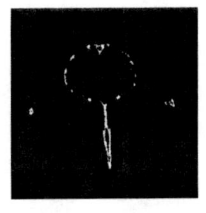

Figure 6.13   Medical image.

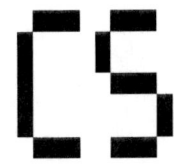

Figure 6.14   Logo image.

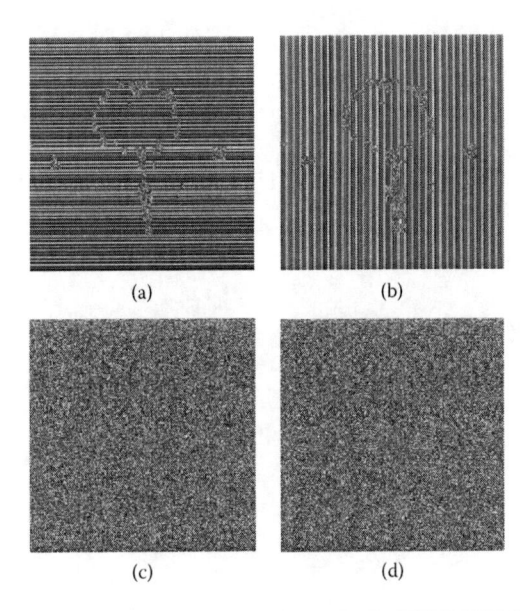

(a)   (b)

(c)   (d)

Figure 6.15   The medical image encrypted with (a) the RC6 algorithm in the ECB mode, (b) the AES in the ECB mode, (c) the RC6 algorithm in the ECB mode with preprocessing, (d) the AES in the ECB mode with preprocessing.

and 6.16, respectively. The histograms of these images are shown in Figures 6.17 and 6.18, respectively. From the obtained results, we notice that the encryption in the ECB mode fails with these images. We notice also that the preprocessing network enhances dramatically the encryption quality of these encrypted images with the ECB mode.

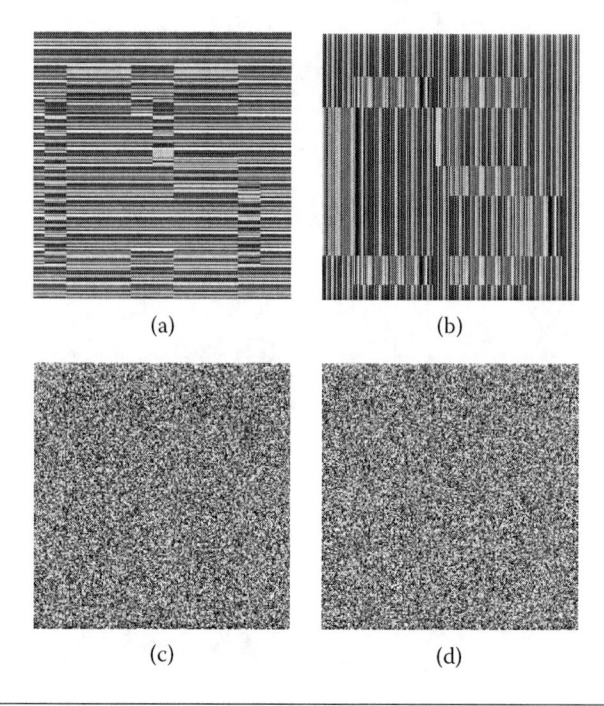

(a)
(b)
(c)
(d)

Figure 6.16 The logo image encrypted with (a) the RC6 algorithm in the ECB mode, (b) the AES in the ECB mode, (c) the RC6 algorithm in the ECB mode with preprocessing, (d) the AES in the ECB mode with preprocessing.

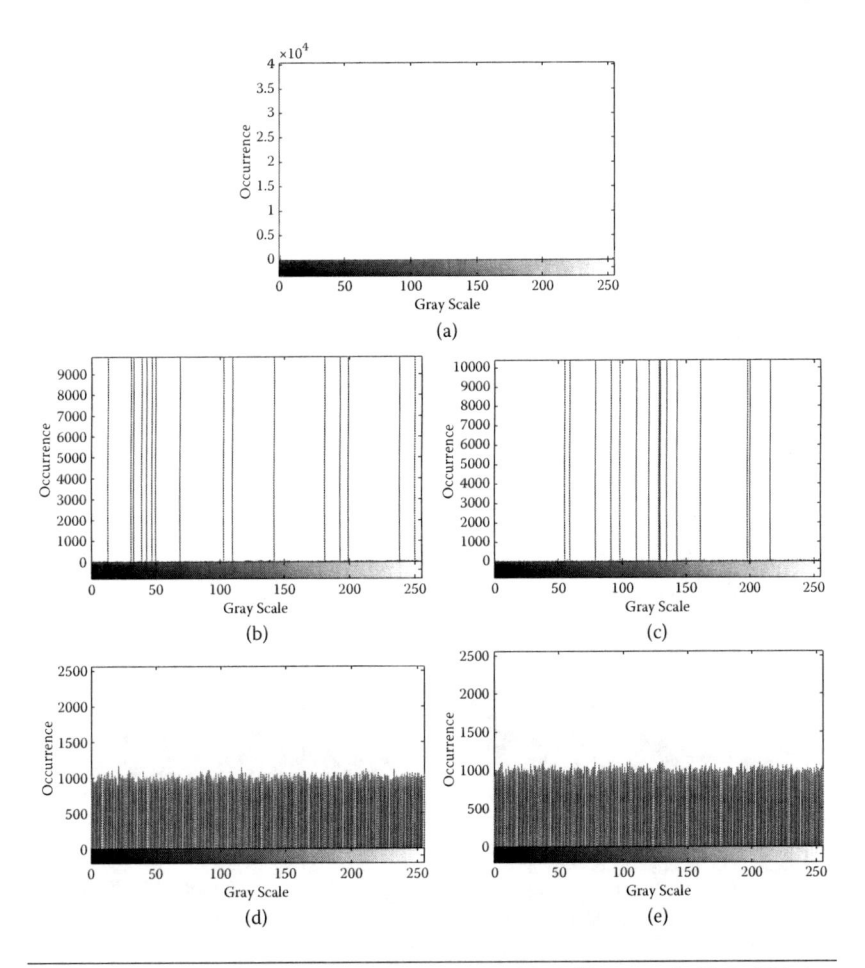

Figure 6.17   The histograms of the medical image: (a) original, (b) encrypted with the RC6 algorithm, (c) encrypted with the AES, (d) encrypted with the RC6 algorithm in the ECB mode with preprocessing, (e) encrypted in AES in the ECB mode with preprocessing.

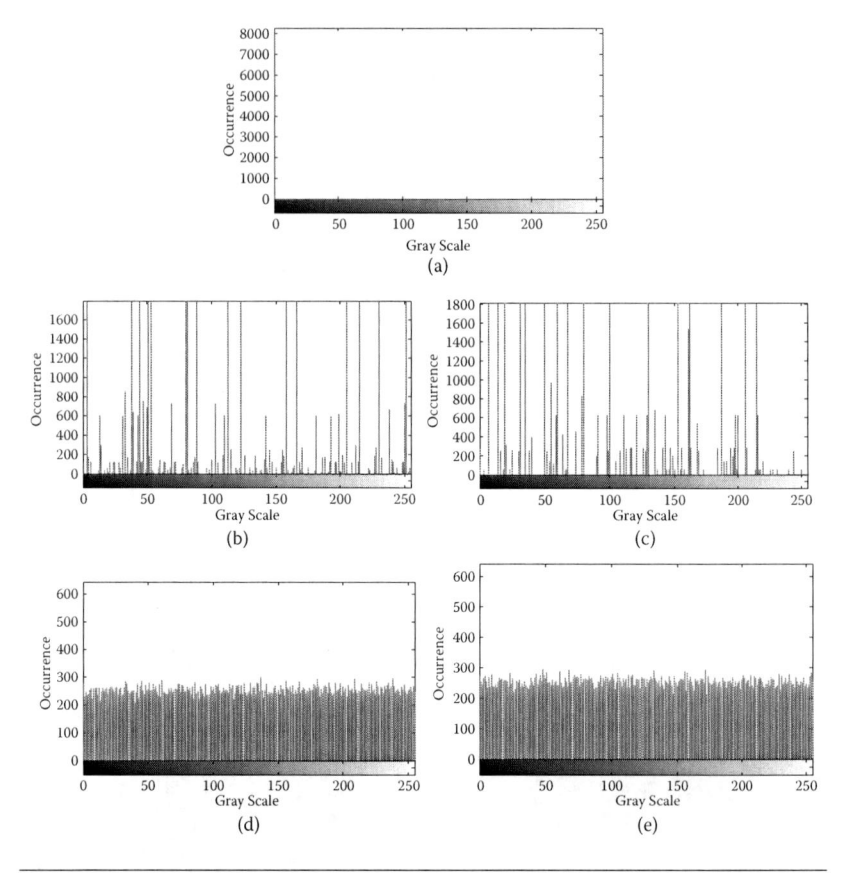

Figure 6.18   The histograms of the logo image: (a) original, (b) encrypted with the RC6 algorithm, (c) encrypted with the AES, (d) encrypted with the RC6 algorithm in the ECB mode with preprocessing, (e) encrypted in AES in the ECB mode with preprocessing.

6.5 Summary

In this chapter, an efficient approach has been presented to solve the problems associated with encryption in the ECB mode. A preprocessing network processes the data before encryption to diffuse the data bytes together prior to the application of the encryption algorithm in the ECB mode. The test results have shown that any encryption algorithm implemented in the ECB mode with the preprocessing network achieves good diffusion characteristics and high encryption quality without losing the parallel processing advantage.

7

ORTHOGONAL FREQUENCY DIVISION MULTIPLEXING

7.1 Introduction

The main objective of this book is to study image encryption from a communication perspective. So, we need to determine if an encryption algorithm fits the communication requirements. It is time now to switch to wireless communication to understand the communication process and its limitations and hence to study the performance of the decryption algorithms after the communication process. We consider multicarrier modulation (MCM), especially orthogonal frequency division multiplexing (OFDM), in our study because it is the new trend in wireless communication systems. MCM is used not only in the physical layer of several wireless network standards such as Institute of Electrical and Electronics Engineers (IEEE) 802.11a, IEEE 802.16a, and HIPERLAN2 (High-Performance Radio Local Area Network Type 2), but also in HDTV (high-definition television) applications that include image communication [111,112]. OFDM overcomes the effects of multipath fading by breaking the signal into several narrow-bandwidth carriers. This results in a low symbol rate reducing the amount of ISI (intersymbol interference). The high tolerance to multipath fading makes OFDM more suited to transmissions with a high data rate in terrestrial environments compared to single-carrier transmissions.

The transmission frequency, receiver velocity, and required multipath tolerance all determine the most suitable transmission mode to use. Doppler spread is caused by rapid changes in the channel response due to movement of the receiver through a multipath environment. It results in random frequency modulation of the OFDM subcarriers, leading to signal degradation. The amount of Doppler spread is proportional to the transmission frequency and the velocity of movement.

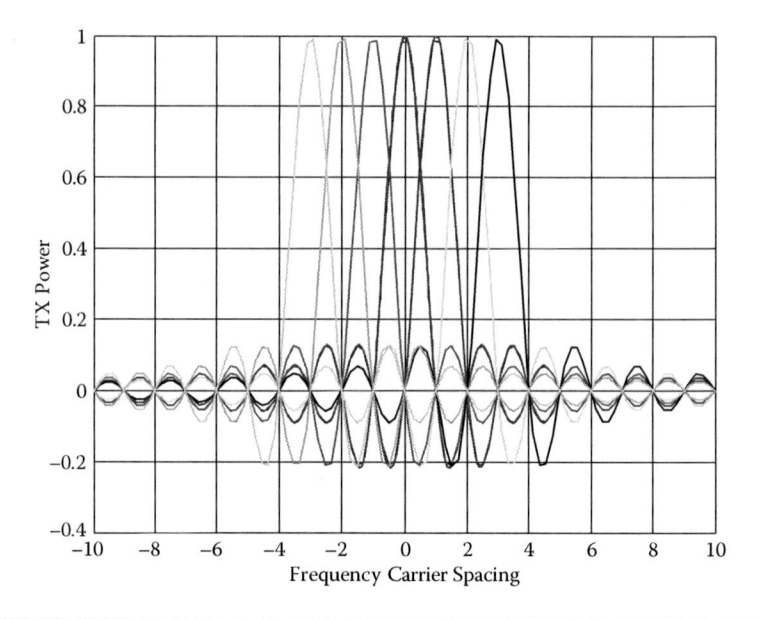

Figure 7.1   Frequency response of the subcarriers in a seven-tone OFDM signal.

The closer the subcarriers are spaced, the more susceptible the OFDM signal is to Doppler spread, so the different transmission modes in digital audio broadcasting (DAB) and HDTV allow a trade-off between the amount of multipath protection (length of the guard period) and the Doppler spread tolerance [113].

The OFDM signal has the spectrum shown in Figure 7.1. In the frequency domain, each OFDM subcarrier has several sinc-shaped frequency responses. The *sinc* shape has a narrow main lobe with many side lobes that decay slowly with the magnitude of the frequency shift away from the center. Each carrier has a peak at the center frequency and nulls evenly spaced with a frequency gap equal to the carrier spacing [114,115].

7.2 Basic Principles of OFDM

All wireless communication systems use a modulation scheme to map the information signal to a form that can be effectively transmitted over the communication channel. A wide range of modulation schemes has been developed, with the most suitable one depending on whether the information signal is an analog waveform or a digital signal.

Common analog modulation schemes include frequency modulation (FM), amplitude modulation (AM), and phase modulation (PM). Common single-carrier modulation schemes for digital communications include amplitude shift keying (ASK), frequency shift keying (FSK), phase shift keying (PSK), and quadrature amplitude modulation (QAM) [115].

OFDM is different from frequency division multiplexing (FDM) in several ways. In conventional broadcasting, each radio station transmits on a different frequency, effectively using FDM to maintain a separation between the stations. There is, however, no coordination or synchronization between each of these stations. With an OFDM transmission system such as DAB or digital video broadcasting (DVB), the information signals from multiple stations are combined into a single multiplexed stream of data. These data are then transmitted using an OFDM ensemble that is made from dense packing of several subcarriers. All the subcarriers within the OFDM signal are time and frequency synchronized to each other, allowing the interference between subcarriers to be carefully controlled. These multiple subcarriers overlap in the frequency domain but cause small effects of intercarrier interference (ICI) due to the orthogonal nature of the modulation. Typically, with FDM the transmission signals need to have a large frequency guard band between channels to prevent interference. This lowers the overall spectral efficiency. However, with OFDM, the orthogonal packing of the subcarriers greatly reduces this guard band, improving the spectral efficiency [116–122].

Each of the carriers in an FDM transmission system can use an analog or digital modulation scheme. There is no synchronization between the transmissions, so one station could transmit using FM and another in a digital form using FSK. In a single OFDM transmission system, all the subcarriers are synchronized to each other, restricting the transmission to digital modulation schemes. OFDM is symbol based and can be thought of as a large number of low-bit-rate carriers transmitted in parallel. All these carriers transmit using synchronized time and frequency, forming a single block of spectrum to ensure that the orthogonal nature of the structure is maintained. Because these multiple carriers form a single OFDM transmission, they are commonly referred to as *subcarriers*, with the term *carrier* reserved for describing the radio-frequency (RF) carrier mixing the signal from baseband. There are several ways of looking at what makes the subcarriers in an OFDM signal orthogonal and why this prevents interference between them [116–119].

7.2.1 Orthogonality

Signals are orthogonal if they are mutually independent of each other. Orthogonality is a property that allows multiple information signals to be transmitted perfectly over a common channel and detected without interference. Loss of orthogonality results in blurring between these information signals and degradation in communications.

Many common multiplexing schemes are inherently orthogonal. Time division multiplexing (TDM) allows transmission of multiple information signals over a single channel by assigning unique time slots to each separate information signal. During each time slot, only the signal from a single source is transmitted, preventing any interference between the multiple information sources. Because of this, TDM is orthogonal in nature [116–119]. In the frequency domain, most FDM systems are orthogonal as each of the separate transmission signals is well spaced out in frequency, preventing interference. The term *OFDM* has been reserved for a special form of FDM. The subcarriers in an OFDM signal are spaced as close as is theoretically possible while maintaining orthogonality between them.

OFDM achieves orthogonality in the frequency domain by allocating each of the separate information signals onto different subcarriers. OFDM signals are made up from a sum of sinusoids, with each one corresponding to a subcarrier. The baseband frequency of each subcarrier is chosen to be an integer multiple of the inverse of the symbol time, resulting in all subcarriers having an integer number of cycles per symbol. As a consequence, the subcarriers are orthogonal to each other [123–130].

Sets of functions are orthogonal to each other if they match the conditions in Equation (7.1). If any two different functions within the set are multiplied and integrated over a symbol period, the result is zero for orthogonal functions. If we look at a matched receiver for one of the orthogonal functions, then the receiver will only see the result for that function. The results from all other functions in the set integrate to zero and thus have no effect [115].

$$\int_0^T s_i(t)s_j(t)\,dt = \begin{cases} c & i=j \\ 0 & i\neq j \end{cases} \tag{7.1}$$

Equation (7.2) shows a set of orthogonal sinusoids that represent the subcarriers for an unmodulated real OFDM signal [115,116].

$$s_k(t) = \begin{cases} \sin(2\pi k f_0 t) & 0 < t < T \quad k=1,2,....N \\ 0 & otherwise \end{cases} \tag{7.2}$$

where f_0 is the subcarrier spacing, N is the number of subcarriers, and T is the symbol period. Because the highest-frequency component is $N f_0$, the transmission bandwidth is also $N f_0$.

7.2.2 Frequency Domain Orthogonality

In the frequency domain, each OFDM subcarrier has a sinc $\sin(x)/x$ frequency response. This is a result of the symbol time corresponding to the inverse of the carrier spacing. As far as the receiver is concerned, each OFDM symbol is transmitted for a fixed time T_{FFT} with no tapering at the ends of the symbol. This symbol time corresponds to the inverse of the subcarrier spacing of $1/T_{FFT}$ Hz. The rectangular, boxcar waveform in the time domain results in a sinc frequency response in the frequency domain. The sinc shape has a narrow main lobe with many side lobes that decay slowly with the magnitude of the frequency difference away from the center. Each carrier has a peak at the center frequency and nulls evenly spaced with a frequency gap equal to the carrier spacing [113].

The orthogonal nature of the transmission is a result of the peak of each subcarrier corresponding to the nulls of all other subcarriers. When this signal is detected using an FFT (fast Fourier transform), the spectrum is not continuous but has discrete samples. If the FFT is time synchronized, the frequency samples of the FFT correspond to just the peaks of the subcarriers; thus, the overlapping frequency region between subcarriers does not affect the receiver. The measured peaks correspond to the nulls for all other subcarriers, resulting in orthogonality between the subcarriers [113–115].

7.3 OFDM System Model

This section briefly reviews the key steps in an OFDM communication system that can be used for the transmission of encrypted images. The first step is the image encryption; the general model of a typical encryption system could be described with the following equation [131]:

$$E(P, K) = C \qquad (7.3)$$

where P is the plainimage, E is the encryption algorithm, K is the encryption key, and C is the cipherimage. The cipherimage is transmitted through the communication channel. At the receiver side, the decryption procedure could be represented by [131]

$$D(C, K') = P' \qquad (7.4)$$

where D is the decryption algorithm, K' is the decryption key (it may or may not be the same as the encryption key K), and P' is the recovered plainimage.

At the transmitter side, the encrypted data are converted into parallel data of N subchannels. Then, the data of each parallel subchannel are modulated using a modulation scheme like PSK or QPSK (quaternary PSK). For QPSK of N subchannels $(d_0, d_1, d_2, \ldots, d_{N-1})$, each d_n is a complex number $d_n = d_{I_n} + jd_{Q_n}$, where d_{I_n} and d_{Q_n} are $\{1, -1\}$ [131–135].

7.3.1 FFT-OFDM

In FFT-OFDM, the modulated data are fed into an inverse FFT (IFFT) circuit, and the OFDM signal is generated. The OFDM splits a high-data-rate sequence into a number of low-rate sequences that are transmitted over a number of subcarriers equal to N. The N subcarriers are chosen to be orthogonal; that is, $f_n = n\Delta f$, where $\Delta f = 1/T_s$, and T_s is the OFDM symbol duration. The resulting signal can be expressed as follows [136]:

$$x(n) = \frac{1}{\sqrt{N}} \sum_{k=0}^{N-1} X(k) e^{\frac{j2\pi kn}{N}}, \quad 0 \leq t \leq T_s \qquad (7.5)$$

where $X(k)$ represents the discrete-time samples.

A guard interval is added at the start of each OFDM symbol to eliminate the ISI, which occurs in multipath channels. An OFDM symbol is extended in a cyclic manner to avoid the ICI. As a result, a channel that is highly frequency selective is transformed into a large set of individual flat fading, non-frequency-selective, narrowband channels. At the receiver, the guard interval is removed, and the time interval $[0, T_s]$ is evaluated.

7.3.2 DCT-OFDM

The structure of the discrete cosine transform (DCT)-OFDM system is similar to that of FFT-OFDM but with the IFFT and the FFT modules replaced by inverse discrete cosine transform (IDCT) and DCT modules, respectively. The main advantage of the DCT lies in its excellent spectral energy compaction property, which makes most of the samples transmitted close to zero, leading to a reduction in the ISI. In addition, it uses only real arithmetic rather than the complex arithmetic used in the FFT. This reduces the signal processing complexity and the in-phase/quadrature imbalance [136,137]. In the DCT-OFDM system, the transmitted signal is given by [138–140]

$$x(n) = \sqrt{\frac{2}{N}} \sum_{k=0}^{N-1} X(k)\beta(k)\cos\left(\frac{\pi k(2n+1)}{2N}\right), \quad p = 0, ..., N-1 \quad (7.6)$$

where $X(k)$ is the kth symbol of the input signal. $\beta(k)$ can be written as follows:

$$\beta(k) = \begin{cases} \dfrac{1}{\sqrt{2}} & k = 0 \\ 1 & k = 1, 2, ..., N-1 \end{cases} \quad (7.7)$$

7.3.3 Discrete Wavelet Transform–OFDM

The OFDM requires a cyclic prefix to remove ISI. This causes overhead, and this overhead may sometimes be too large for the system to be effective. The use of the wavelet transform reduces the ISI and ICI [138, 141,142]. The structure of the discrete wavelet transform

(DWT)-OFDM system is similar to that of FFT-OFDM but with the IFFT and the FFT modules replaced by indirect discrete wavelet transform (IDWT) and DWT modules, respectively. The transmitted signal with DWT-OFDM is given by [138]

$$x(n) = \sum_{k=0}^{N} X(k)\varphi(t - kT) \qquad (7.8)$$

where $\varphi(t)$ is the wavelet basis function.

Further advantages of the DWT-OFDM are as follows:

1. It requires less overhead as it does not require a cyclic perfix (CP).
2. It does not require a pilot tone, which takes about 8% of the subbands.
3. DWT-OFDM is inherently robust to ISI and ICI.

The DWT is explained in the following section.

7.3.4 Discrete Wavelet Transform

Wavelets have become a popular tool in most signal-processing and communications applications. The conventional DWT may be regarded as equivalent to filtering the input signal with a bank of band-pass filters whose impulse responses are all approximately given by scaled versions of a mother wavelet. The scaling factor between adjacent filters is usually 2:1, leading to octave bandwidths and center frequencies that are one octave apart [138,139]. The outputs of the filters are usually maximally decimated so that the number of DWT output samples equals the number of input samples, and the transform is invertible as shown in Figure 7.2.

7.3.4.1 Implementation of the DWT

The DWT is normally implemented by digital filters as shown for the one-dimensional (1D) case in Figure 7.2. The art of finding a good wavelet lies in the design of the set of filters H_0, H_1, G_0, and G_1 to achieve various trade-offs between spatial and frequency domain characteristics while satisfying the perfect reconstruction (PR) condition [142]. In Figure 7.2,

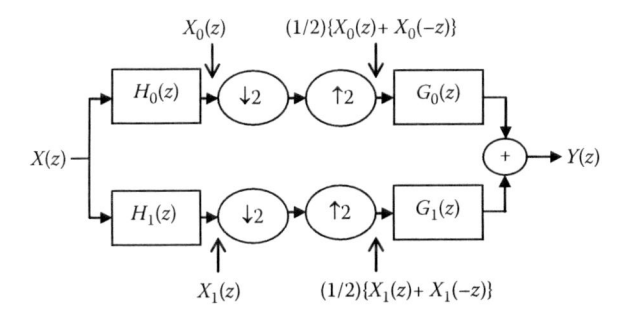

Figure 7.2   The two-band decomposition-reconstruction wavelet filter bank.

the process of decimation and interpolation by 2:1 at the output of H_0 and H_1 effectively sets all odd samples of these signals to zero.

For the low-pass branch, this is equivalent to multiplying $x_0(n)$ by $\frac{1}{2}\left(1+(-1)^n\right)$. Hence, $X_0(z)$ is converted to $\{X_0(z) + X_0(-z)\}$. Similarly, $X_1(z)$ is converted to $\frac{1}{2}\{X_1(z)+X_1(-z)\}$.

Thus, the expression for $Y(z)$ is given by [142]

$$Y(z)=\frac{1}{2}\{X_0(z)+X_0(-z)\}G_0(z)+\frac{1}{2}\{X_1(z)+X_1(-z)\}G_1(z)$$

$$=\frac{1}{2}X(z)\{H_0(z)G_0(z)+H_1(z)G_1(z)\}$$

$$+\frac{1}{2}X(-z)\{H_0(-z)G_0(z)+H_1(-z)G_1(z)\} \tag{7.9}$$

The first PR condition requires aliasing cancellation and forces this term in $X(-z)$ to be zero. Hence, $\{H_0(-z)G_0(z) + H_1(-z)G_1(z)\} = 0$, which can be achieved if [142]

$$H_1(z) = z^{-a}G_0(-z) \text{ and } G_1(z) = z^{a}H_0(-z) \tag{7.10}$$

where a must be odd (usually $a = \pm1$).

The second PR condition is that the transfer function from $X(z)$ to $Y(z)$ should be unity [142]:

$$\{H_0(z)G_0(z) + H_1(z)G_1(z)\} = 2 \tag{7.11}$$

If we define a product filter $P(z) = H_0(z)G_0(z)$ and substitute from Equation (7.10) into Equation (7.11), then the PR condition becomes [142]

$$H_0(z)G_0(z) + H_1(z)G_1(z) = P(z) + P(-z) = 2 \qquad (7.12)$$

This needs to be true for all z, and because the odd powers of z in $P(z)$ cancel with those in $P(-z)$, it requires that $p_0 = 1$ and $p_k = 0$ for all n even and nonzero. The polynomial $P(z)$ should be a zero-phase polynomial to minimize distortion. In general, $P(z)$ is of the following form [142]:

$$P(z) = \ldots + p_5 z^5 + p_3 z^3 + p_1 z + 1 + p_1 z^{-1} + p_3 z^{-3} + p_5 z^{-5} + \ldots \qquad (7.13)$$

The design method for the PR filters can be summarized in the following steps [142]:

1. Choose p_1, p_3, p_5, ... to give a zero-phase polynomial $P(z)$ with good characteristics.
2. Factorize $P(z)$ into $H_0(z)$ and $G_0(z)$ with similar low-pass frequency responses.
3. Calculate $H_1(z)$ and $G_1(z)$ from $H_0(z)$ and $G_0(z)$.

To simplify this procedure, we can use the following relation:

$$P(z) = P_t(Z) = 1 + P_{t,1}Z + P_{t,3}Z^3 + P_{t,5}Z^5 + \ldots \qquad (7.14)$$

where

$$Z = \frac{1}{2}\left(z + z^{-1}\right) \qquad (7.15)$$

7.3.4.2 Haar Wavelet Transform   The Haar wavelet is the simplest type of wavelet. In discrete form, Haar wavelets are related to a mathematical operation called the Haar transform. The Haar transform serves as a prototype for all other wavelet transforms. Like all wavelet transforms, the Haar transform decomposes a discrete signal into two subsignals of half its length. One subsignal is a running average or trend; the other subsignal is a running difference or fluctuation.

The Haar wavelet uses the simplest possible $P_t(Z)$ with a single zero at $Z = -1$. It is represented as follows [142]:

$$P_t(Z) = 1 + Z \text{ and } Z = \frac{1}{2}\left(z + z^{-1}\right) \qquad (7.16)$$

Thus,

$$P(z) = \frac{1}{2}\left(z + 2 + z^{-1}\right) = \frac{1}{2}(z+1)\left(1 + z^{-1}\right)$$

$$= G_0(z)H_0(z) \qquad (7.17)$$

We can find $H_0(z)$ and $G_0(z)$ as follows:

$$H_0(z) = \frac{1}{2}\left(1 + z^{-1}\right) \qquad (7.18)$$

$$G_0(z) = (z + 1) \qquad (7.19)$$

Using Equation (7.10) with $a = 1$, we obtain

$$G_1(z) = zH_0(-z) = \frac{1}{2}z\left(1 - z^{-1}\right) = \frac{1}{2}(z - 1) \qquad (7.20)$$

$$H_1(z) = z^{-1}G_0(-z) = z^{-1}(-z + 1) = (z^{-1} - 1) \qquad (7.21)$$

7.4 Guard Interval Insertion

The guard interval insertion is an important step in all OFDM systems. The purpose of this step is to avoid the ISI. Figure 7.3 illustrates the process of guard interval insertion [143–146]. Also, the guard period may consist of two sections: zero-amplitude transmission data and a cyclic extension of the transmitted symbol. We consider all these types of guard interval and their effects on the transmitted encrypted images.

Figure 7.3 Guard interval insertion in OFDM.

7.5 Communication Channels

7.5.1 Additive White Gaussian Noise Channel

For an additive white Gaussian noise (AWGN) channel, the received signal $r(t)$ is expressed as [141]

$$r(t) = s(t) + n(t) \tag{7.22}$$

where $s(t)$ is the transmitted signal, $r(t)$ is the received signal, and $n(t)$ is the AWGN with power spectral density given by [141]

$$\Phi(f) = N_{0/2} \ [\text{W/Hz}] \tag{7.23}$$

where N_0 is a constant that is often called the noise power density.

7.5.2 Fading Channel

The path between the transmitter and receiver is characterized by various obstacles and reflections, which lead to a fading effect at the receiver as shown in Figure 7.4. We can take the following expression to simulate the incoming signal at the receiver in the Raleigh fading scenario [141]:

$$f(t) = x(t) + j \cdot y(t) \tag{7.24}$$

$$= \left[\sqrt{\frac{2}{N_1 + 1}} \sum_{n=1}^{N_1} \sin\left(\frac{\pi n}{N_1}\right) \cos\left\{ 2\pi f_d \cos\left(\frac{2\pi n}{N_1}\right) t \right\} \right.$$

$$\left. + \frac{1}{\sqrt{N_1 + 1}} \cos(2\pi f_d t) \right]$$

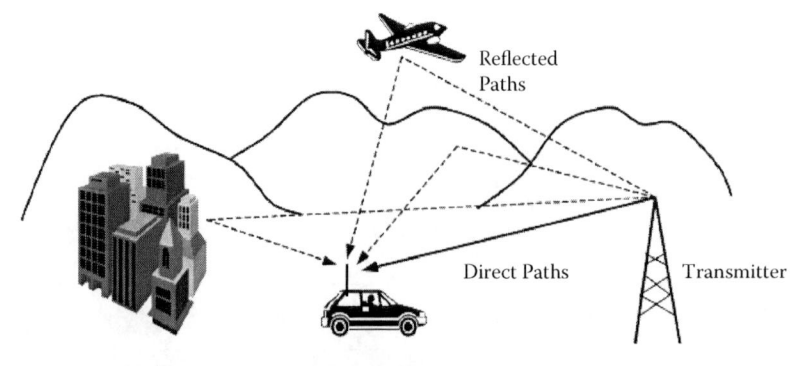

Figure 7.4   A wireless channel with multipath propagation.

$$+j\sqrt{\frac{2}{N_1}}\sum_{n=1}^{N_1}\sin\left(\frac{\pi n}{N_1}\right)\cos\left\{2\pi f_d\cos\left(\frac{2\pi n}{N_1}\right)t\right\} \qquad (7.25)$$

where the wave number of the incoming wave is N_0, and N_1 is given by

$$N_1 = 1/2((N_0/2) - 1) \qquad (7.26)$$

7.6 Channel Estimation and Equalization

Channel estimation and equalization constitute a major challenge in current and future communication systems. The equalizers utilized to compensate for the ISI can be classified as linear and nonlinear equalizers [147]. Linear equalizers are simple linear filter structures that try to invert the channel in the sense that the product of the transfer functions of the channel and the equalizer fulfills a certain criterion. This criterion can either be achieving a completely flat transfer function of the channel filter concatenation or minimizing the mean square error at the filter output [148]. One of the most popular linear equalizers is the zero-forcing (ZF) equalizer. The ZF solution can be written as [148]

$$\mathbf{W}_{ZF} = (\mathbf{H}^H\mathbf{H})^{-1}\mathbf{H}^H \qquad (7.27)$$

where $\mathbf{H}$ is the matrix of the channel transfer function.

The advantage of the ZF equalizer is that the statistics of the additive noise and source data are not required.

Channel estimation can be performed with pilot symbol estimation (PSE). With an OFDM system, the wideband channel is divided into a number of narrowband channels. Thus, channel estimation can be performed by inserting pilot symbols with a known modulation scheme into the transmitted signal. Based on these pilot symbols, the receiver can measure the channel transfer function for each subcarrier using interpolation techniques [113].

The mathematical model for the PSE and equalization can be represented as follows [144]:

$$\begin{pmatrix} i_1 \\ q_1 \end{pmatrix} = A \begin{pmatrix} i_0 \\ q_0 \end{pmatrix} \qquad (7.28)$$

where i_1 is the received in-phase symbol, q_1 is the quadrature received symbol, i_0 is the transmitted in-phase symbol, q_0 is the quadrature transmitted symbol, and $\mathbf{A}$ is the transition matrix of the fading environment that is given by [130]

$$\mathbf{A} = \begin{pmatrix} a_{11} & a_{12} \\ a_{21} & a_{22} \end{pmatrix} \tag{7.29}$$

The channel matrix is Toplitz with $a_{11} = a_{22}$ and $a_{12} = -a_{21}$. To perform channel equalization, we need to estimate the matrix $\mathbf{A}^{-1}$ as follows [130]:

$$\mathbf{A}^{-1} = \frac{1}{\sqrt{a_{11}^2 + a_{12}^2}} \begin{pmatrix} a_{11} & a_{12} \\ -a_{12} & a_{11} \end{pmatrix} \tag{7.30}$$

Because the transmitted and received symbols are known, we can estimate the channel coefficients using Equations (7.28) and (7.30) as follows [130]:

$$a_{11} = a_{22} = \frac{1}{\sqrt{i_1 + q_1}} \left(i_0 i_1 + q_0 q_1 \right) \tag{7.31}$$

$$a_{21} = -a_{12} = \frac{1}{\sqrt{i_1 + q_1}} \left(q_0 i_1 + i_0 q_1 \right) \tag{7.32}$$

Once the channel is estimated, it can be used for equalization with unknown transmitted symbols [113]. Equalization is not the only problem encountered in wireless communication of images with OFDM. There are some other limitations that may affect the quality of transmitted images, especially if they are encrypted. Some of these problems are addressed in the next chapter.

8
OFDM LIMITATIONS

8.1 Introduction

OFDM has been adopted in the European digital audio and video broadcasting radio system and is being investigated for broadband indoor wireless communications. Standards such as High-Performance Radio Local-Area Network (HIPERLAN2), Institute of Electrical and Electronics Engineers (IEEE) 802.11a, and IEEE 802.11g have emerged to support services based on the Internet Protocol (IP). Such systems are based on OFDM and are designed to operate in the 5-GHz band [147].

Unfortunately, OFDM communication systems have two primary problems: the high sensitivity to carrier frequency offsets (CFOs) and the high peak-to-average power ratio (PAPR). The sensitivity to CFOs breaks the subcarriers' orthogonality, and the high PAPR requires system components with a wide linear range to accommodate for the signal variations. Otherwise, nonlinear distortion, which results in a loss of subcarrier orthogonality and hence a degradation in the system performance, occurs [148–152].

Researchers have proposed various methods to combat the intercarrier interference (ICI in OFDM systems. The existing approaches that have been developed to reduce ICI can be categorized into frequency domain equalization, time domain windowing, and self-cancellation (SC) schemes. In addition, statistical approaches have been explored to estimate and cancel ICI. In this chapter, the effects of CFO and PAPR problems are studied with solutions to these problems and to reduce their effects on the transmitted encrypted images.

8.2 Analysis of Intercarrier Interference

The main problem of OFDM is its susceptibility to small differences in frequency at the transmitter and receiver, normally referred to as frequency offset. This frequency offset can be caused by Doppler shift due to relative motion between the transmitter and receiver or by differences between the frequencies of the local oscillators at the transmitter and receiver. The frequency offset is modeled as a multiplicative factor introduced in the channel, as shown in Figure 8.1) [153–155].

The received signal is given by

$$y(n) = i(n)\exp\left(\frac{j2\pi n\varepsilon}{N}\right) + w(n) \qquad (8.1)$$

where ε is the normalized frequency offset and is given by ΔfNT_s. Δf is the frequency difference between the transmitted and received carrier frequencies, and T_s is the subcarrier symbol period. $w(n)$ is the additive white Gaussian noise (AWGN) introduced in the channel.

The effect of this frequency offset on the received symbol stream can be understood by considering the received symbol $Y(k)$ on the kth subcarrier [151–153].

$$Y(k) = X(k)S(0) + \sum_{l=0, l\neq k}^{N-1} X(l)S(l-k) + W(k) \qquad (8.2)$$

where $k = 0, 1, \ldots, N-1$; N is the total number of subcarriers; $X(k)$ is the transmitted symbol for the kth subcarrier; $W(k)$ is the fast Fourier transform (FFT) of $w(n)$; and $S(l-k)$ are the complex coefficients for the ICI components in the received signal. The ICI components are

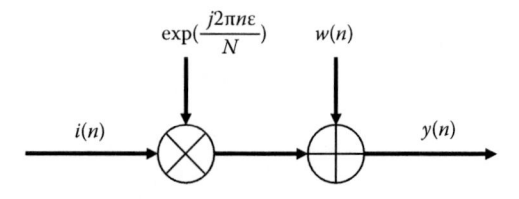

Figure 8.1 Frequency offset model.

the interfering signals transmitted on subcarriers other than the kth subcarrier. The complex coefficients are given by [151–153]

$$S(l-k) = \frac{\sin(\pi(l+\varepsilon-k))}{N\sin(\pi(l+\varepsilon-k)/N)} \exp(j\pi(1-\frac{1}{N})(l+\varepsilon-k)) \quad (8.3)$$

8.3 CFO in FFT-OFDM System Model

In FFT-OFDM, the modulated symbols are grouped into blocks, each containing N symbols, and the inverse FFT (IFFT) is performed. The resulting signal after the IFFT can be expressed as follows [138,151–153]:

$$x(n) = \frac{1}{\sqrt{N}} \sum_{k=0}^{N-1} X(k) e^{\frac{j2\pi nk}{T_s}}, \quad 0 \le n \le N-1 \quad (8.4)$$

where N is the number of subcarriers. $X(k)$ represents the kth modulated symbol.

At the end of the transmitter, a guard interval is inserted between symbols with hybrid guard interval or zero padding to eliminate the intersymbol interference (ISI). The resulting signal is then transmitted through the wireless channel.

At the receiver side, the padded zeros or the hybrid guard interval are removed from the received signal. The received signal in the presence of CFO is given by

$$r(n) = x(n) e^{\frac{j2\pi m\varepsilon}{N}} + w(n) \quad (8.5)$$

where $\varepsilon = \Delta f T_s$ is the normalized CFO, Δf is the CFO, T_s is the OFDM symbol duration, and $w(n)$ is AWGN. After that, the received signal is transformed into the frequency domain via the FFT as follows:

$$R(k) = \sum_{n=0}^{N-1} r(n) e^{\left(\frac{-2\pi nk}{N}\right)} \quad (8.6)$$

Substituting Equation (8.3) into Equation (8.4), we obtain [151–153]

$$R(k) = \sum_{n=0}^{N-1} x(n)e^{\left(\frac{2\pi m\varepsilon}{N}\right)}e^{\left(\frac{-2\pi nk}{N}\right)} + W(k)$$

$$= \frac{1}{N}\sum_{l=0}^{N-1} X(l)\sum_{n=0}^{N-1} e^{\left(\frac{2\pi n(l+\varepsilon\ -k)}{N}\right)} + W(k) \tag{8.7}$$

where $W(k)$ is the kth frequency domain sample of the noise. We can expand $\dfrac{1}{N}\sum_{n=0}^{N-1} e^{\left(\frac{j2\pi n(l-\varepsilon\ -k)}{N}\right)}$ using the geometric series as follows [151–153]:

$$\frac{1}{N}\sum_{n=0}^{N-1} e^{\left(\frac{j2\pi n(l-\varepsilon-k)}{N}\right)} = \frac{1}{N}\frac{1-e^{(j2\pi(l+\varepsilon-k))}}{1-e^{\left(\frac{j2\pi(l+\varepsilon-k)}{N}\right)}}$$

$$= \frac{1}{N}e^{(j\pi(l+\varepsilon-k)(1-1/N))}\frac{\sin(\pi(l-k+\varepsilon\ ))}{\sin(\pi(l-k-\varepsilon)/N)} \tag{8.8}$$

We obtain [151–153]

$$R(k) = \sum_{l=0}^{N-1} X(l)S(l-k+\varepsilon\ ) + W(k) \tag{8.9}$$

where $S(l - k + \varepsilon)$ represents the complex coefficients for the interference components in the received signal. $S(l - k + \varepsilon)$ is given by [151–153]

$$S\ (l-k+\varepsilon\ ) = e^{(j\pi(l+\varepsilon-k)(1-1/N))}\frac{\sin(\pi(l-k+\varepsilon))}{M\sin(\pi(l-k-\varepsilon)/N)} \tag{8.10}$$

8.4 CFO in DCT-OFDM System Model

The main advantage of the discrete cosine transform (DCT) lies in its excellent spectral energy compaction property, which makes most of the samples transmitted close to zero, leading to a reduction in the ISI. In addition, it uses only real arithmetic rather than the complex arithmetic used in the FFT. This reduces the signal-processing complexity, and the in-phase/quadrature imbalance [150,151]. In the DCT-OFDM system, the transmitted signal is given by [136,137,139–143]

$$x(n) = \sqrt{\frac{2}{N}} \sum_{k=0}^{N-1} X(k)\beta(k)\cos\left(\frac{\pi k(2n+1)}{2N}\right), \quad p = 0,\dots, N-1 \quad (8.11)$$

where $X(k)$ is the kth symbol of the input signal. $\beta(k)$ can be written as follows:

$$\beta(k) = \begin{cases} \dfrac{1}{\sqrt{2}} & k = 0 \\ 1 & k = 1, 2, \dots\dots, N-1 \end{cases} \quad (8.12)$$

In the DCT-OFDM system, the complex coefficients for the interference components in the received signal can be expressed as follows [151–153]:

$$\begin{aligned} S(l,k,\zeta) = \frac{1}{2N}\beta_k\beta_l &\left[\phi(l+k-\zeta)+\phi(l-k-\zeta)\right. \\ &+\phi(l+k+\zeta)+\phi(l-k+\zeta)\big] \\ &+ j\big[\Gamma(l+k-\zeta)+\Gamma(l-k-\zeta) \\ &-\Gamma(l+k+\zeta)-\Gamma(l-k+\zeta)\big] \end{aligned} \quad (8.13)$$

where $\zeta = 2T\Delta f$ is the normalized CFO for the DCT-OFDM system [151–153].

$$\Gamma(x) = \frac{\sin\left(\dfrac{\pi x}{2}\right)^2}{\sin\left(\dfrac{\pi x}{2N}\right)} \tag{8.14}$$

$$\phi(x) = \frac{\sin\left(\dfrac{\pi x}{2}\right)\cos\left(\dfrac{\pi x}{2}\right)}{\sin\left(\dfrac{\pi x}{2N}\right)} \tag{8.15}$$

8.5 CFO in DWT-OFDM System Model

The transmitted signal with discrete wavelet transform (DWT)-OFDM is given by [138]

$$x(n) = \sum_{k=0}^{N} X(k)\varphi(t - kT) \tag{8.16}$$

where $\varphi(t)$ is the wavelet basis functors. Some attempts have been made to predict the CFO effect on DWT-OFDM, but, as of yet, no closed-form expression has been presented for this task.

8.6 CFO Compensation

There are many techniques that were developed to compensate for the CFO in multicarrier communication systems. In this chapter, the compensation process is carried out in the time domain. Mathematically, for the OFDM system, the received sequence is multiplied by a time domain sequence $e^{-\frac{j2\pi n\varepsilon}{N}}$ before the FFT/DCT/DWT processing. The compensated signal can be formulated as follows for AWGN [151–153]:

$$r_c(n) = e^{-\frac{j2\pi n\varepsilon}{N}} r(n) = x\,(n) + e^{-\frac{j2\pi n\varepsilon}{N}} w(n) \tag{8.17}$$

In fading channels, as we know, estimation of the channel transfer function is needed. The channel transfer function estimate may be computed using any algorithm that gives reliable estimates. In our design, the zero-forcing (ZF) channel estimator was applied to obtain the initial channel estimate when the first OFDM symbol of the superframe is received. This symbol represents the pilot symbol known to the receiver. After receiving the first OFDM symbol, the estimator switches to the tracking mode. The channel estimates are refined and tracked according to the gradient algorithm.

8.7 Simulation Parameters

In our simulations, we carried out the encryption of the Cameraman image with the previously discussed encryption algorithms. Then, the encrypted images were transformed into a binary format and used for OFDM modulation. Both AWGN and Raleigh fading channels were considered in the simulations. A normalized CFO was used in some of the simulation experiments. PAPR reduction techniques were used in some other experiments. In all experiments, 128 subcarriers and quaternary phase shift keying (QPSK) were used. The simulation parameters are summarized in Table 8.1.

To evaluate the quality of the decrypted images at the receiver, we used the peak signal-to-noise ratio (PSNR) between the original image and the decrypted image, which is defined as follows:

$$PSNR = 10\log_{10}\left(\frac{255^2}{MSE}\right) \tag{8.18}$$

where the mean square error (MSE) is defined as

$$MSE = \frac{1}{U^2}\sum_{i=1}^{U}\sum_{j=1}^{U}\left[f(i,j)-\hat{f}(i,j)\right]^2 \tag{8.19}$$

Table 8.1 Simulation Parameters

	PARAMETER	DESCRIPTION
Transmitter	System bandwidth	64 MHz
	Modulation type	QPSK
	cyclic prefix length	16 samples
	N	128 subcarriers
	Subcarrier spacing	500 kHz
	Type of transform	FFT, DCT, or DWT
Channel	Channel model	Raleigh fading
Receiver	PSE and equalizer	ZF equalization
Encryption algorithm	Type of encryption	Chaotic encryption, chaotic-OFB encryption

where $f(i, j)$ is the original image of dimensions $U \times U$ and $\hat{f}(i, j)$ is the decrypted image.

Figures 8.2 and 8.3 illustrate the PSNR performance versus the E_b/N_0 for the FFT-OFDM, the DCT-OFDM, and the DWT-OFDM systems with chaotic and chaotic-output feedback (OFB) encryption algorithms in a frequency-selective channel, respectively. Without CFO compensation and with pilot symbol estimation (PSE), it was shown that CFO disrupted the PSNR performance of all OFDM systems; with CFO compensation, the PSNR performances of all systems were better than those without CFO compensation. These figures also show that the PSNR performances of all encryption algorithms with the FFT-OFDM system were the best compared with those of the DCT-OFDM and the DWT-OFDM systems.

Figure 8.4 gives the output of the FFT-OFDM system for the transmission of the original Cameraman image over an AWGN channel at $N = 128$ and $G = 32$.

Figure 8.5 gives the output of the FFT-OFDM system for the transmission of the original Cameraman image over a fading channel at $N = 128$ and $G = 32$.

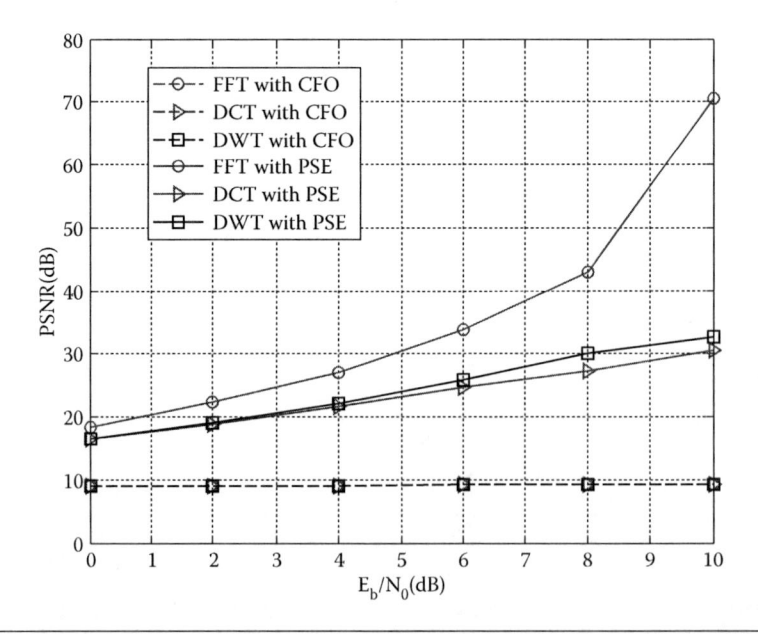

Figure 8.2   PSNR versus E_b/N_0 for chaotic-encrypted image transmission with OFDM system over a Raleigh channel.

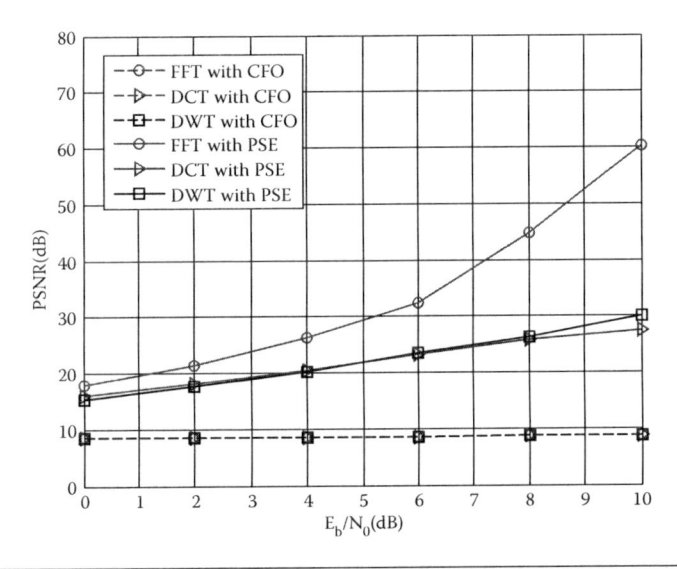

Figure 8.3 PSNR versus E_b/N_0 for chaotic-OFB-encrypted image transmission with OFDM system over a Raleigh fading channel.

(a) E_b/N_0 = 0 dB
PSNR = 14.6993

(b) E_b/N_0 = 5 dB
PSNR = 23.7838

(c) E_b/N_0 = 10 dB
PSNR = 48.0593

(d) E_b/N_0 = 15 dB
PSNR > 60

Figure 8.4 Original image with FFT-OFDM over an AWGN channel.

(a) $E_b/N_0 = 0$ dB
PSNR = 17.8214

(b) $E_b/N_0 = 5$ dB
PSNR = 29.4547

(c) $E_b/N_0 = 10$ dB
PSNR = 53.8719

(d) $E_b/N_0 = 15$ dB
PSNR > 60

Figure 8.5   Original image with FFT-OFDM over a fading channel.

Figure 8.6 gives the output of the FFT-OFDM system for the transmission of the chaotic-encrypted image over an AWGN channel at $N = 128$ and $G = 32$.

Figure 8.7 gives the output of the FFT-OFDM system for the transmission of the chaotic-encrypted image over a fading channel at $N = 128$ and $G = 32$.

Figure 8.8 gives the output of the DCT-OFDM system for the transmission of the original Cameraman image over an AWGN channel at $N = 128$ and $G = 32$.

Figure 8.9 gives the output of the DCT-OFDM system for the transmission of the original Cameraman image over a fading channel at $N = 128$ and $G = 32$.

(a) $E_b/N_0 = 0$ dB
PSNR = 15.1527

(b) $E_b/N_0 = 5$ dB
PSNR = 25.1709

(c) $E_b/N_0 = 10$ dB
PSNR = 50.5529

(d) $E_b/N_0 = 15$ dB
PSNR >60

Figure 8.6   Chaotic-encrypted image with FFT-OFDM over an AWGN channel.

Figure 8.10 gives the output of the DCT-OFDM system for the transmission of the chaotic-encrypted image over an AWGN channel at $N = 128$ and $G = 32$.

Figure 8.11 gives the output of the DCT-OFDM system for the transmission of the chaotic-encrypted image over a fading channel at $N = 128$ and $G = 32$.

Figure 8.12 gives the output of the DWT-OFDM system for the transmission of the original Cameraman image over an AWGN channel at $N = 128$ and $G = 32$.

Figure 8.13 gives the output of the DWT-OFDM system for the transmission of the original Cameraman image over a fading channel at $N = 128$ and $G = 32$.

(a) $E_b/N_0 = 0$ dB
PSNR = 18.3597

(b) $E_b/N_0 = 5$ dB
PSNR = 30.0828

(c) $E_b/N_0 = 10$ dB
PSNR = 53.8881

(d) $E_b/N_0 = 15$ dB
PSNR >60

Figure 8.7   Chaotic-encrypted image with FFT-OFDM over a fading channel.

Figure 8.14 gives the output of the DWT-OFDM system for the transmission of the chaotic-encrypted image over an AWGN channel at $N = 128$ and $G = 32$.

Figure 8.15 gives the output of the DWT-OFDM system for the transmission of the chaotic-encrypted image over a fading channel at $N = 128$ and $G = 32$.

Table 8.2 summarizes these results.

8.8 Effect of PAPR

One of the main limitations of OFDM is the high PAPR. A signal with large peaks can be obtained by the constructive superposition of subcarriers. The PAPR is linearly dependent on the number of

(a) $E_b/N_0 = 0$ dB
PSNR = 15.7172

(b) $E_b/N_0 = 5$ dB
PSNR = 26.3188

(c) $E_b/N_0 = 10$ dB
PSNR = 53.5993

(d) $E_b/N_0 = 15$ dB
PSNR >60

Figure 8.8 Original image with DCT-OFDM over an AWGN channel.

subcarriers. High-power peaks make certain demands on power amplifiers and analog-to-digital (A/D) and digital-to-analog (D/A) converters. Large peaks are distorted nonlinearly due to the power amplifier imperfections, and intermodulation products appear. They can be interpreted as ICI and out-of-band radiation.

This high PAPR results in significant in-band distortion and out-of-band radiation when the signal passes through a nonlinear device. The in-band distortion increases the bit error rate (BER), and the out-of-band radiation results in unacceptable adjacent channel interference (ACI) [152]. Without the use of any PAPR reduction technique, the efficiency of power consumption at the transmitter becomes very poor. Because OFDM signals are modulated independently in each subcarrier, the combined OFDM signals are likely to have large peak powers at certain instances. The peak power increases as the number of subcarriers increases. The peak power is generally

(a) $E_b/N_0 = 0$ dB
PSNR = 16.7843

(b) $E_b/N_0 = 5$ dB
PSNR = 23.2367

(c) $E_b/N_0 = 10$ dB
PSNR = 27.9452

(d) $E_b/N_0 = 15$ dB
PSNR = 29.3708

Figure 8.9   Original image with DCT-OFDM over a fading channel.

evaluated in terms of the PAPR. The PAPR of the transmitted OFDM signal is defined as

$$PAPR = \frac{P_{peak}}{P_{avg}} = \frac{\max\left(\left|\bar{x}_m\right|^2\right)}{1/N\sum_{m=0}^{N}\left|\bar{x}_m\right|^2}, m = 0, 1, \ldots\ldots, N-1 \quad (8.20)$$

In Equation (8.20), the numerator represents the maximum envelope power, and the denominator represents the average power.

The cumulative distribution function (CDF) of the PAPR is one of the most frequently used performance metrics for PAPR performance before and after applying PAPR reduction techniques. In the literature, the complementary CDF (CCDF) is commonly used instead of

(a) E_b/N_0 = 0 dB
PSNR = 15.0481

(b) E_b/N_0 = 5 dB
PSNR = 24.4613

(c) E_b/N_0 = 10 dB
PSNR = 53.4971

(d) E_b/N_0 = 15 dB
PSNR >60

Figure 8.10 Chaotic-encrypted image with DCT-OFDM over an AWGN channel.

the CDF itself. The CCDF of the PAPR denotes the probability that the PAPR of a data block exceeds a given threshold. In [153], a simple approximate expression was derived for the CCDF of the PAPR of a multicarrier signal; the real and imaginary values of $x(t)$ follow a Gaussian distribution. The CDF of the amplitude z of an OFDM signal sample is given by

$$F(z) = 1 - e^{-z} \tag{8.21}$$

We need to derive the CDF of the PAPR for an OFDM data block. Assuming that the signal samples are mutually independent and uncorrelated, the CDF of the PAPR for an OFDM data block can be found as

$$P(PAPR \leq z) = F(z)^N = (1 - e^{-z})^N \tag{8.22}$$

(a) $E_b/N_0 = 0$ dB
PSNR = 16.3858

(b) $E_b/N_0 = 5$ dB
PSNR = 23.0375

(c) $E_b/N_0 = 10$ dB
PSNR = 30.3995

(d) $E_b/N_0 = 15$ dB
PSNR = 33.2591

Figure 8.11 Chaotic-encrypted image with DCT-OFDM over a fading channel.

The assumption made that the signal samples are mutually independent and uncorrelated is not true when oversampling is applied. Also, this expression is not accurate for a small number of subcarriers because the Gaussian assumption does not hold in this case [156–159]. It was shown that the PAPR of an oversampled signal for N subcarriers is approximated by the distribution for αN subcarriers without oversampling, where α is larger than 1 [160]. In other words, the effect of oversampling is approximated by adding a certain number of extra signal samples.

The CDF of the PAPR of an oversampled signal is thus given by

$$P(PAPR \leq z) \approx (1 - e^{-z})^{\alpha N} \qquad (8.23)$$

(a) $E_b/N_0 = 0$ dB
PSNR = 14.6839

(b) $E_b/N_0 = 5$ dB
PSNR = 23.7368

(c) $E_b/N_0 = 10$ dB
PSNR = 45.9748

(d) $E_b/N_0 = 15$ dB
PSNR >60

Figure 8.12 Original image with DWT-OFDM over an AWGN channel.

It was found that $\alpha = 2.3$ is a good approximation for four-times oversampled OFDM signals [161]. According to Equation (8.23), the probability that the PAPR exceeds $PAPR_0$ is given by

$$CCDF = \Pr[PAPR > PAPR_0] = 1 - (1 - e^{PAPR_0})^{\alpha N} \qquad (8.24)$$

8.9 PAPR Reduction Methods

Several methods have been presented in the literature for PAPR reduction of OFDM signals. In this book, we concentrate on three such methods: clipping, companding, and hybrid clipping and companding. These are three simple PAPR reduction methods. Our objective of is to show the effect of PAPR reduction methods on encrypted image transmission.

(a) $E_b/N_0 = 0$ dB
PSNR = 17.5731

(b) $E_b/N_0 = 5$ dB
PSNR = 24.9482

(c) $E_b/N_0 = 10$ dB
PSNR = 32.6435

(d) $E_b/N_0 = 15$ dB
PSNR = 37.7109

Figure 8.13   Original image with DWT-OFDM over a fading channel.

8.9.1 The Clipping Method

Clipping is the simplest method that can limit the amplitude of the signal to some desired threshold value. It is a nonlinear process and may cause in-band distortion and out-of-band radiation. This leads to degradation in the BER performance [162,163]. In the clipping method, when the amplitude exceeds a certain threshold, it is hard clipped, while the phase is saved. Namely, when we assume the phase of the baseband OFDM signal $\bar{x}_m$ is ϕ_m and the threshold is A, the output signal after clipping will be given as follows:

$$\bar{x}_m = \left\{ \begin{array}{ll} Ae^{j\phi_m} & |\bar{x}_m| > A \\ \bar{x}_m & |\bar{x}_m| \leq A \end{array} \right\} \tag{8.25}$$

(a) $E_b/N_0 = 0$ dB
PSNR = 14.7090

(b) $E_b/N_0 = 5$ dB
PSNR = 23.5544

(c) $E_b/N_0 = 10$ dB
PSNR = 44.6552

(d) $E_b/N_0 = 15$ dB
PSNR >60

Figure 8.14   Chaotic-encrypted image with DWT-OFDM over an AWGN channel.

The clipping ratio (CR) is a useful means to represent the clipping level. It is the ratio between the maximum power of the clipped signal and the average power of the unclipped signal. If the IFFT/inverse direct cosine transform (IDCT)/inverse direct wavelet transform (IDWT) output signal is normalized, the unclipped signal power is 1. If we clip the IFFT/IDCT/IDWT output at level A, then the CR is $(A)^2 / 1 = (A)^2$.

8.9.2 The Companding Method

The companding transform uses a compander to reduce the signal amplitude. Such an approach can effectively reduce the PAPR with less computational complexity [164]. The corresponding transmitter and receiver need a compander and an expander, respectively, which

(a) $E_b/N_0 = 0$ dB
PSNR = 16.3689

(b) $E_b/N_0 = 5$ dB
PSNR = 23.8108

(c) $E_b/N_0 = 10$ dB
PSNR = 32.7138

(d) $E_b/N_0 = 15$ dB
PSNR = 37.5838

Figure 8.15 Chaotic-encrypted image with DWT-OFDM over a fading channel.

of course slightly increases the hardware cost. In simulation experiments, compression is performed according to the well-known μ-law. The compression process can be expressed as follows [164,165]:

$$x_c = V_{max} \frac{\ln[1 + \mu|\bar{x}_m|/V_{max}]}{\ln[1 + \mu]} \mathrm{sgn}[\bar{x}_m] \tag{8.26}$$

where

$$V_{max} = \max(|\bar{x}_m|) \quad m = 0,1,.....N-1$$

where μ is the companding coefficient, x_c is the companded sample, and $\bar{x}_m$ is the original sample. The expansion process is simply the inverse of Equation (8.26):

$$\hat{\bar{x}}_m = \frac{V_{max}}{\mu}[\exp(\frac{|x_c|\ln(1+\mu)}{V_{max}} - 1)]\mathrm{sgn}(x_c)$$

Table 8.2 Summarized Results

E_b/N_0	$E_b/N_0 = 0$ dB	$E_b/N_0 = 5$ dB	$E_b/N_0 = 10$ dB
ORIGINAL + AWGN CHANNEL			
PSNR with FFT	14.6993	23.7838	48.0593
PSNR with DCT	15.7172	26.3188	53.5993
PSNR with DWT	14.6839	23.7368	45.9748
ORIGINAL + FADING CHANNEL			
PSNR with FFT	17.8214	29.4547	53.8719
PSNR with DCT	16.7843	23.2367	27.9452
PSNR with DWT	17.5731	24.9482	32.6435
CHAOTIC + AWGN CHANNEL			
PSNR with FFT	15.1527	25.1709	50.5529
PSNR with DCT	15.0481	24.4613	53.4971
PSNR with DWT	14.7090	23.5544	44.6552
CHAOTIC + FADING CHANNEL			
PSNR with FFT	18.3597	30.0828	53.8881
PSNR with DCT	16.3858	23.0375	30.3995
PSNR with DWT	16.3689	23.8108	32.7138
DES + AWGN CHANNEL			
PSNR with FFT	8.3766	11.1743	36.4777
PSNR with DCT	8.3762	11.1133	34.4818
PSNR with DWT	8.3954	11.2270	32.2373
DES + FADING CHANNEL			
PSNR with FFT	9.6637	16.7520	41.4821
PSNR with DCT	9.0390	11.7067	16.5200
PSNR with DWT	8.4729	10.7940	17.4856
AES + AWGN CHANNEL			
PSNR with FFT	8.4026	9.4551	31.5464
PSNR with DCT	8.4211	9.4006	30.5205
PSNR with DWT	8.4049	9.3407	28.8218
AES + FADING CHANNEL			
PSNR with FFT	8.6932	14.1673	36.8018
PSNR with DCT	8.5829	10.2435	14.1012
PSNR with DWT	8.3737	9.2508	13.7209
RC6 + AWGN CHANNEL			
PSNR with FFT	8.3916	9.4165	30.9051
PSNR with DCT	8.3961	9.3528	33.1261
PSNR with DWT	8.4344	9.3419	29.1988
RC6 + FADING CHANNEL			
PSNR with FFT	8.7540	13.8893	40.0734
PSNR with DCT	8.5691	10.2276	14.1490
PSNR with DWT	8.3983	9.2497	13.9828

AES, Advanced Encryption Standard; DES, Data Encryption Standard.

where $\hat{\overline{x}}_m$ is the estimated sample after expansion.

8.9.3 The Hybrid Clipping-and-Companding Method

The hybrid clipping-and-companding method comprises clipping followed by μ-law companding. By exploiting the clipping in the first step, the hybrid scheme can reduce the PAPR of OFDM signals with slight BER degradation. Moreover, the companding in the second step further reduces the PAPR of the OFDM signals. Consequently, we expect that the hybrid scheme effectively reduces the PAPR with slight BER degradation, while the complexity of the system is slightly increased.

A complete image transmission system with OFDM comprises a PAPR reduction stage and an error-correction coding stage to reduce the effect of PAPR reduction at the receiver. A convolutional code with rate 1/2, constraint length 7, and octal generator polynomial (133,171) is adopted in the simulation experiments.

8.10 Simulation Experiments of PAPR Reduction Methods

Several experiments were carried out to test the effect of the PAPR reduction methods on the process of encrypted image transmission with OFDM. The compared PAPR reduction methods are the clipping, companding, and hybrid clipping-and-companding methods.

The performance of the process of chaotic image transmission with encrypted (FFT/DCT/DWT)-OFDM systems having $N = 128$ and $G = 32$ over an AWGN is shown in Figures 8.16 to 8.18. From these figures, it is clear that the PSNR performance of the chaotic-encrypted image transmission with FFT/DCT/DWT-OFDM is better with $\mu = 0.1$ than that with $\mu = 4$ at $E_b/N_0 \geq 10$ dB. At high SNR values, it is possible to receive the chaotic encrypted images with a very large PSNR if error-correction codes are utilized.

The performance of the process of encrypted image transmission using the chaotic-OFB algorithm with (FFT/DCT/DWT)-OFDM systems over an AWGN is shown in Figures 8.19 to 8.21. From these figures, it is clear that the PSNR performance of the chaotic-OFB-encrypted image transmission with FFT/DCT/DWT-OFDM is better with $\mu = 0.1$ than with $\mu = 4$ at $E_b/N_0 \geq 10$ dB. At high SNR

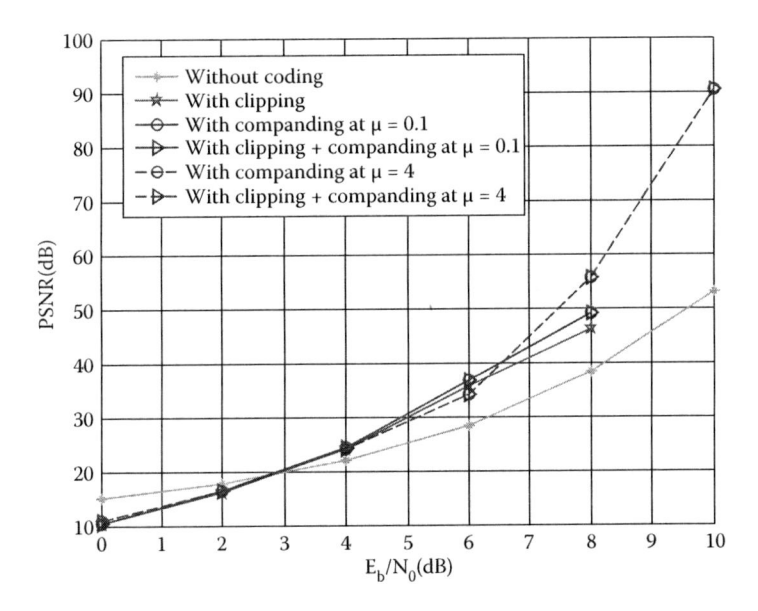

Figure 8.16 PSNR versus E_b/N_0 for chaotic-encrypted image transmission with FFT-OFDM system over an AWGN channel. Convolutional coding is applied with all cases except the first one.

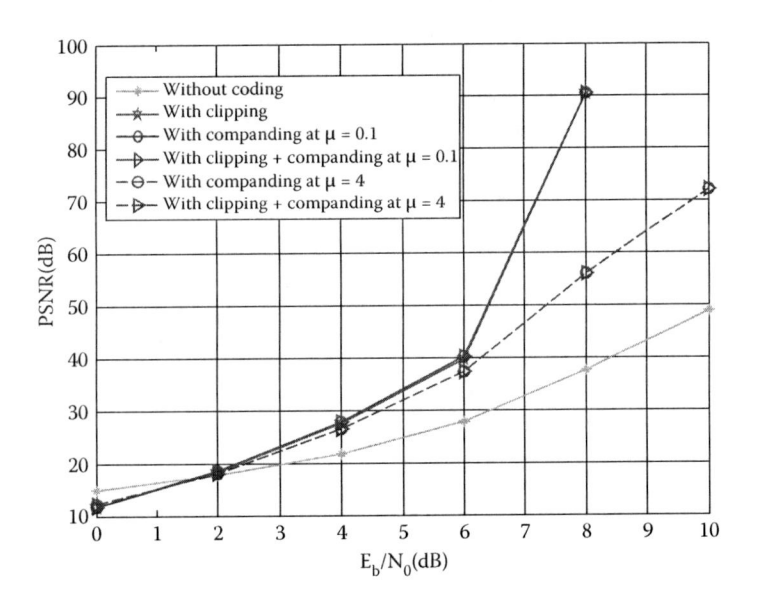

Figure 8.17 PSNR versus E_b/N_0 for chaotic-encrypted image transmission with DCT-OFDM system over an AWGN channel. Convolutional coding is applied with all cases except the first one.

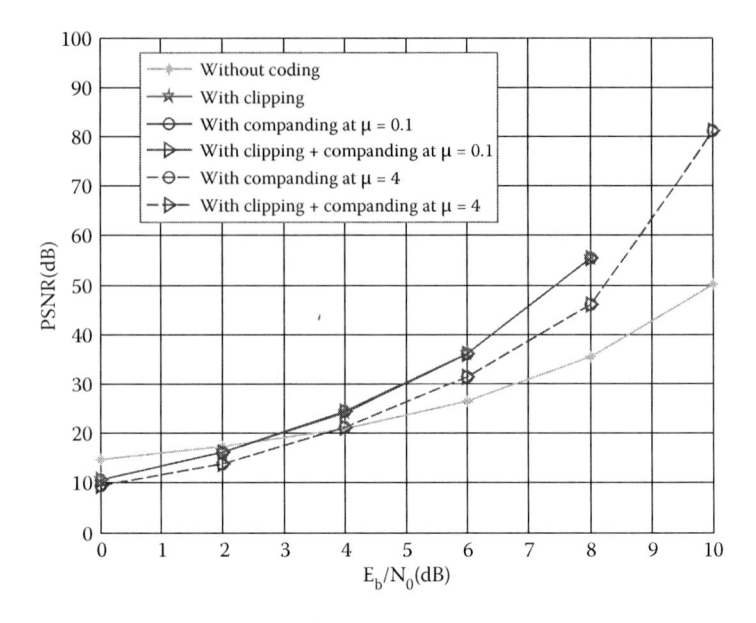

Figure 8.18 PSNR versus E_b/N_0 for chaotic-encrypted image transmission with DWT-OFDM system over an AWGN channel. Convolutional coding is applied with all cases except the first one.

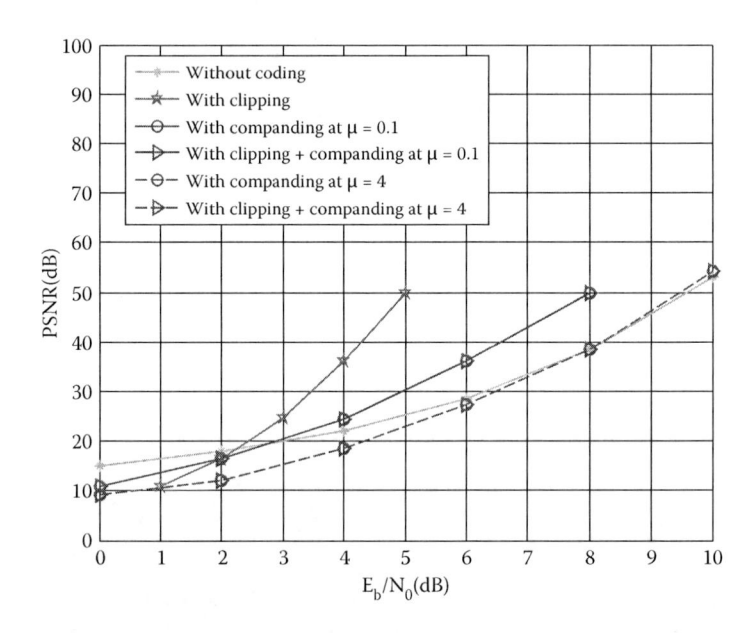

Figure 8.19 PSNR versus E_b/N_0 for chaotic-OFB-encrypted image transmission with FFT-OFDM system over an AWGN channel. Convolutional coding is applied with all cases except the first one.

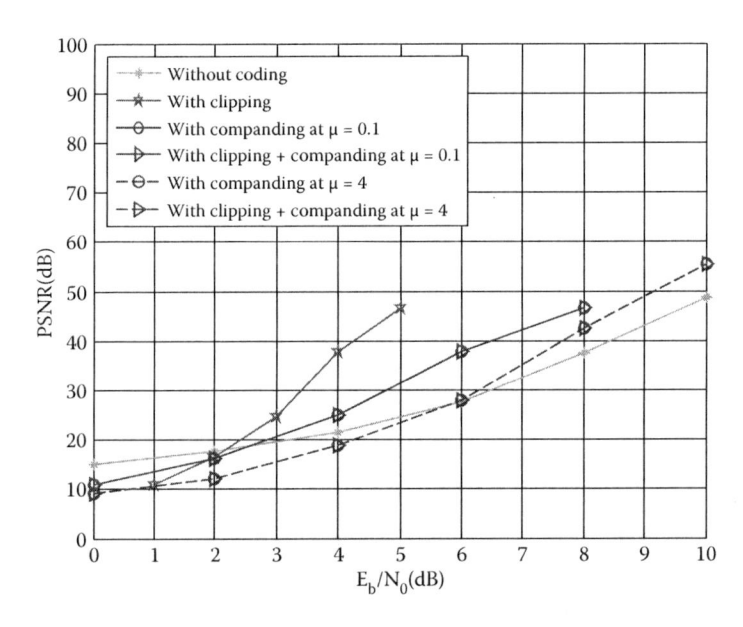

Figure 8.20 PSNR versus E_b/N_0 for chaotic-OFB-encrypted image transmission with DCT-OFDM system over an AWGN channel. Convolutional coding is applied with all cases except the first one.

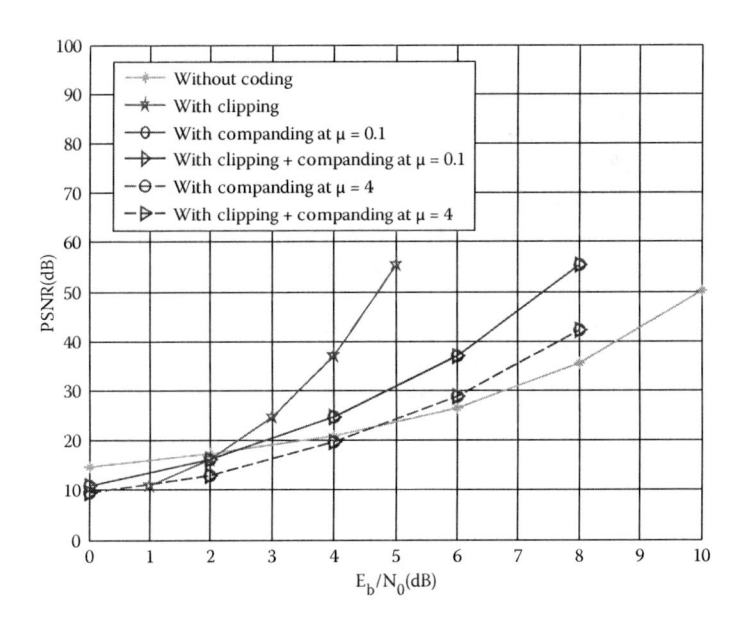

Figure 8.21 PSNR versus E_b/N_0 for chaotic-OFB-encrypted image transmission with DWT-OFDM system over an AWGN channel. Convolutional coding is applied with all cases except the first one.

values, it is possible to receive the chaotic-OFB-encrypted images with a very large PSNR if error-correction codes are utilized. From all obtained results, we can see that PAPR reduction technique effects can be mitigated to a great extent with error-correcting codes such as the convolutional code. Another conclusion is that chaotic-OFB encryption preserves its rank as the best encryption algorithm from the immunity to noise perspective even in the presence of PAPR reduction. A simple clipping, companding, or hybrid clipping-and-companding method may be feasible for PAPR reduction in the case of encrypted image transmission.

8.11 Sampling Clock Offset

The detection of OFDM symbols cannot be done properly without reliable clock synchronization. One synchronization step consists of estimating the OFDM symbol timing, which is the delay between the transmitted and the received OFDM symbols. In a certain number of applications where these symbols are short, estimating this delay is enough. However, as soon as the number of samples per OFDM symbol (or, equivalently, the number of subcarriers) becomes large, the frequency offset between the transmitter sampling clock and the receiver sampling clock in its free oscillation mode also has to be considered. Indeed, this offset leads to a sampling delay that drifts linearly in time over the OFDM symbol. Without any compensation, this drift hampers the receiver performance as soon as the product of the relative clock frequency offset with the number of subcarriers becomes nonnegligible in comparison with one [180]. For instance, in very-high-speed digital subscriber line (VDSL) transmissions, these two quantities can respectively reach 10^{-4} and 4096 [167], making the clock frequency offset compensation mandatory. As another example, power line transmissions (PLTs) in the band (1 MHz, 20 MHz) [168] show similar behavior with respect to this phenomenon.

As it is well known, the part of the OFDM symbol that enters the FFT device at the receiver comes after a cyclic prefix. As the prefix has a length comparable to the channel impulse response length, it is precisely when the channel is long that a long duration has to be chosen for the useful part of the OFDM symbol to reduce the impact of the cyclic prefix on the spectral efficiency. It is therefore worth considering the problem of the joint estimation of the clock frequency offset and of the channel impulse response, particularly when the observation window has to be rather large [166–168].

8.11.1 System Model

Let us consider the reception of one standard OFDM block that has passed through a nonflat fading channel. After removing the guard interval, the observation window size is $T_o = NT$ where N is the number of subcarriers, T is the sampling period at the transmitter, and $1/T_o$ is the spacing between two adjacent subcarriers. Consequently, the continuous-time received signal $y_N^{(a)}(t)$ is written as follows:

$$y_N^{(a)}(t) = \sum_{k \in Z} d_{N,k} g^{(a)}(t - kT) + v^{(a)}(t) \qquad (8.27)$$

where $(d_{N,k})_{k=0,1,\dots,N-1}$ represents the output of the N-fold IFFT device of the transmitter. This OFDM symbol is devoted to training and therefore is assumed to be known at the receiver. As usual, N is a power of 2. The unknown impulse response $g^{(a)}(t)$ represents the complete channel that includes the transmit filter, the propagation channel, and the receiver low-pass filter. Finally. $v^{(a)}(t)$ is an additive noise independent of the data. Because of the oscillators' imperfection, the transmitter and receiver clocks are not synchronized.

Therefore, $y_N^{(a)}(t)$ is sampled at $(1 + \delta)T$ instead of T, where δ is an unknown offset lying in the known interval $[-\delta_{max}, \delta_{max}]$. The parameter δ_{max} is related to the precision of the oscillators used in the transmission chain. The asymmetric digital subscriber line (ASDL)/VDSL norms [167], for instance, recommend that δ_{max} be equal to 10^{-4}. The discrete time received signal $y_N(n) = y_N^{(a)}(n(1+\delta)T)$ is then written

$$y_N(n) = \sum_{m \in Z} d_{N,n-m} g^{(a)}(mT + n\delta T) + v(n) \qquad (8.28)$$

where $v(n) = v^{(a)}(n(1+\delta)T)$ is assumed white Gaussian circular with zero mean and known variance $\sigma^2 = E[|v(n)|^2]$. As usual, $g^{(a)}(lT)$ is assumed time limited with the time support included in $[0, LT]$ where L is a known integer. We thus write $g_l = g^{(a)}(lT)$ for $l = 0, \dots, L-1$.

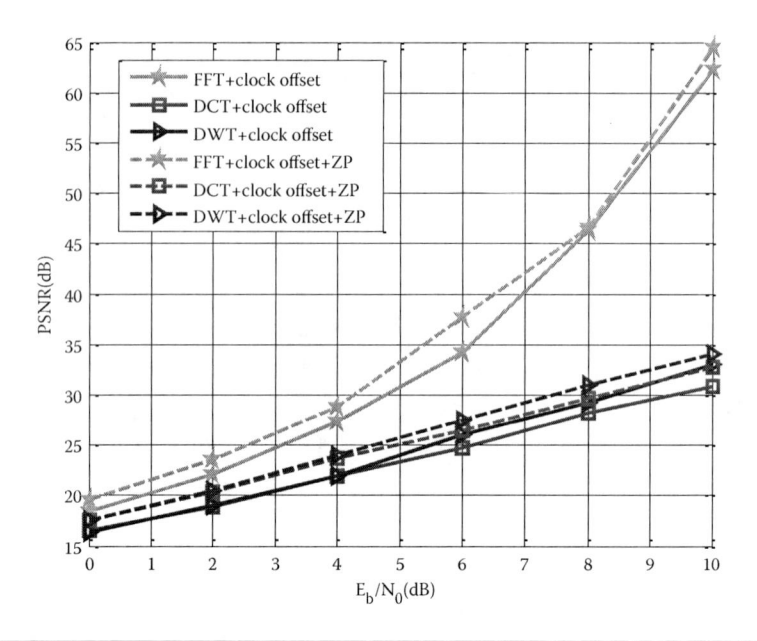

Figure 8.22 PSNR versus E_b/N_0 for chaotic-encrypted image transmission with OFDM system over a Raleigh fading channel.

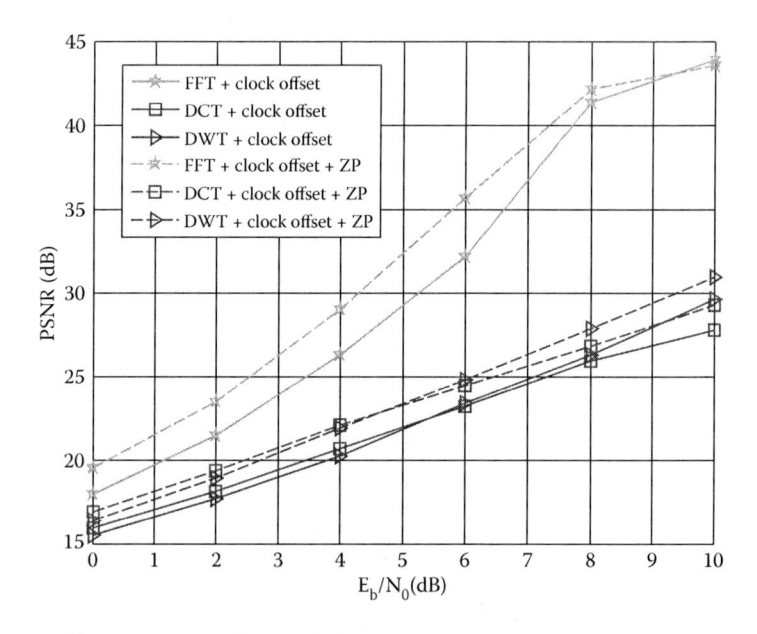

Figure 8.23 PSNR versus E_b/N_0 for chaotic-OFB-encrypted image transmission with OFDM system over a Raleigh fading channel.

The performance of the process of original image transmission with the chaotic and chaotic-OFB algorithm for (FFT/DCT/DWT)-OFDM systems over a Raleigh fading channel is shown in Figures 8.22 and 8.23. From these figures, it is clear that the PSNR performance of both chaotic and chaotic-OFB-encrypted image transmission with FFT/DCT/DWT-OFDM is better with applying the zero padding (ZP) guard and PSE method. At high E_b/N_0 values, it is possible to receive the chaotic-OFB-encrypted images with a very large PSNR if error-correction codes are utilized. From all obtained results, we can see that sampling clock offset reduction technique effects can be mitigated to a great extent. Another conclusion is that the chaotic-OFB encryption preserves its rank as the best encryption algorithm from the immunity to noise point of view even in the presence of sampling clock offset.

9
SIMULATION EXAMPLES

9.1 Simulation Parameters

In our simulations, we carried out the encryption of the Cameraman image with the previously discussed encryption algorithms. Then, the encrypted images were transformed into a binary format and used for orthogonal frequency division multiplexing (OFDM) modulation. Both additive white Gaussian noise (AWGN) and Rayleigh fading channels were considered in the simulations. A normalized carrier frequency offset (CFO) of 0.1 was used in some of the simulation experiments. PAPR reduction techniques were used in some other experiments. In all experiments, 128 subcarriers and quaternary phase shift keying (QPSK) were used. The simulation parameters are summarized in Table 9.1.

To evaluate the quality of the decrypted images at the receiver, we used the peak signal-to-noise ratio (PSNR) between the original image and the decrypted image. Figures 9.1 to 9.3 illustrate the variation of the PSNR of the decrypted image with the signal-to-noise ratio (SNR) in the channel (E_b/N_0) for the fast Fourier transform (FFT)-OFDM system with different encryption algorithms in an AWGN channel for different numbers of subcarriers. These figures showed that as the number of subcarriers is increased, the quality of the decrypted images is improved.

In this section, several experiments compare between the encryption algorithms and select the most suitable system for the transmission of encrypted images. To better illustrate the results, each encryption algorithm was studied for a different number of subchannels as shown in Figures 9.4 to 9.6 without equalization and Figures 9.7 to 9.9 with zero-forcing (ZF) equalization. From these figures, it is clear that equalization enhances the system performance dramatically. The best-obtained results from an encryption algorithm were for the chaotic encryption algorithm because it is a permutation-based ciphering algorithm.

Table 9.1   Simulation Parameters

	PARAMETER	DESCRIPTION
Transmitter	System bandwidth	64 MHz
	Modulation type	QPSK
	CP length	16 samples
	N	128 subcarriers
	Subcarrier spacing	500 kHz
	Type of transform	FFT, DCT, or DWT
Channel	Channel model	AWGN or Rayleigh fading
Receiver	PSE and equalizer	ZF equalization
Encryption algorithm	Type of encryption	Chaotic encryption, DES, AES, or RC6
	Key of DES	(00010203040506)16
	Key of RC6	(000102030405060708090a0b0c0d0e0f)16
	Key of AES	(000102030405060708090a0b0c0d0e0f)16

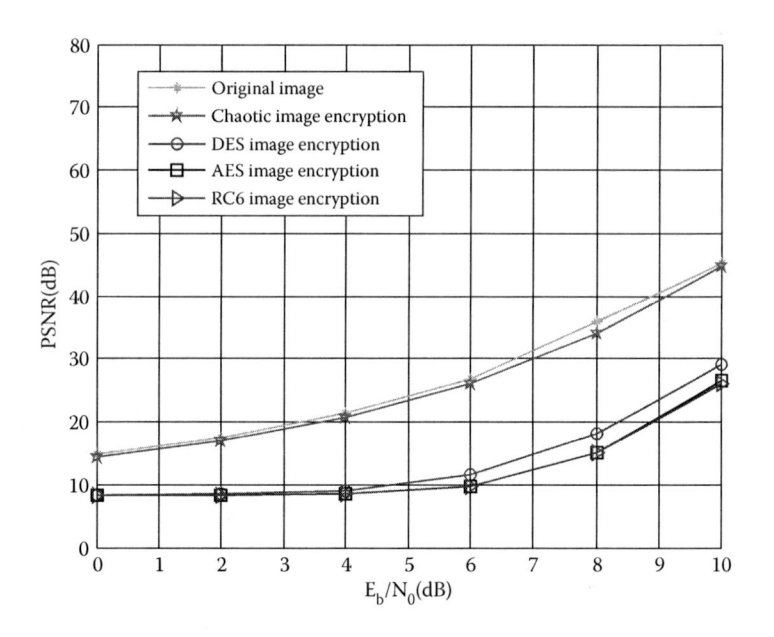

Figure 9.1   PSNR vs. E_b/N_0 in the FFT-OFDM system for an AWGN channel with $N = 64$.

The Data Encryption Standard (DES), the Advanced Encryption Standard (AES), and the RC6 encryption algorithms provided lower PSNR performance for the FFT-OFDM system compared with the chaotic algorithm because these algorithms have a diffusion mechanism, which reduces the noise immunity. On the other hand, the chaotic algorithm is more robust to noise. From these

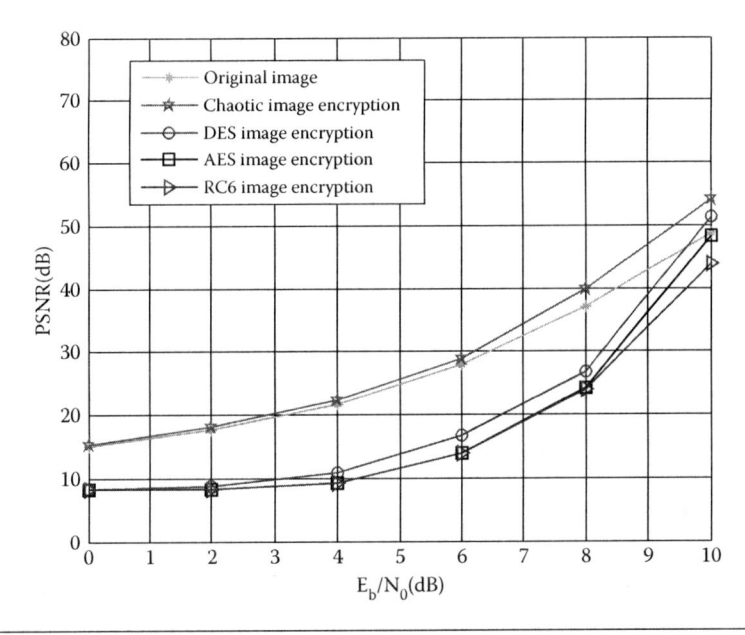

Figure 9.2 PSNR versus E_b/N_0 in the FFT-OFDM system for an AWGN channel with $N = 512$.

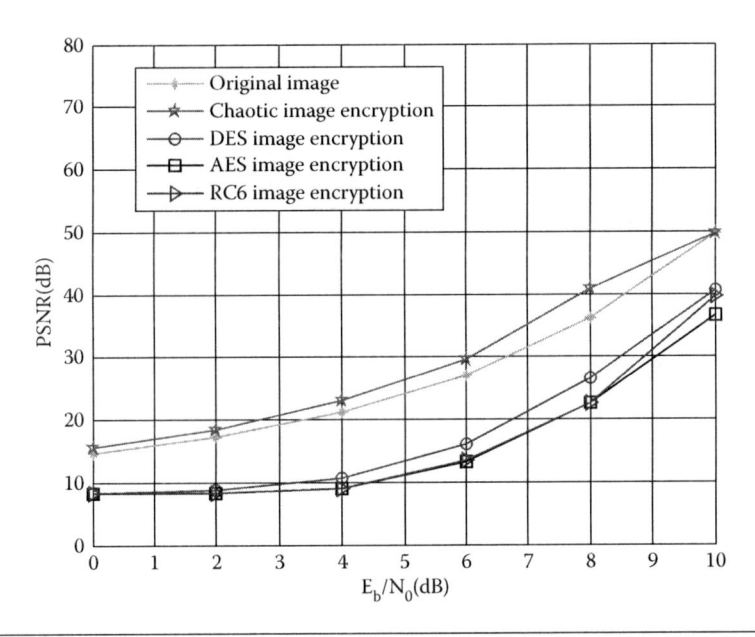

Figure 9.3 PSNR versus E_b/N_0 in the FFT-OFDM system for an AWGN channel with $N = 1024$.

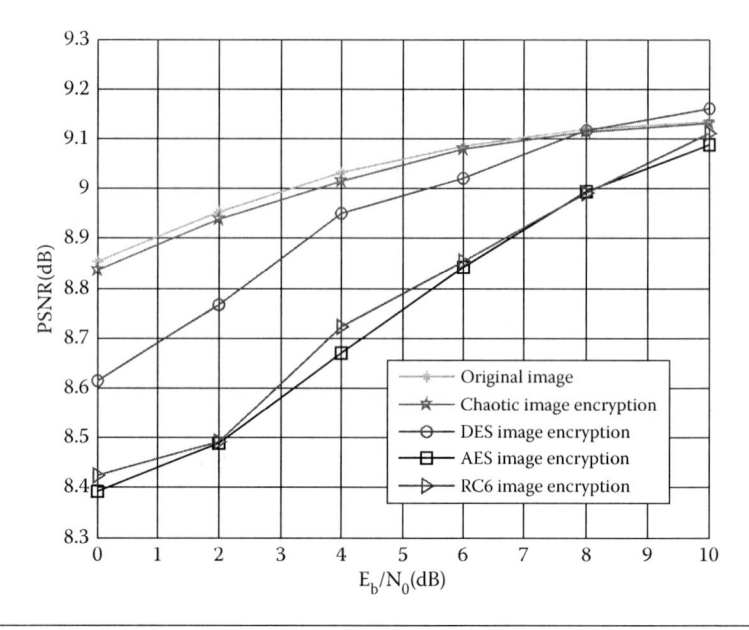

Figure 9.4   PSNR versus E_b/N_0 in the FFT-OFDM system for a Raleigh channel with $N = 64$ before PSE and equalization.

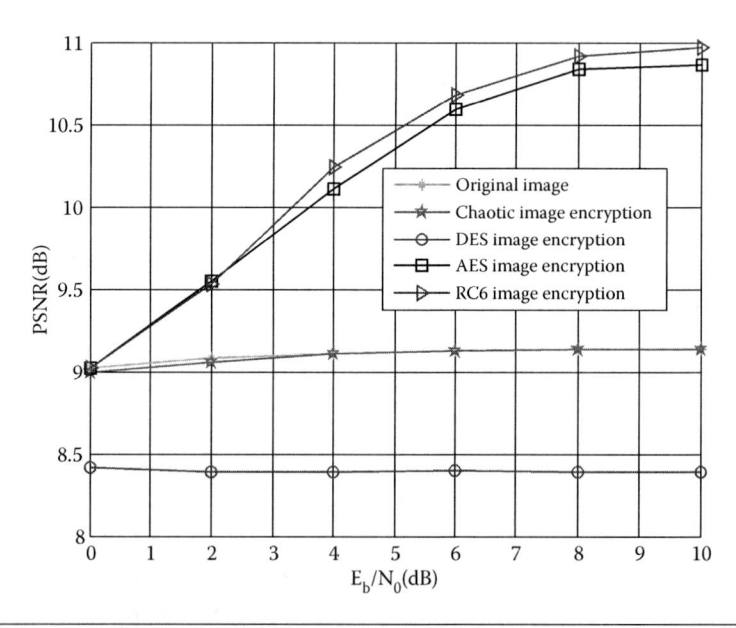

Figure 9.5   PSNR versus E_b/N_0 in the FFT-OFDM system for a Raleigh channel with $N = 512$ before PSE and equalization.

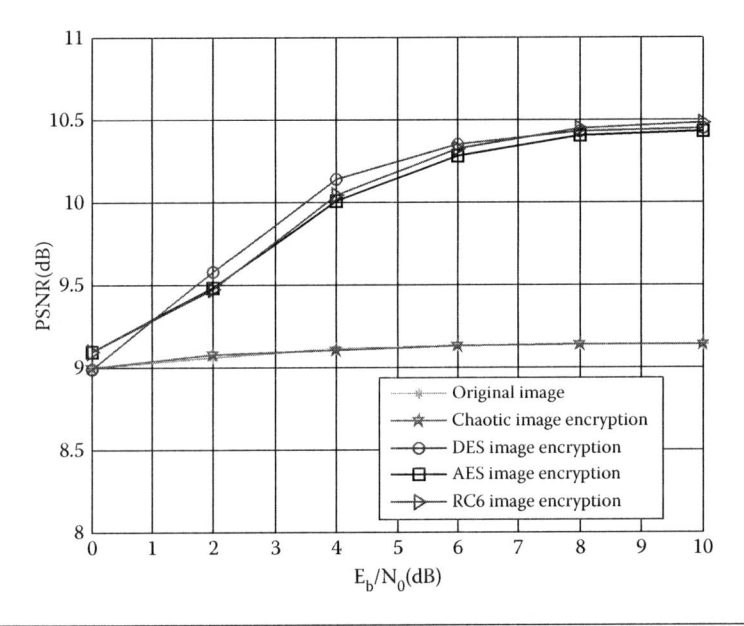

Figure 9.6 PSNR versus E_b/N_0 in the FFT-OFDM system for a Raleigh channel with $N = 1024$ before PSE and equalization.

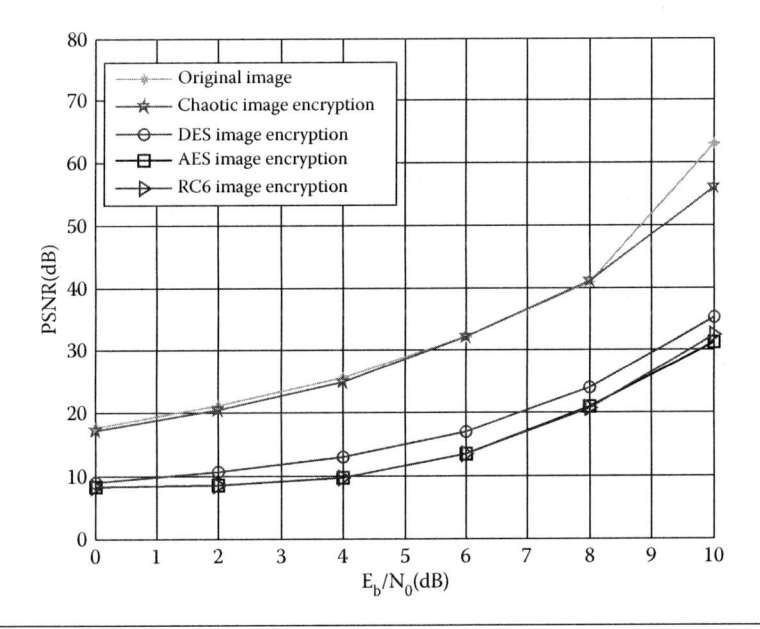

Figure 9.7 PSNR versus E_b/N_0 in the FFT-OFDM system for a Raleigh channel with $N = 64$ after PSE and equalization.

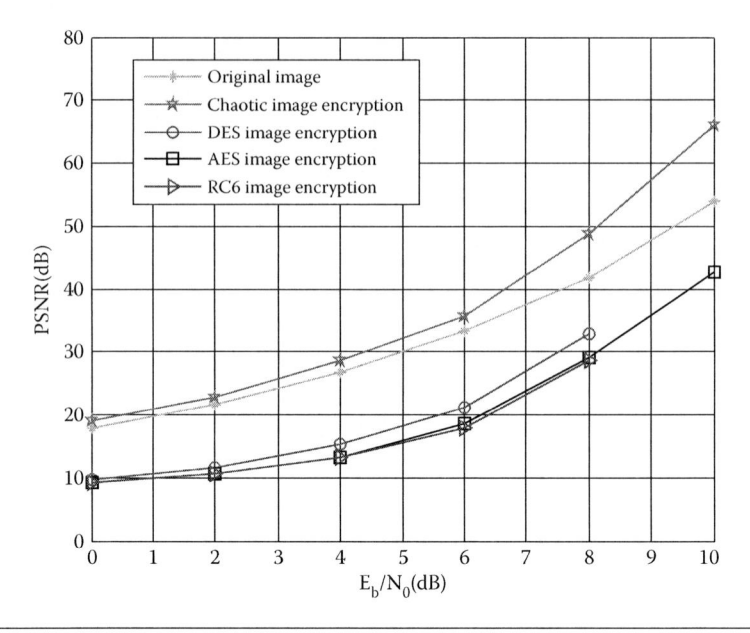

Figure 9.8   PSNR versus E_b/N_0 in the FFT-OFDM system for a Raleigh channel with $N = 512$ after PSE and equalization.

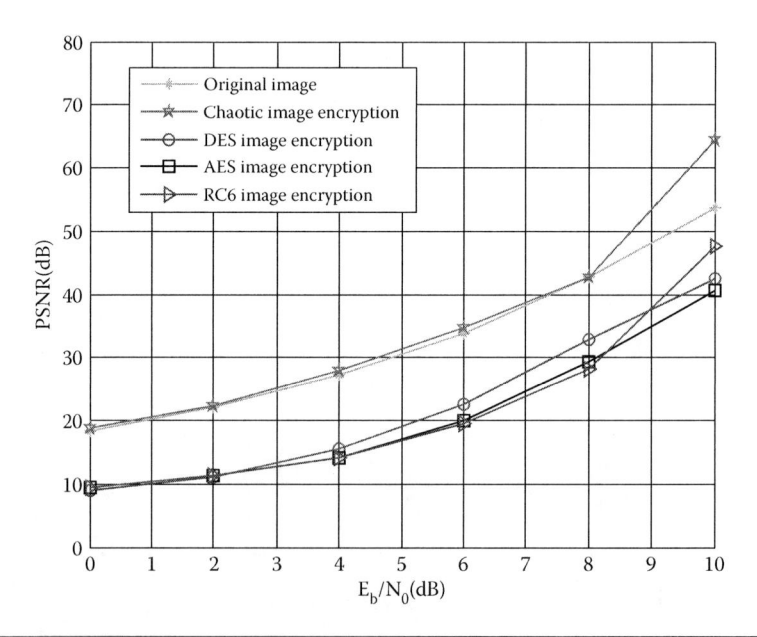

Figure 9.9   PSNR versus E_b/N_0 in the FFT-OFDM system for a Raleigh channel with $N = 1024$ after PSE and equalization.

figures, it is clear that equalization enhances the system performance dramatically. The best-obtained results from an encryption algorithm were for the chaotic encryption algorithm because it is a permutation-based ciphering algorithm.

The effect of the guard interval length on the system was studied; Figures 9.10 to 9.12 illustrate the PSNR performance versus the E_b/N_0

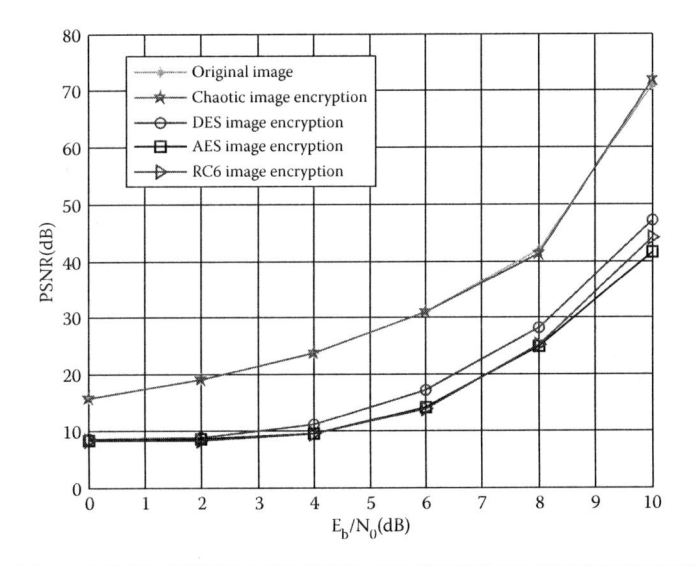

Figure 9.10   PSNR versus E_b/N_0 in the FFT-OFDM system for an AWGN channel with $G = 0$.

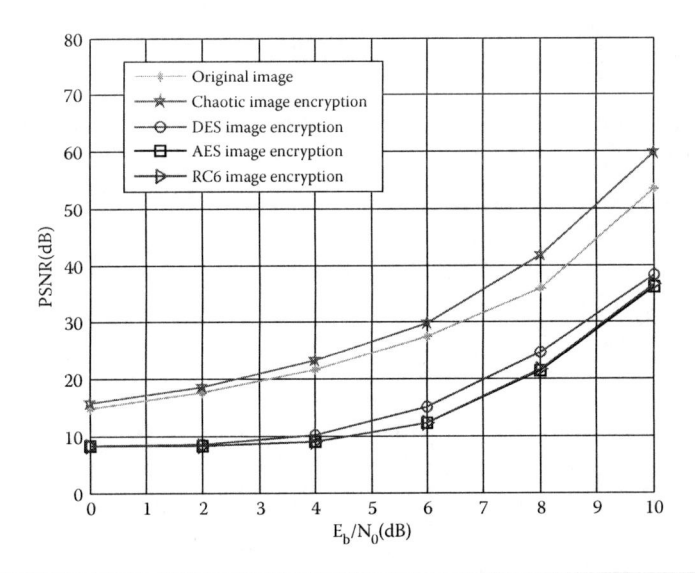

Figure 9.11   PSNR versus E_b/N_0 in the FFT-OFDM system for an AWGN channel with $G = 16$.

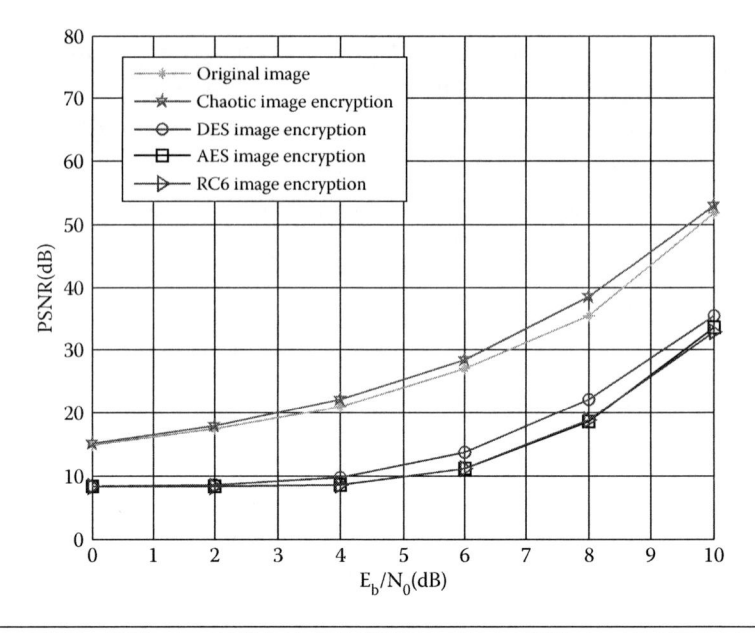

Figure 9.12   PSNR versus E_b/N_0 in the FFT-OFDM system for an AWGN channel with $G = 32$.

for the FFT-OFDM systems with different encryption algorithms in an AWGN channel. The guard intervals utilized ranged from 0 to 32 samples. From these figures, it is clear that the PSNR performances of all encryption algorithms with the FFT-OFDM system at $G = 0$ were the best compared with those with $G = 16$ and 32. The chaotic-encryption gives the best PSNR values.

Figures 9.13 to 9.15 illustrate the variation of the PSNR with the E_b/N_0 for the FFT-OFDM with $G = [0, 16, 32]$, with different encryption algorithms in a frequency-selective channel. Without pilot symbol estimation (PSE), it was shown that fading disrupts the PSNR performance of FFT-OFDM systems regardless of the encryption algorithm. However, Figures 9.16 to 9.18 show that, with PSE, the PSNR performances of all systems were better than those without PSE.

The DES, AES, and RC6 provide lower PSNR performance for the FFT-OFDM system than the chaotic algorithm because these algorithms have a diffusion mechanism, which highly reduces the noise immunity. On the other hand, the chaotic algorithm is more robust to noise.

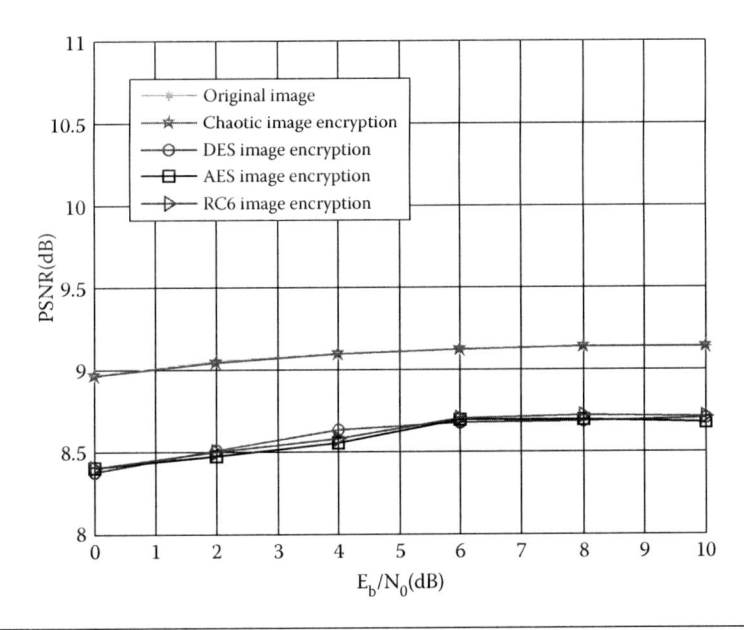

Figure 9.13 PSNR versus E_b/N_0 in the FFT-OFDM system for a Raleigh channel with $G = 0$ without PSE and equalization.

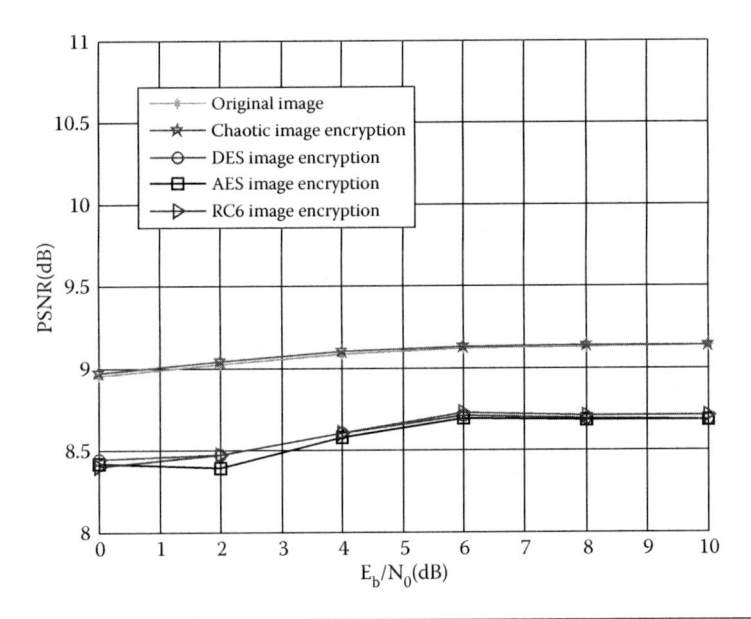

Figure 9.14 PSNR versus E_b/N_0 in the FFT-OFDM system for a Raleigh channel with $G = 16$ without PSE and equalization.

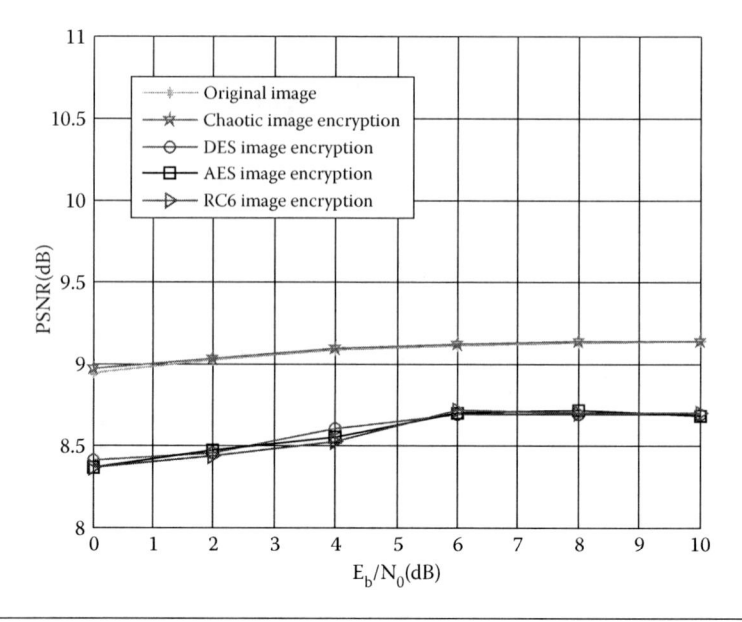

Figure 9.15   PSNR versus E_b/N_0 in the FFT-OFDM system for a Raleigh channel with $G = 32$ without PSE and equalization.

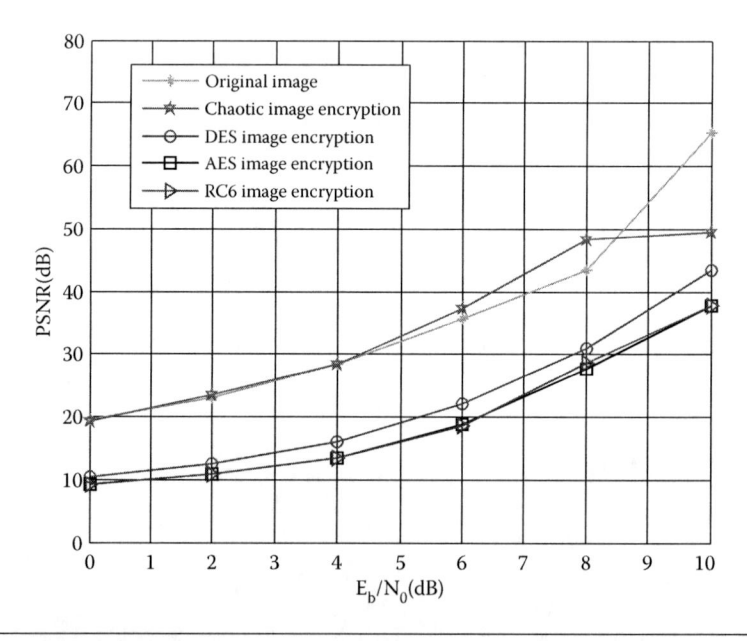

Figure 9.16   PSNR versus E_b/N_0 in the FFT-OFDM system for a Raleigh channel with $G = 0$ with PSE and equalization.

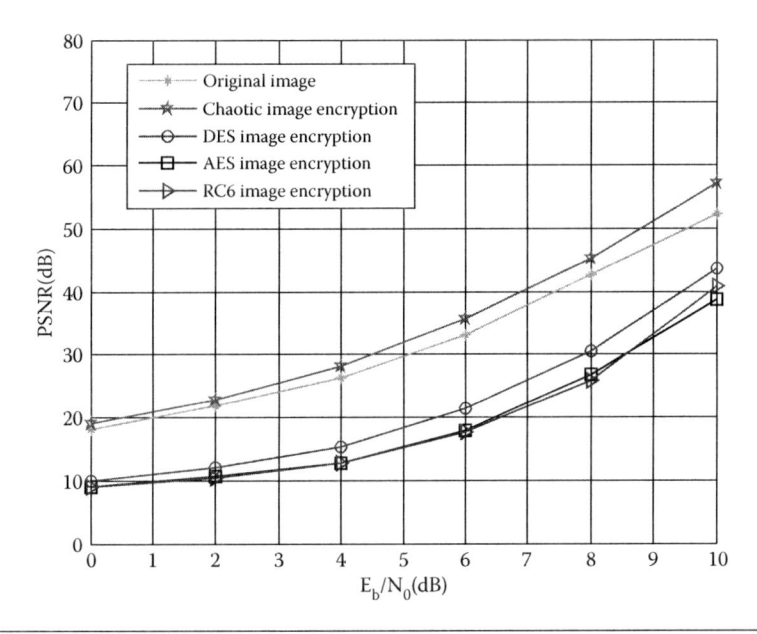

Figure 9.17 PSNR versus E_b/N_0 in the FFT-OFDM system for a Raleigh channel with $G = 16$ with PSE and equalization.

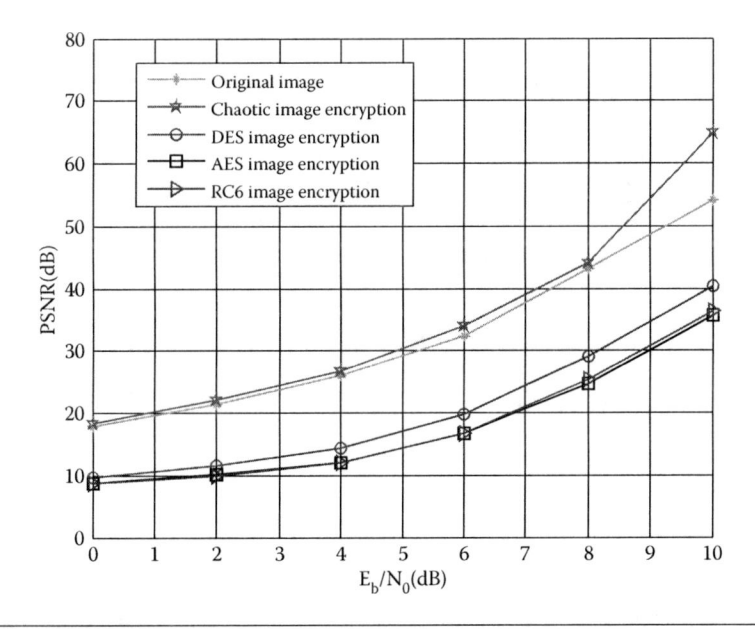

Figure 9.18 PSNR versus E_b/N_0 in the FFT-OFDM system for a Raleigh channel with $G = 32$ with PSE and equalization.

The different guard intervals were studied. A comparison study between the hybrid guard interval (HGI), which is composed of a cyclic prefix (CP) and zero samples, and the zero padding (ZP) guard interval with the different encryption algorithms in a frequency-selective channel is illustrated in Figures 9.19 and 9.20. CFO compensation was performed. It is shown that the PSNR performance of the ZP-OFDM systems is slightly better than that of the HGI-OFDM systems.

Figure 9.21 gives the output of the FFT-OFDM system for the transmission of the original Cameraman image over an AWGN channel at $N = 128$ and $G = 32$.

Figure 9.22 gives the output of the FFT-OFDM system for the transmission of the original Cameraman image over a fading channel at $N = 128$ and $G = 32$.

Figure 9.23 gives the output of the FFT-OFDM system for the transmission of the chaotic-encrypted image over an AWGN channel at $N = 128$ and $G = 32$.

Figure 9.24 gives the output of the FFT-OFDM system for the transmission of the chaotic-encrypted image over a fading channel at $N = 128$ and $G = 32$.

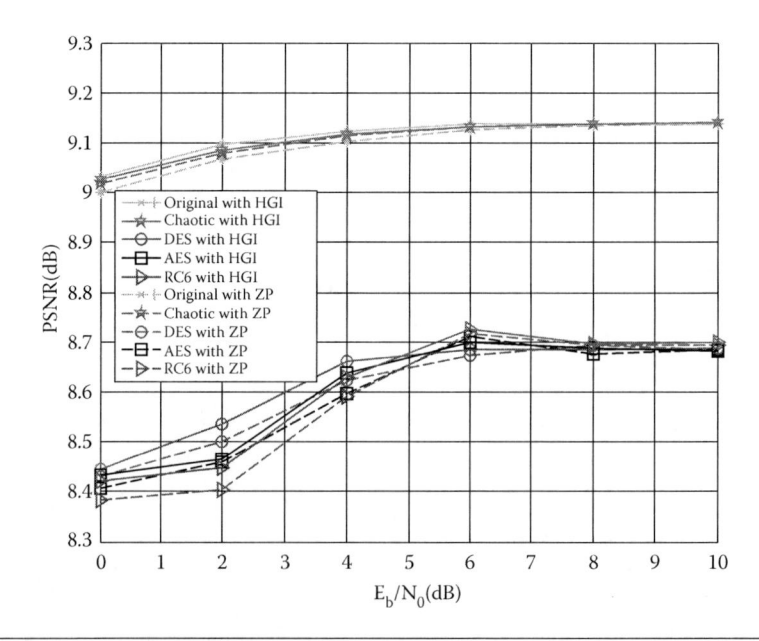

Figure 9.19 PSNR versus E_b/N_0 in the FFT-OFDM system with ZP and hybrid G for a Raleigh channel with $N = 128$ and $G = 32$ without PSE and equalization.

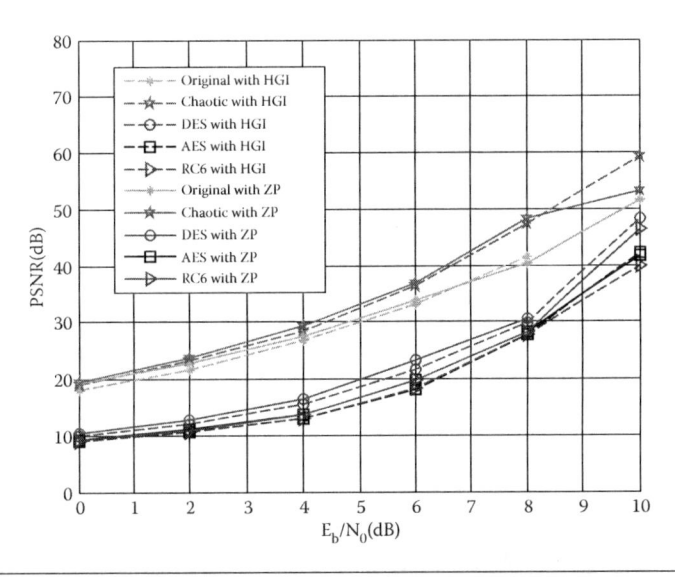

Figure 9.20   PSNR versus E_b/N_0 in the FFT-OFDM system with ZP and HGI for a Raleigh channel with $N = 128$ and $G = 32$ with PSE and equalization.

(a) $E_b/N_0 = 0$ dB
PSNR = 14.6993

(b) $E_b/N_0 = 5$ dB
PSNR = 23.7838

(c) $E_b/N_0 = 10$ dB
PSNR = 48.0593

(d) $E_b/N_0 = 15$ dB
PSNR >60

Figure 9.21   Original image with FFT-OFDM over an AWGN channel.

(a) $E_b/N_0 = 0$ dB
PSNR = 17.8214

(b) $E_b/N_0 = 5$ dB
PSNR = 29.4547

(c) $E_b/N_0 = 10$ dB
PSNR = 53.8719

(d) $E_b/N_0 = 15$ dB
PSNR >60

Figure 9.22　Original image with FFT-OFDM over a fading channel.

Figure 9.25 gives the output of the FFT-OFDM system for the transmission of the DES-encrypted image over an AWGN channel at $N = 128$ and $G = 32$.

Figure 9.26 gives the output of the FFT-OFDM system for the transmission of the DES-encrypted image over a fading channel at $N = 128$ and $G = 32$.

Figure 9.27 gives the output of the FFT-OFDM system for the transmission of the AES-encrypted image over an AWGN channel at $N = 128$ and $G = 32$.

Figure 9.28 gives the output of the FFT-OFDM system for the transmission of the AES-encrypted image over a fading channel at $N = 128$ and $G = 32$.

(a)E_b/N_0 = 0 dB
PSNR = 15.1527

(b) E_b/N_0 = 5 dB
PSNR = 25.1709

(c) E_b/N_0 = 10 dB
PSNR = 50.5529

(d) E_b/N_0 = 15 dB
PSNR >60

Figure 9.23 Chaotic-encrypted image with FFT-OFDM over an AWGN channel.

Figure 9.29 gives the output of the FFT-OFDM system for the transmission of the RC6-encrypted image over an AWGN channel at N = 128 and G = 32.

Figure 9.30 gives the output of the FFT-OFDM system for the transmission of the RC6-encrypted image over a fading channel at N = 128 and G = 32.

Figures 9.31 to 9.33 illustrate the variation of the PSNR of a decrypted image with the SNR in the channel (E_b/N_0) for the DCT-OFDM system with different encryption algorithms in an AWGN channel for different numbers of subcarriers. These figures show that as the number of subcarriers is increased, the quality of the decrypted images improved.

(a) E_b/N_0 = 0 dB
PSNR = 18.3597

(b) E_b/N_0 = 5 dB
PSNR = 30.0828

(c) E_b/N_0 = 10 dB
PSNR = 53.8881

(d) E_b/N_0 = 15 dB
PSNR >60

Figure 9.24   Chaotic-encrypted image with FFT-OFDM over a fading channel.

To better illustrate the results, each encryption algorithm was studied for a different number of subchannels as shown in Figures 9.34 to 9.36 without equalization and Figures 9.37 to 9.39 with ZF equalization. From these figures, it is clear that equalization enhances the system performance dramatically. The best-obtained results from an encryption algorithm were for the chaotic encryption algorithm because it is a permutation-based ciphering algorithm.

The effect of the guard interval length on the system is studied in Figures 9.40 to 9.42, which illustrate the PSNR performance versus the E_b/N_0 for the DCT-OFDM systems with different encryption algorithms in an AWGN channel. The guard intervals utilized range from 0 to 32 samples. From these figures, it is clear that the PSNR performances of all encryption algorithms with the DCT-OFDM

(a) $E_b/N_0 = 0$ dB
PSNR = 8.3766

(b) $E_b/N_0 = 5$ dB
PSNR = 11.1743

(c) $E_b/N_0 = 10$ dB
PSNR = 36.4777

(d) $E_b/N_0 = 15$ dB
PSNR >60

Figure 9.25   DES-encrypted image with FFT-OFDM over an AWGN channel.

system at $G = 0$ are the best compared with those with $G = 16$ and 32. The chaotic-encrypted image gives the best PSNR values.

Figures 9.43 to 9.45 illustrate the variation of the PSNR with the E_b/N_0 for the DCT-OFDM with $G = [0, 16, 32]$, with different encryption algorithms in a frequency-selective channel. Without PSE, it is shown that fading disrupts the PSNR performance of the DCT-OFDM system, regardless of the encryption algorithm. However, Figures 9.46 to 9.48 show that, with PSE, the PSNR performances of all systems were better than those without PSE.

The different types of guard intervals were studied. A comparison study between the HGI, which is composed of a CP and zero samples, and the ZP guard interval with the different encryption algorithms in a frequency-selective channel is illustrated in Figures 9.49 and 9.50.

(a) $E_b/N_0 = 0$ dB
PSNR = 9.6637

(b) $E_b/N_0 = 5$ dB
PSNR = 16.7520

(c) $E_b/N_0 = 10$ dB
PSNR = 41.4821

(d) $E_b/N_0 = 15$ dB
PSNR >60

Figure 9.26   DES-encrypted image with FFT-OFDM over a fading channel.

CFO compensation was performed. It is shown that the PSNR performance of the ZP-OFDM system was slightly better than that of the HGI-OFDM system.

Figure 9.51 gives the output of the DCT-OFDM system for the transmission of the original Cameraman image over an AWGN channel at $N = 128$ and $G = 32$.

Figure 9.52 gives the output of the DCT-OFDM system for the transmission of the original Cameraman image over a fading channel at $N = 128$ and $G = 32$.

Figure 9.53 gives the output of the DCT-OFDM system for the transmission of the chaotic-encrypted image over an AWGN channel at $N = 128$ and $G = 32$.

(a) $E_b/N_0 = 0$ dB
PSNR = 8.4026

(b) $E_b/N_0 = 5$ dB
PSNR = 9.4551

(c) $E_b/N_0 = 10$ dB
PSNR = 31.5464

(d) $E_b/N_0 = 15$ dB
PSNR >60

Figure 9.27   AES-encrypted image with FFT-OFDM over an AWGN channel.

Figure 9.54 gives the output of the DCT-OFDM system for the transmission of the chaotic-encrypted image over a fading channel at $N = 128$ and $G = 32$.

Figure 9.55 gives the output of the DCT-OFDM system for the transmission of the DES-encrypted image over an AWGN channel at $N = 128$ and $G = 32$.

Figure 9.56 gives the output of the DCT-OFDM system for the transmission of the DES-encrypted image over a fading channel at $N = 128$ and $G = 32$.

Figure 9.57 gives the output of the DCT-OFDM system for the transmission of the AES-encrypted image over an AWGN channel at $N = 128$ and $G = 32$.

(a) E_b/N_0 = 0 dB
PSNR = 8.6932

(b) E_b/N_0 = 5 dB
PSNR = 14.1673

(c) E_b/N_0 = 10 dB
PSNR = 36.8018

(d) E_b/N_0 = 15 dB
PSNR >60

Figure 9.28   AES-encrypted image with FFT-OFDM over a fading channel.

Figure 9.58 gives the output of the DCT-OFDM system for the transmission of the AES-encrypted image over a fading channel at N = 128 and G = 32.

Figure 9.59 gives the output of the DCT-OFDM system for the transmission of the RC6-encrypted image over an AWGN channel at N = 128 and G = 32.

Figure 9.60 gives the output of the DCT-OFDM system for the transmission of the RC6-encrypted image over a fading channel at N = 128 and G = 32.

Figures 9.61 to 9.63 illustrate the variation of the PSNR of a decrypted image with the SNR in the channel (E_b/N_0) for the DWT-OFDM system with different encryption algorithms in an AWGN channel for different numbers of subcarriers. These figures show that

(a) E_b/N_0 = 0 dB
PSNR = 8.3916

(b) E_b/N_0 = 5 dB
PSNR = 9.4165

(c) E_b/N_0 = 10 dB
PSNR = 30.9051

(d) E_b/N_0 = 15 dB
PSNR >60

Figure 9.29   RC6-encrypted image with FFT-OFDM over an AWGN channel.

as the number of subcarriers is increased, the quality of the decrypted images is improved.

Each encryption algorithm was studied for a different number of subchannels as shown in Figures 9.64 to 9.66 without equalization and Figures 9.67 to 9.69 with ZF equalization. From these figures, it is clear, as obtained in the previous results, that equalization enhanced system performance. The best-obtained results from an encryption algorithm were for the chaotic encryption algorithm because it is a permutation-based ciphering algorithm. The DES, the AES, and the RC6 encryption algorithms provides worse PSNR performance for the DWT-OFDM system compared with the chaotic algorithm because these algorithms have a diffusion mechanism that reduces the

(a) $E_b/N_0 = 0$ dB
PSNR = 8.7540

(b) $E_b/N_0 = 5$ dB
PSNR = 13.8893

(c) $E_b/N_0 = 10$ dB
PSNR = 40.0734

(d) $E_b/N_0 = 15$ dB
PSNR >60

Figure 9.30   RC6-encrypted image with FFT-OFDM over a fading channel.

noise immunity. On the other hand, the chaotic algorithm is more robust to noise.

The effect of the guard interval length on the system was studied; Figures 9.70 to 9.72 illustrate the PSNR performance versus the E_b/N_0 for the DWT-OFDM systems with different encryption algorithms in an AWGN channel. The guard intervals utilized ranged from 0 to 32 samples. From these figures, it is clear that the PSNR performances of all encryption algorithms with the DWT-OFDM system at $G = 0$ were the best compared with those with $G = 16$ and 32. The chaotic encryption gives the best PSNR values.

Figures 9.73 to 9.75 illustrate the variation of the PSNR with the E_b/N_0 for the DWT-OFDM with $G = [0, 16, 32]$, with different encryption algorithms in a frequency-selective channel. Without

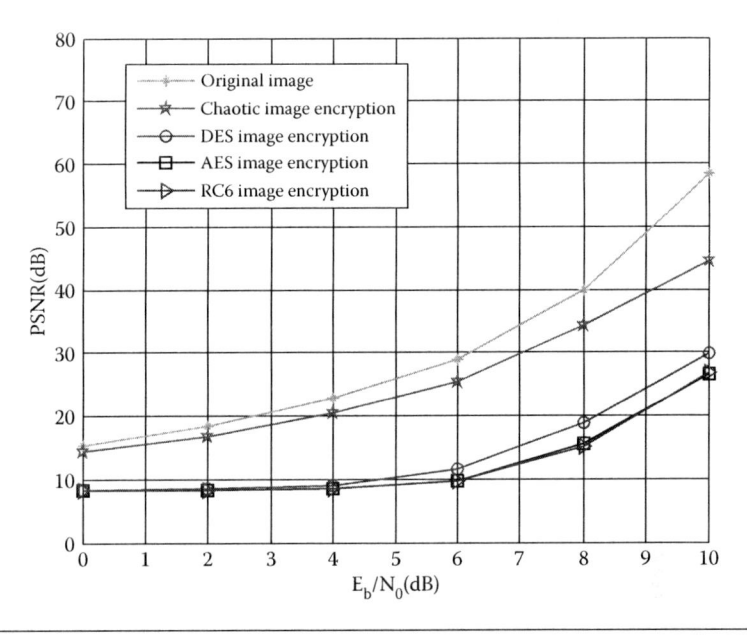

Figure 9.31 PSNR versus E_b/N_0 in the DCT-OFDM system for an AWGN channel with $N = 64$.

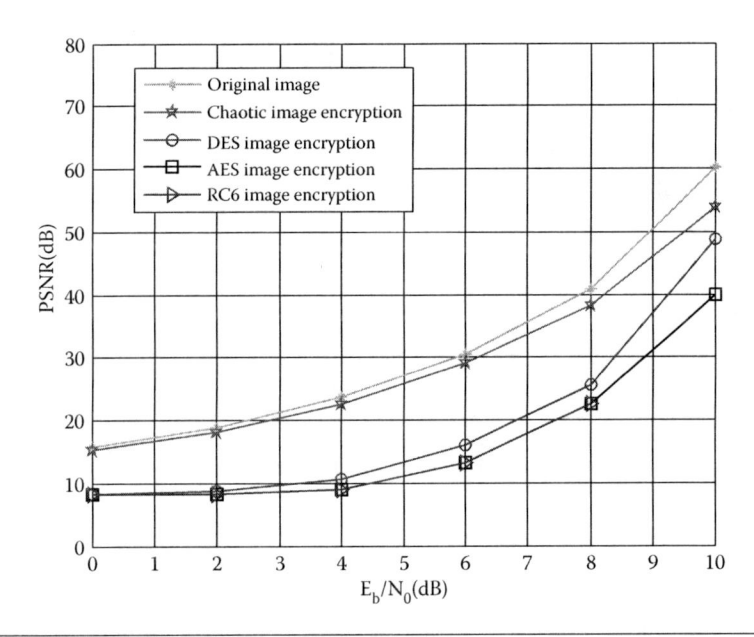

Figure 9.32 PSNR versus E_b/N_0 in the DCT-OFDM system for an AWGN channel with $N = 512$.

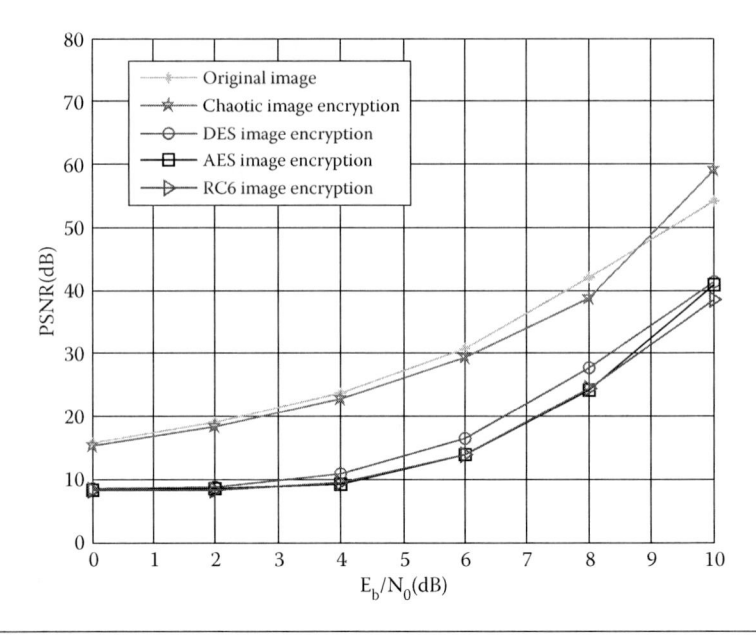

Figure 9.33   PSNR versus E_b/N_0 in the DCT-OFDM system for an AWGN channel with $N = 1024$.

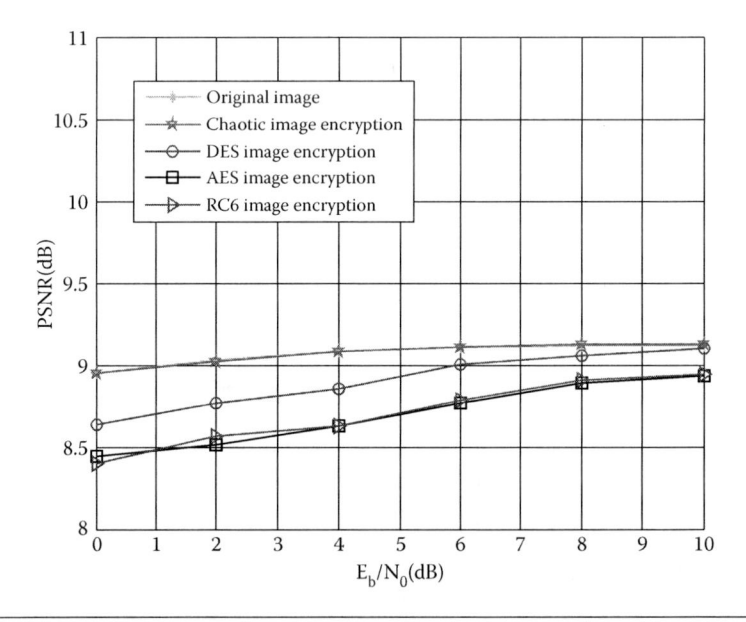

Figure 9.34   PSNR versus E_b/N_0 in the DCT-OFDM system for a Raleigh channel with $N = 64$ without PSE and equalization.

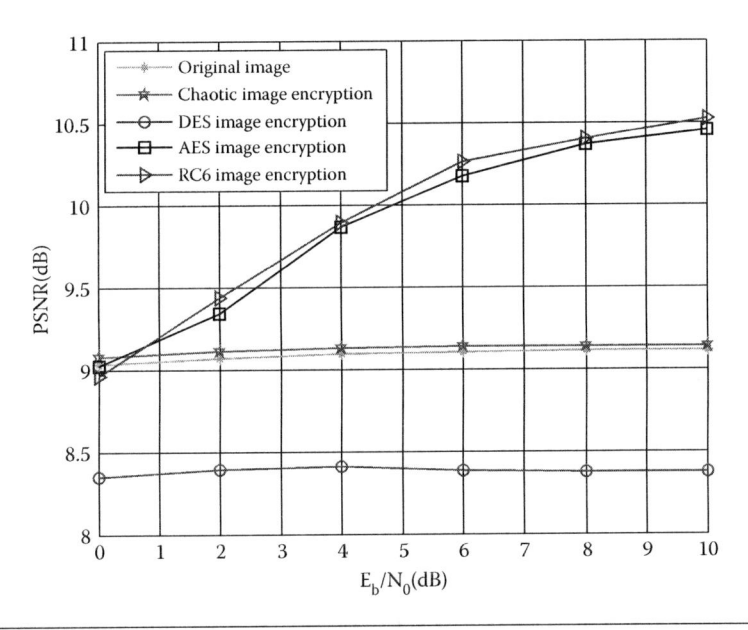

Figure 9.35 PSNR versus E_b/N_0 in the DCT-OFDM system for a Raleigh channel with $N = 512$ without PSE and equalization.

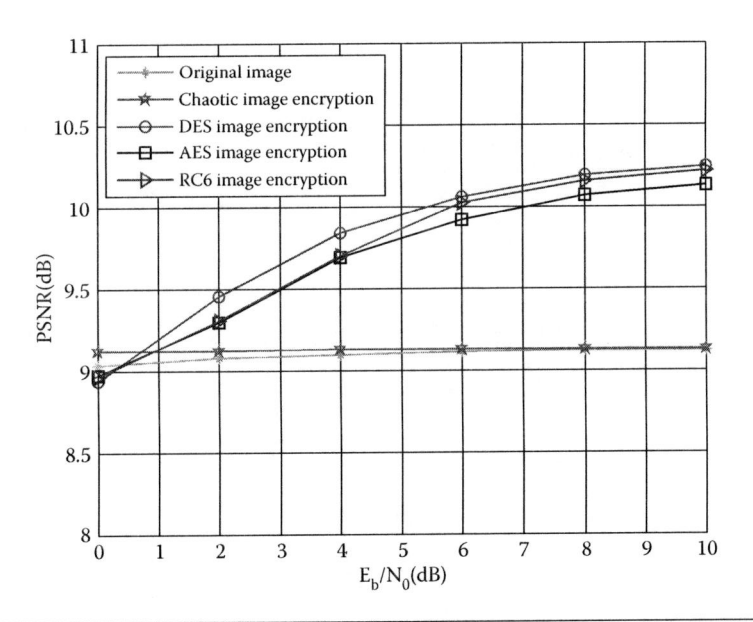

Figure 9.36 PSNR versus E_b/N_0 in the DCT-OFDM system for a Raleigh channel with $N = 1024$ without PSE and equalization.

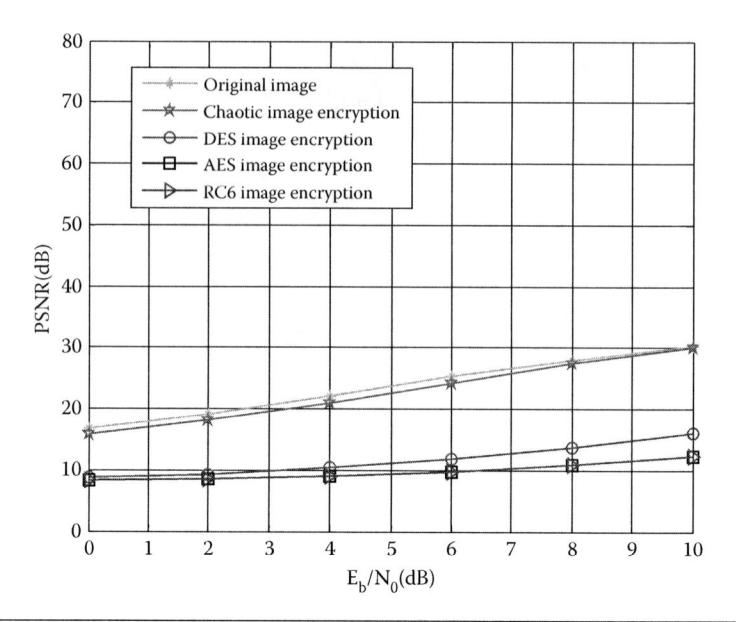

Figure 9.37 PSNR versus E_b/N_0 in the DCT-OFDM system for a Raleigh channel with $N = 64$ with PSE and equalization.

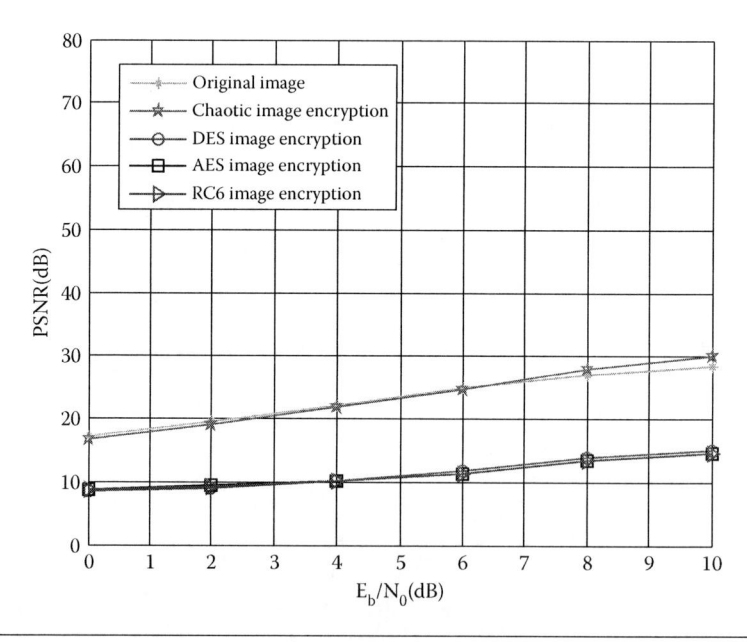

Figure 9.38 PSNR versus E_b/N_0 in the DCT-OFDM system for a Raleigh channel with $N = 512$ with PSE and equalization.

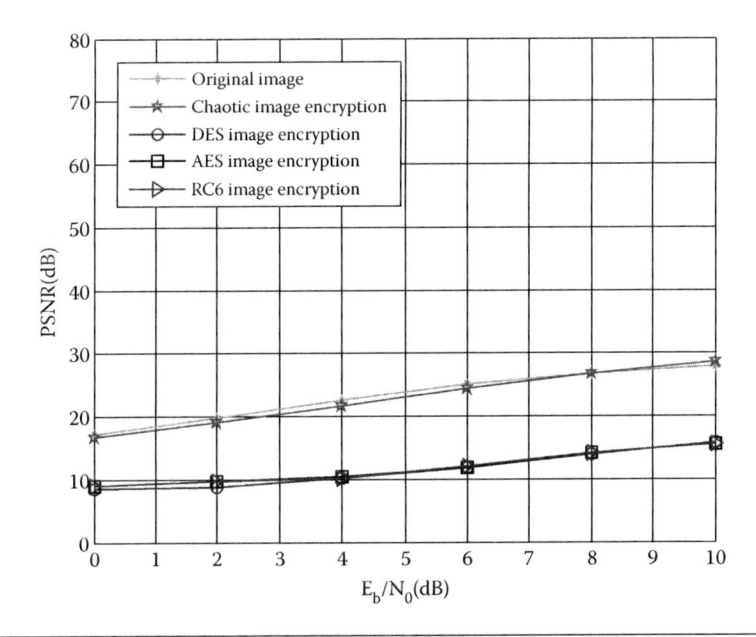

Figure 9.39   PSNR versus E_b/N_0 in the DCT-OFDM system for a Raleigh channel with $N = 1024$ with PSE and equalization.

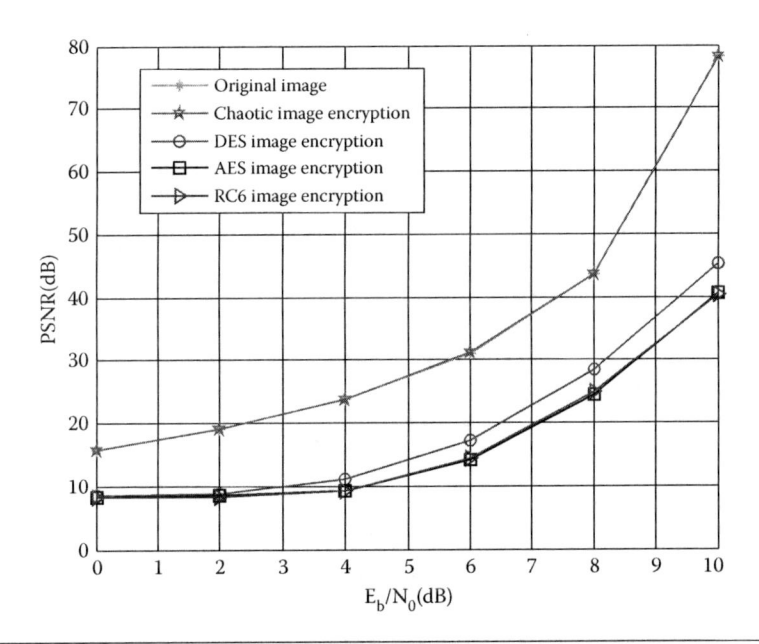

Figure 9.40   PSNR versus E_b/N_0 in the DCT-OFDM system for an AWGN channel with $G = 0$.

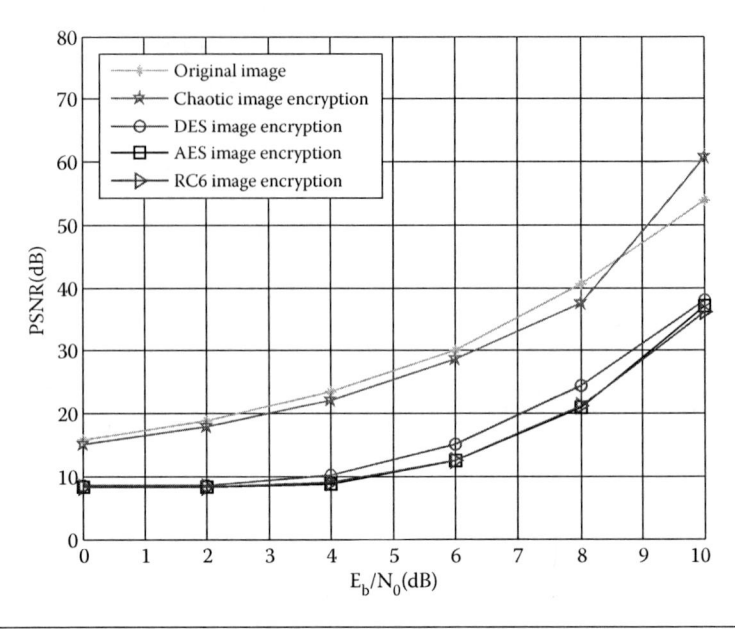

Figure 9.41   PSNR versus E_b/N_0 in the DCT-OFDM system for an AWGN channel with $G = 16$.

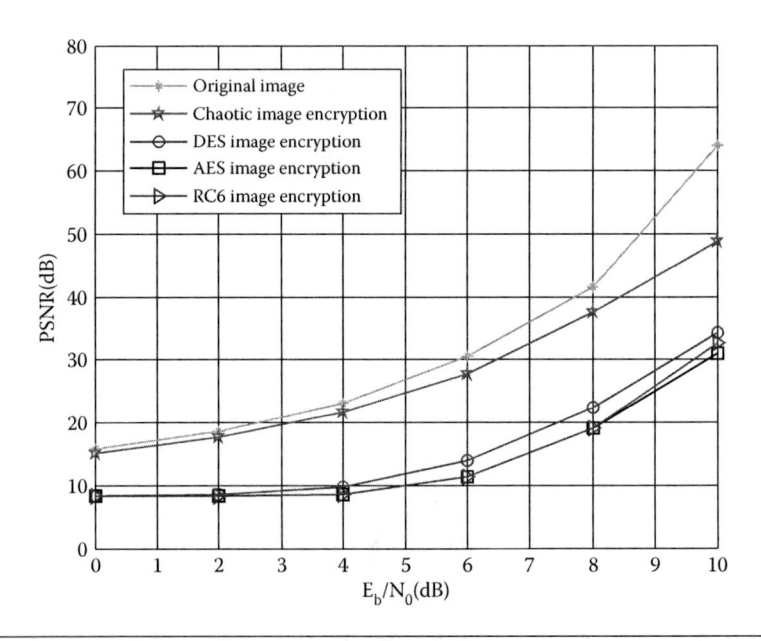

Figure 9.42   PSNR versus E_b/N_0 in the DCT-OFDM system for an AWGN channel with $G = 32$.

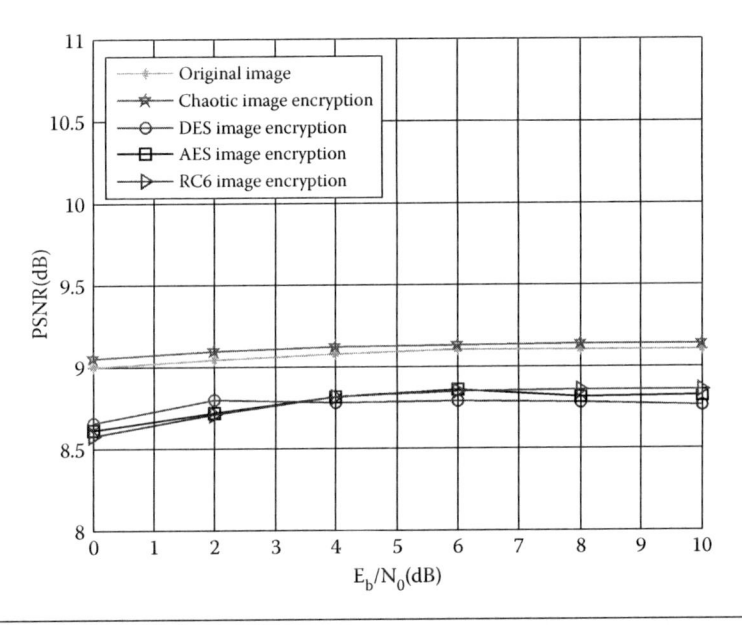

Figure 9.43 PSNR versus E_b/N_0 in the DCT-OFDM system for a Raleigh channel with $G = 0$ without PSE and equalization.

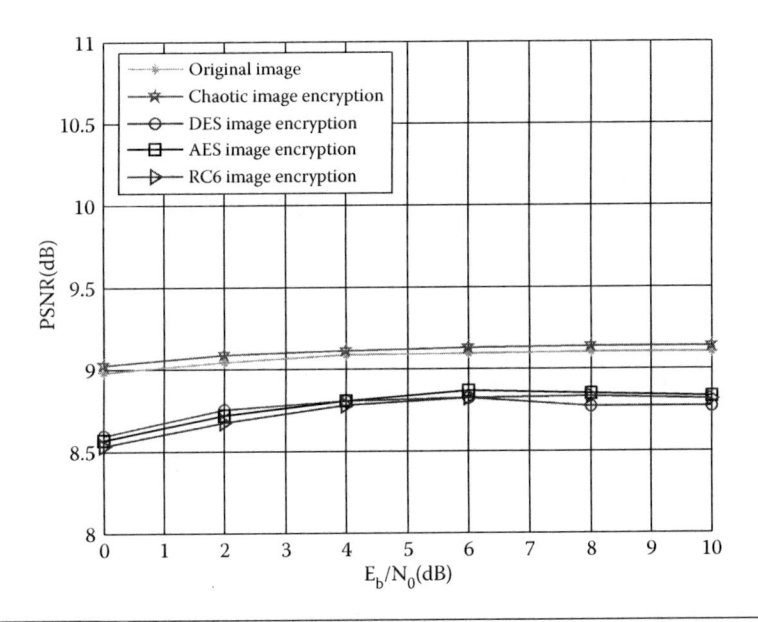

Figure 9.44 PSNR versus E_b/N_0 in the DCT-OFDM system for a Raleigh channel with $G = 16$ without PSE and equalization.

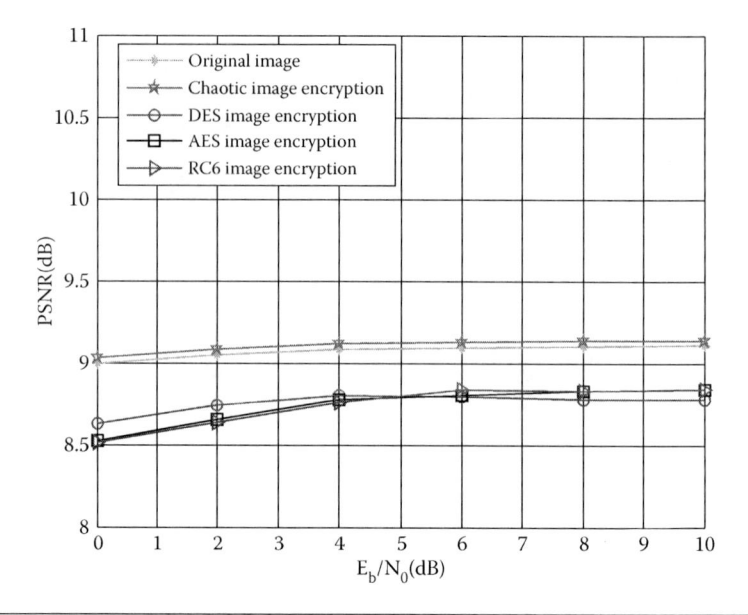

Figure 9.45 PSNR versus E_b/N_0 in the DCT-OFDM system for a Raleigh channel with $G = 32$ without PSE and equalization.

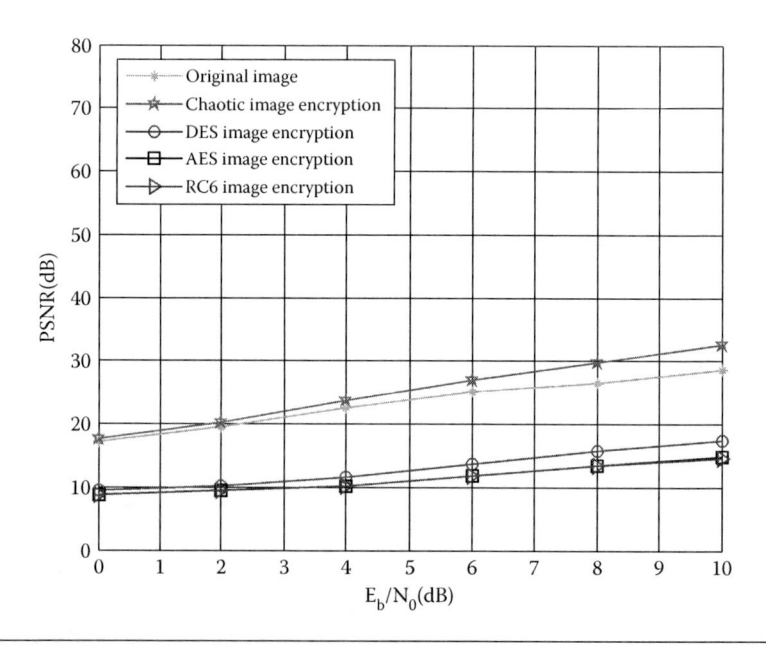

Figure 9.46 PSNR versus E_b/N_0 in the DCT-OFDM system for a Raleigh channel with $G = 0$ with PSE and equalization.

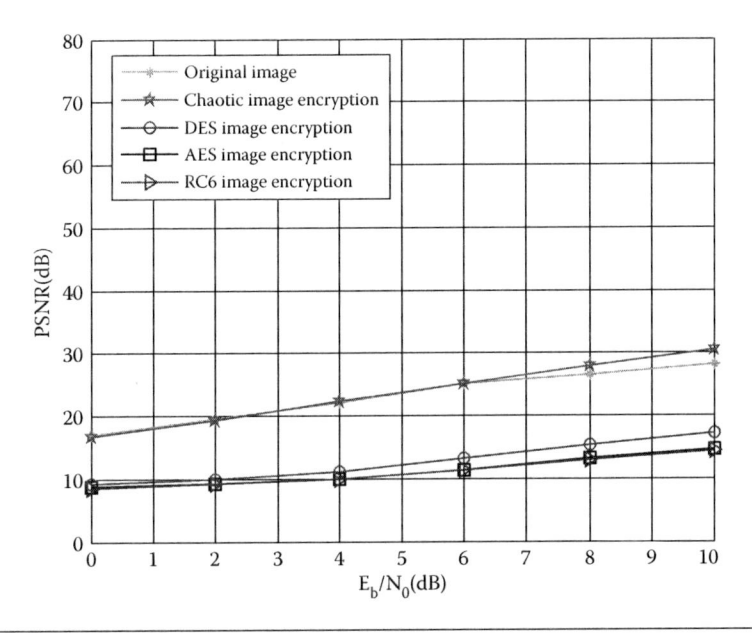

Figure 9.47 PSNR versus E_b/N_0 in the DCT-OFDM system for a Raleigh channel with $G = 16$ with PSE and equalization.

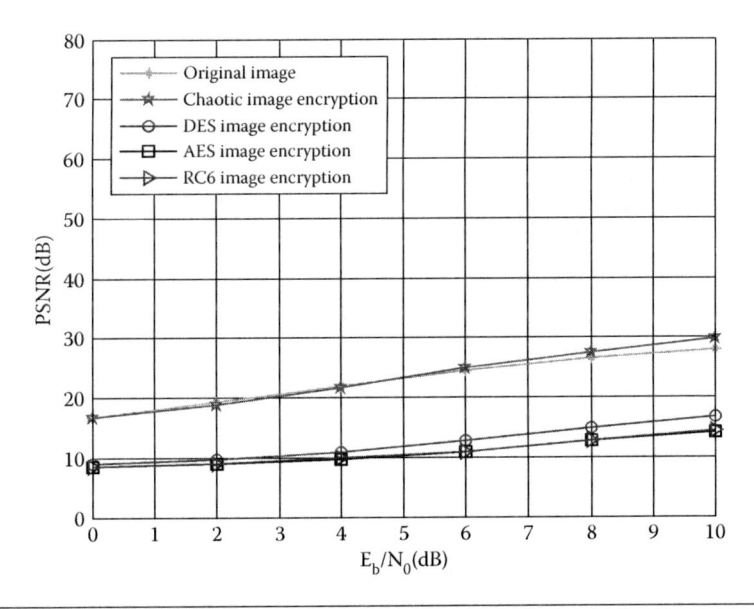

Figure 9.48 PSNR versus E_b/N_0 in the DCT-OFDM system for a Raleigh channel with $G = 32$ with PSE and equalization.

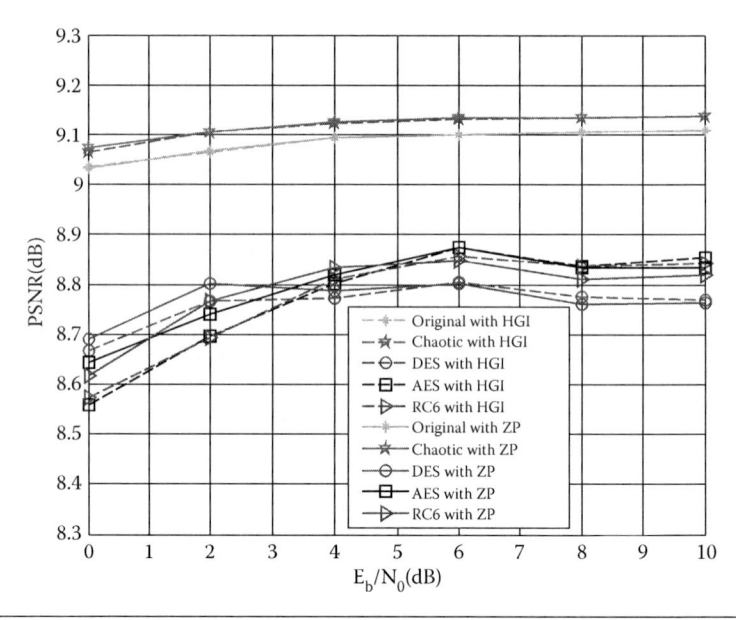

Figure 9.49   PSNR versus E_b/N_0 in the DCT-OFDM system with ZP and HGI for a Raleigh channel with $N = 128$ and $G = 32$ without PSE and equalization.

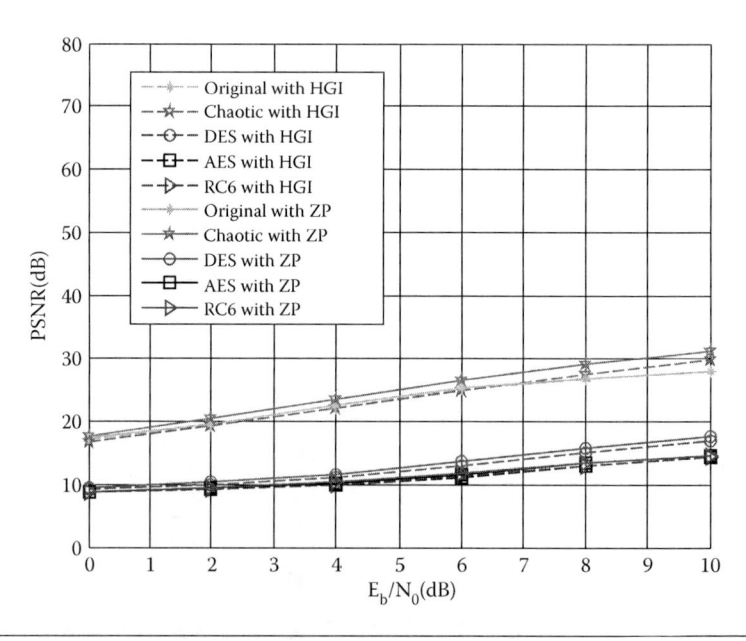

Figure 9.50   PSNR versus E_b/N_0 in the DCT-OFDM system with ZP and HGI for a Raleigh channel with $N = 128$ and $G = 32$ with PSE and equalization.

(a) $E_b/N_0 = 0$ dB
PSNR = 15.7172

(b) $E_b/N_0 = 5$ dB
PSNR = 26.3188

(c) $E_b/N_0 = 10$ dB
PSNR = 53.5993

(d) $E_b/N_0 = 15$ dB
PSNR >60

Figure 9.51 Original image with DCT-OFDM over an AWGN channel.

PSE and equalization, it is shown that fading disrupts the PSNR performance of the DWT-OFDM systems regardless of the encryption algorithm. However, Figures 9.76 to 9.78 show that, with PSE and equalization, the PSNR performances of all systems were better than those without PSE and equalization.

The DES, AES, and RC6 provide lower PSNR performance for the DWT-OFDM system than the chaotic algorithm because these algorithms have a diffusion mechanism that highly reduces the noise immunity. On the other hand, the chaotic algorithm was more robust to noise.

The different types of guard intervals were studied. A comparison study between the HGI and the ZP guard interval with the different encryption algorithms in a frequency-selective channel is illustrated

(a) $E_b/N_0 = 0$ dB
PSNR = 16.7843

(b) $E_b/N_0 = 5$ dB
PSNR = 23.2367

(c) $E_b/N_0 = 10$ dB
PSNR = 27.9452

(d) $E_b/N_0 = 15$ dB
PSNR = 29.3708

Figure 9.52 Original image with DCT-OFDM over a fading channel.

in Figures 9.79 and 9.80. CFO compensation was performed. The PSNR performance of the ZP-OFDM systems was slightly better than that of the HGI-OFDM systems.

Figure 9.81 gives the output of the DWT-OFDM system for the transmission of the original Cameraman image over an AWGN channel at $N = 128$ and $G = 32$.

Figure 9.82 gives the output of the DWT-OFDM system for the transmission of the original Cameraman image over a fading channel at $N = 128$ and $G = 32$.

Figure 9.83 gives the output of the DWT-OFDM system for the transmission of the chaotic-encrypted image over an AWGN channel at $N = 128$ and $G = 32$.

(a) $E_b/N_0 = 0$ dB
PSNR = 15.0481

(b) $E_b/N_0 = 5$ dB
PSNR = 24.4613

(c) $E_b/N_0 = 10$ dB
PSNR = 53.4971

(d) $E_b/N_0 = 15$ dB
PSNR >60

Figure 9.53 Chaotic-encrypted image with DCT-OFDM over an AWGN channel.

Figure 9.84 gives the output of the DWT-OFDM system for the transmission of the chaotic-encrypted image over a fading channel at $N = 128$ and $G = 32$.

Figure 9.85 gives the output of the DWT-OFDM system for the transmission of the DES-encrypted image over an AWGN channel at $N = 128$ and $G = 32$.

Figure 9.86 gives the output of the DWT-OFDM system for the transmission of the DES-encrypted image over a fading channel at $N = 128$ and $G = 32$.

Figure 9.87 gives the output of the DWT-OFDM system for the transmission of the AES-encrypted image over an AWGN channel at $N = 128$ and $G = 32$.

(a) E_b/N_0 = 0 dB
PSNR = 16.3858

(b) E_b/N_0 = 5 dB
PSNR = 23.0375

(c) E_b/N_0 = 10 dB
PSNR = 30.3995

(d) E_b/N_0 = 15 dB
PSNR = 33.2591

Figure 9.54 Chaotic-encrypted image with DCT-OFDM over a fading channel.

Figure 9.88 gives the output of the DWT-OFDM system for the transmission of the AES-encrypted image over a fading channel at N = 128 and G = 32.

Figure 9.89 gives the output of the DWT-OFDM system for the transmission of the RC6-encrypted image over an AWGN channel at N = 128 and G = 32.

Figure 9.90 gives the output of the DWT-OFDM system for the transmission of the RC6-encrypted image over a fading channel at N = 128 and G = 32.

Table 9.2 summarizes these results.

(a) $E_b/N_0 = 0$ dB
PSNR = 8.3762

(b) $E_b/N_0 = 5$ dB
PSNR = 11.1133

(c) $E_b/N_0 = 10$ dB
PSNR = 34.4818

(d) $E_b/N_0 = 15$ dB
PSNR >60

Figure 9.55 DES-encrypted image with DCT-OFDM over an AWGN channel.

9.2 Simulation Experiments in the Presence of CFO

Figure 9.91 illustrates the variation of the PSNR with the E_b/N_0 for the FFT-OFDM, DCT-OFDM, and DWT-OFDM systems with different encryption algorithms. A symbol rate of 250,000 symbols per second, $\varepsilon = 0.1$, $N = 128$, and AWGN channel are assumed. This figure shows that the PSNR performance of all systems with different encryption algorithms deteriorates due to the presence of the CFO because the CFO disrupts the orthogonality between subcarriers and gives rise to the intercarrier interference (ICI), which leads to performance degradation. This figure also shows that the CFO compensation process can avoid the impact of the CFO and produce high-quality

(a) $E_b/N_0 = 0$ dB
PSNR = 9.0390

(b) $E_b/N_0 = 5$ dB
PSNR = 11.7067

(c) $E_b/N_0 = 10$ dB
PSNR = 16.5200

(d) $E_b/N_0 = 15$ dB
PSNR = 19.9308

Figure 9.56 DES-encrypted image with DCT-OFDM over a fading channel.

decrypted images. On the other hand, without CFO compensation, it is clear that the CFO degrades the PSNR performance of the DES, AES, and RC6 for all OFDM systems.

Figures 9.92 to 9.97 illustrate the PSNR performance versus the E_b/N_0 for the FFT-OFDM, DCT-OFDM, and DWT-OFDM systems with different encryption algorithms in a frequency-selective channel. Without CFO compensation, it is shown that CFO disrupted the PSNR performance of all OFDM systems regardless of the encryption algorithm. However, with CFO compensation, the PSNR performances of all systems are better than those without CFO compensation. These figures also show that the PSNR performances of all encryption algorithms with the FFT-OFDM system

(a) $E_b/N_0 = 0$ dB
PSNR = 8.4211

(b) $E_b/N_0 = 5$ dB
PSNR = 9.4006

(c) $E_b/N_0 = 10$ dB
PSNR = 30.5205

(d) $E_b/N_0 = 15$ dB
PSNR >60

Figure 9.57 AES-encrypted image with DCT-OFDM over an AWGN channel.

were the best compared with those of the DCT-OFDM and DWT-OFDM systems.

The DES, AES, and RC6 provide lower PSNR performance for all OFDM systems compared with the chaotic algorithm because these algorithms have a diffusion mechanism that reduces the noise immunity. On the other hand, the chaotic algorithm is more robust to noise.

9.3 Simulation Experiments for Enhanced Algorithms

The performance of the chaotic algorithm with its modes (ECB, CFB, CBC, and OFB) for (FFT/DCT/DWT)-OFDM systems was compared. With $N = 128$ and $G = 32$ over an AWGN, the PSNR

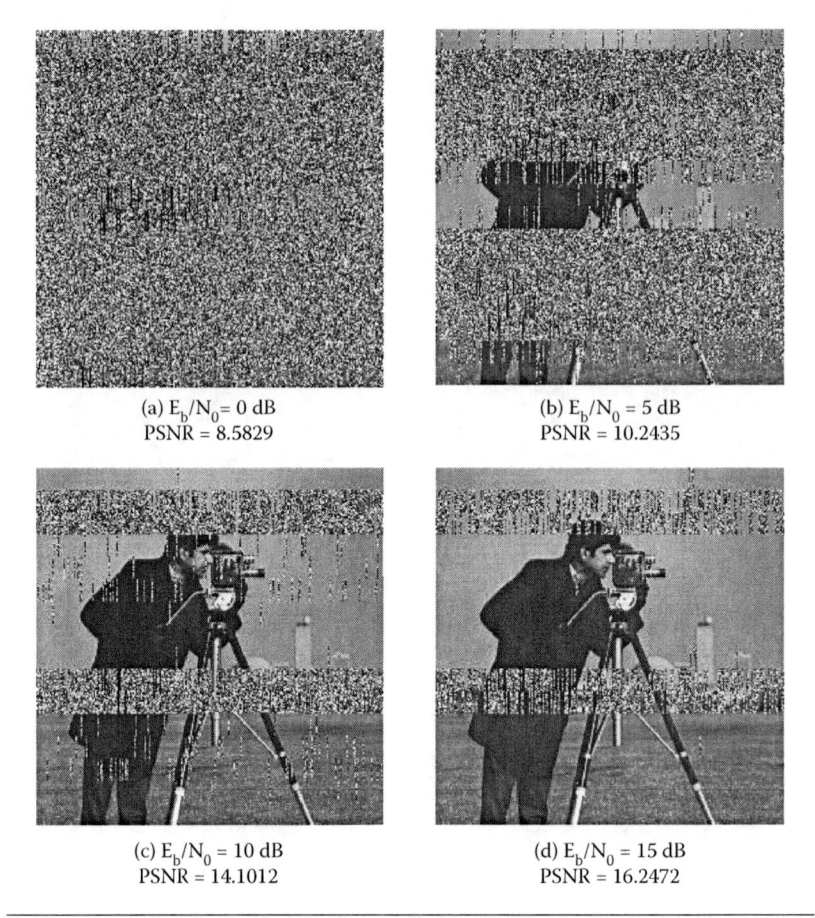

(a) E_b/N_0 = 0 dB
PSNR = 8.5829

(b) E_b/N_0 = 5 dB
PSNR = 10.2435

(c) E_b/N_0 = 10 dB
PSNR = 14.1012

(d) E_b/N_0 = 15 dB
PSNR = 16.2472

Figure 9.58   AES-encrypted image with DCT-OFDM over a fading channel.

performance of the chaotic-ECB, chaotic-CFB, chaotic-CBC, and chaotic-OFB algorithms for all OFDM systems start to increase at moderate-to-high E_b/N_0 values. In Figure 9.98, the PSNR of the chaotic-ECB algorithm is also the best compared to those of all other modes of encryption algorithms with FFT-OFDM. From Figure 9.99, the PSNR of the chaotic-(OFB and ECB) algorithms is also the best compared to those of all other modes of encryption algorithms with DCT-OFDM. From Figure 9.100, the PSNR of the chaotic-OFB algorithm is the best compared to that of all other modes of encryption algorithms with DWT-OFDM.

The performance of the chaotic algorithm with its modes (ECB, CFB, CBC, and OFB) for (FFT/DCT/DWT)-OFDM systems was compared. With N = 128 and G = 32 over a fading channel,

(a) $E_b/N_0 = 0$ dB
PSNR = 8.3961

(b) $E_b/N_0 = 5$ dB
PSNR = 9.3528

(c) $E_b/N_0 = 10$ dB
PSNR = 33.1261

(d) $E_b/N_0 = 15$ dB
PSNR >60

Figure 9.59 RC6-encrypted image with DCT-OFDM over an AWGN channel.

the PSNR performance of the chaotic-ECB, chaotic-CFB, chaotic-CBC, and chaotic-OFB algorithms for all OFDM systems start to increase at moderate-to-high E_b/N_0 values. In Figures 9.101 to 9.103, the PSNR of the chaotic-CBC algorithm is the best compared to those of all other modes of encryption with FFT/DCT/DWT-OFDM without CFO. From Figures 9.104 to 9.106, the PSNR of the chaotic-CBC algorithm is also the best compared to those of all other modes of encryption algorithms with FFT-OFDM systems with CFO.

A comparison study between the CP-OFDM systems and the ZP-OFDM systems with the different chaotic encryption modes in a frequency-selective channel without PSE is illustrated in

(a) $E_b/N_0 = 0$ dB
PSNR = 8.5691

(b) $E_b/N_0 = 5$ dB
PSNR = 10.2276

(c) $E_b/N_0 = 10$ dB
PSNR = 14.1490

(d) $E_b/N_0 = 15$ dB
PSNR = 16.1107

Figure 9.60   RC6-encrypted image with DCT-OFDM over a fading channel.

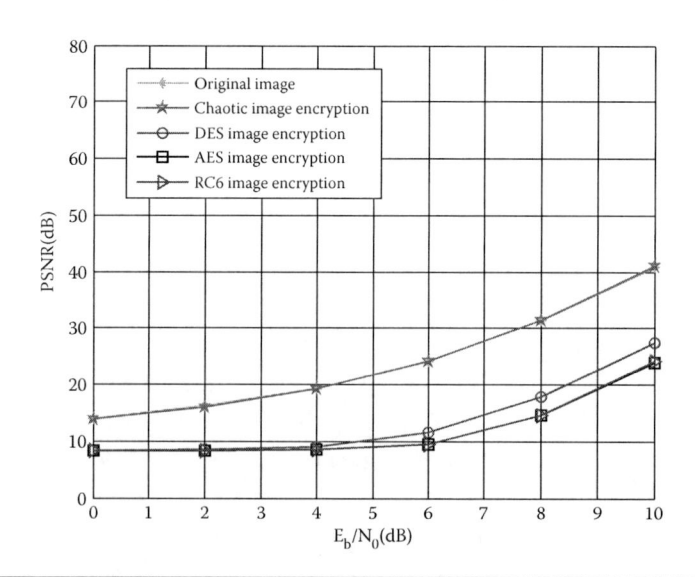

Figure 9.61   PSNR versus E_b/N_0 in the DWT-OFDM system for an AWGN channel with $N = 64$.

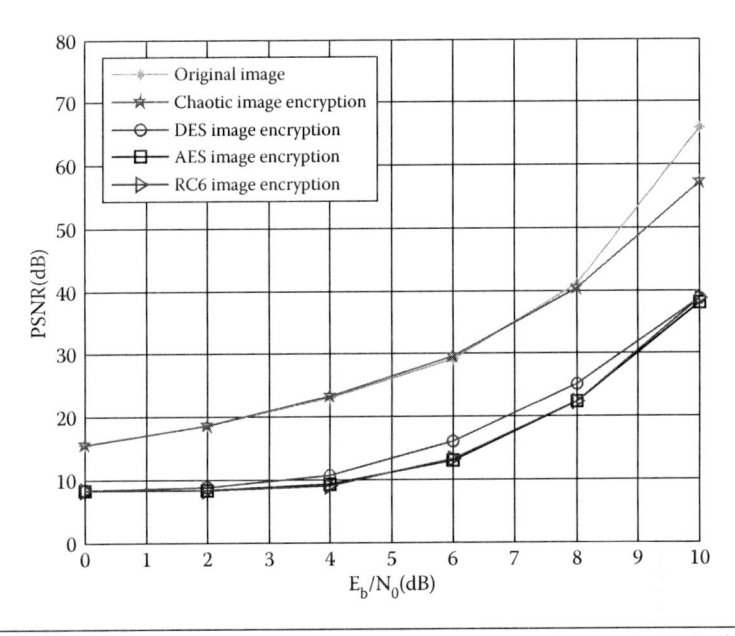

Figure 9.62 PSNR versus E_b/N_0 in the DWT-OFDM system for an AWGN channel with $N = 512$.

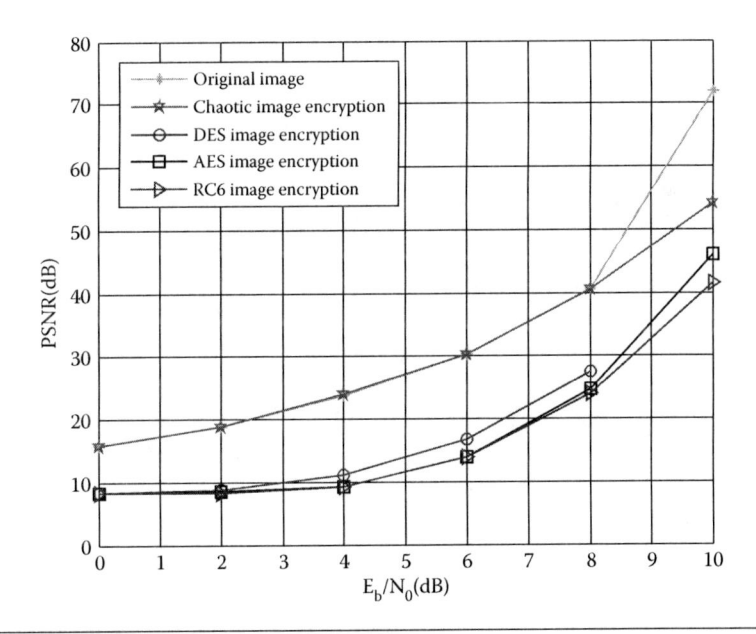

Figure 9.63 PSNR versus E_b/N_0 in the DWT-OFDM system for an AWGN channel with $N = 1024$.

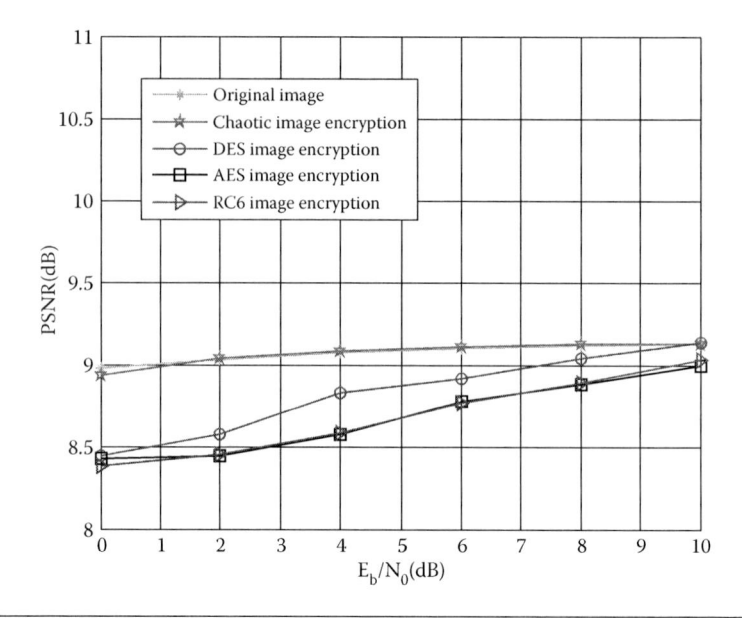

Figure 9.64   PSNR versus E_b/N_0 in the DWT-OFDM system for a Raleigh channel with $N = 64$ without PSE and equalization.

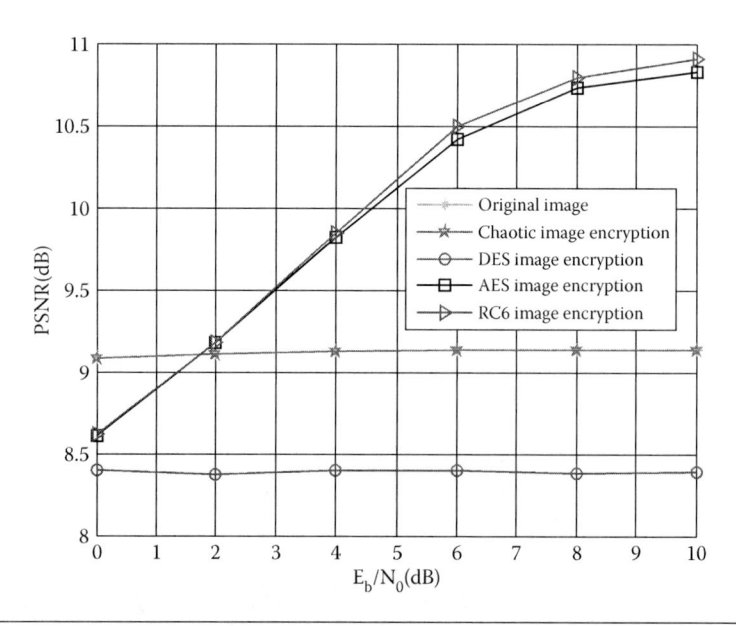

Figure 9.65   PSNR versus E_b/N_0 in the DWT-OFDM system for a Raleigh channel with $N = 512$ without PSE and equalization.

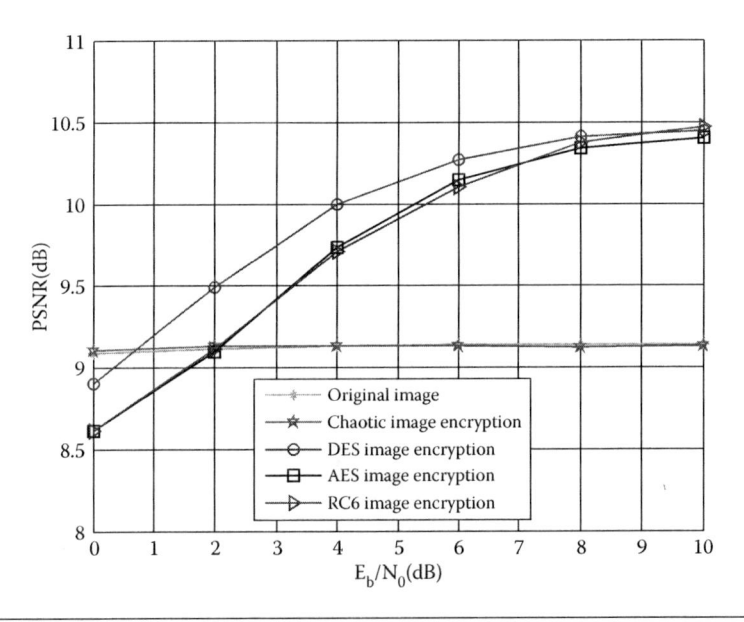

Figure 9.66 PSNR versus E_b/N_0 in the DWT-OFDM system for a Raleigh channel with $N = 1024$ without PSE and equalization.

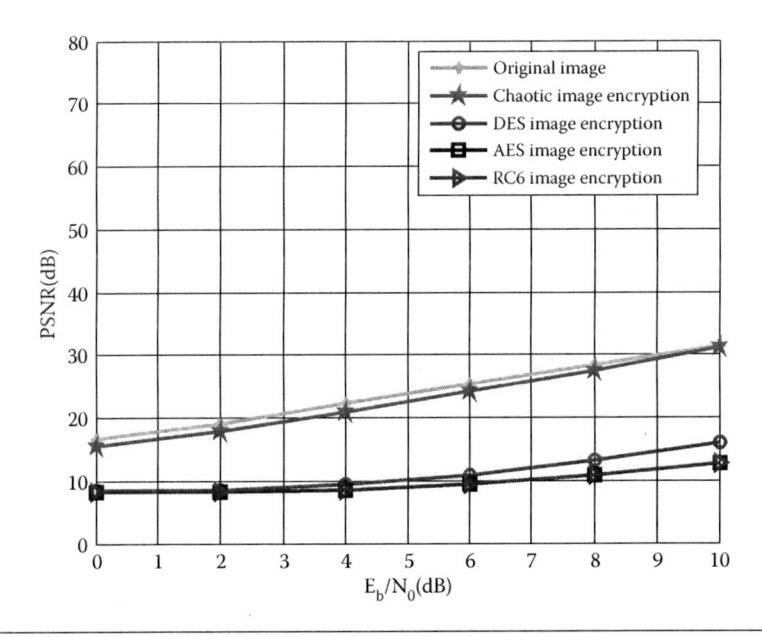

Figure 9.67 PSNR versus E_b/N_0 in the DWT-OFDM system for a Raleigh channel with $N = 64$ with PSE and equalization.

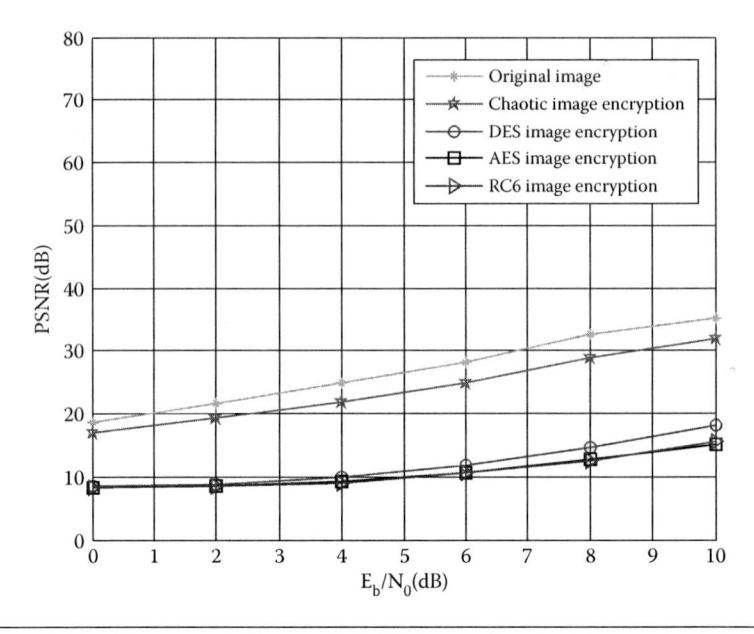

Figure 9.68   PSNR versus E_b/N_0 in the DWT-OFDM system for a Raleigh channel with $N = 512$ with PSE and equalization.

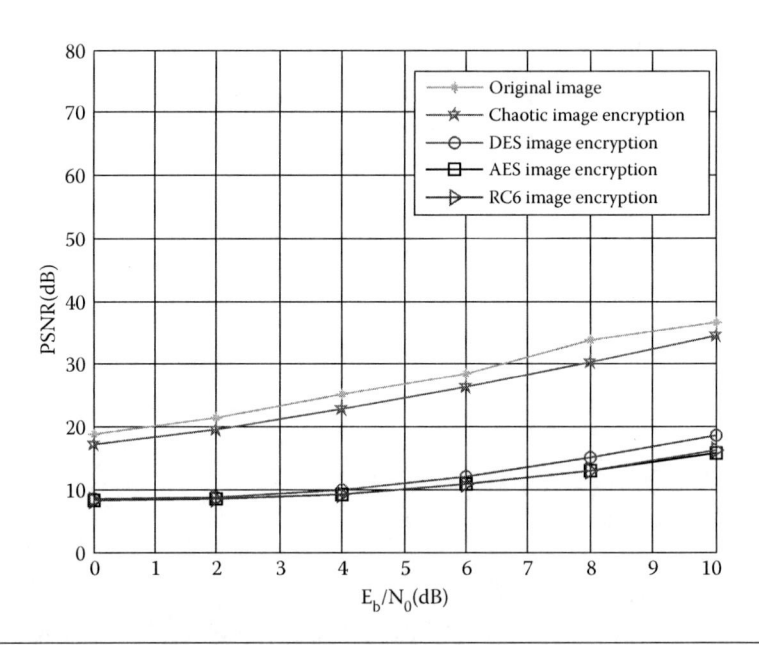

Figure 9.69   PSNR versus E_b/N_0 in the DWT-OFDM system for a Raleigh channel with $N = 1024$ with PSE and equalization.

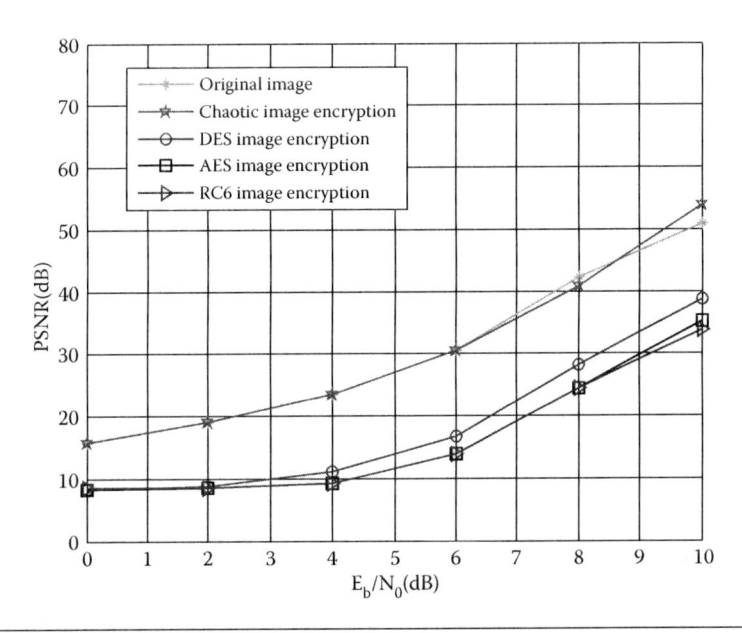

Figure 9.70 PSNR versus E_b/N_0 in the DWT-OFDM system for an AWGN channel with $G = 0$.

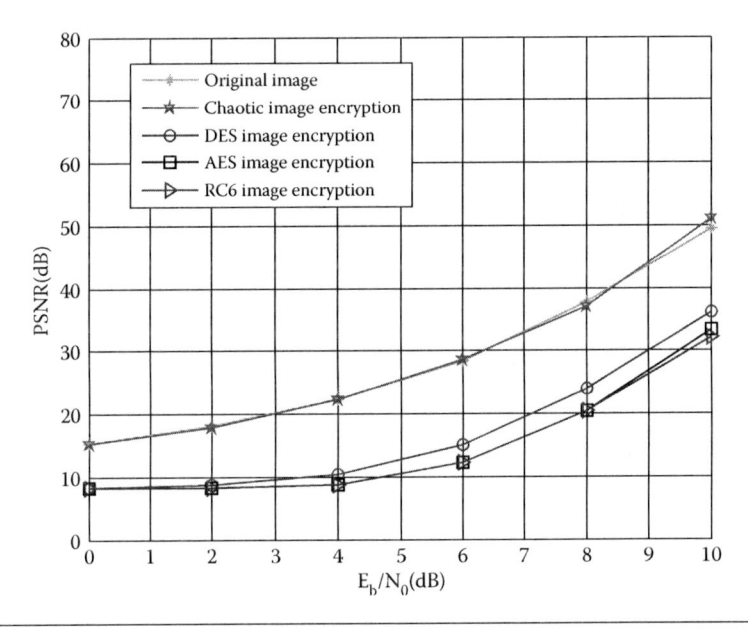

Figure 9.71 PSNR versus E_b/N_0 in the DWT-OFDM system for an AWGN channel with $G = 16$.

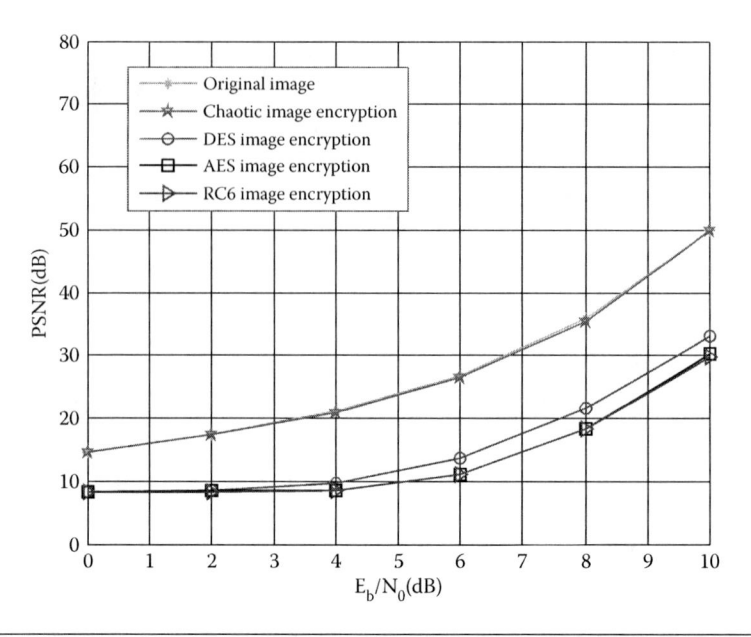

Figure 9.72   PSNR versus E_b/N_0 in the DWT-OFDM system for an AWGN channel with $G = 32$.

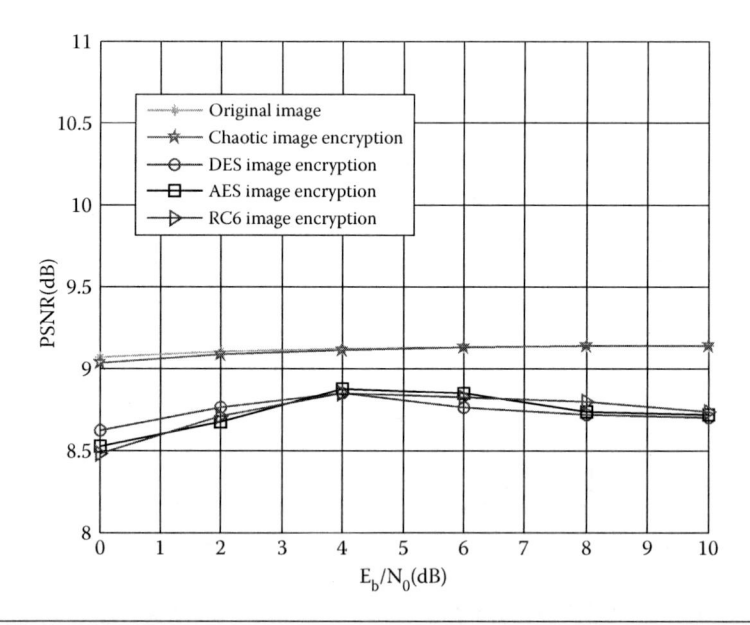

Figure 9.73   PSNR versus E_b/N_0 in the DWT-OFDM system for a Raleigh channel with $G = 0$ without PSE and equalization.

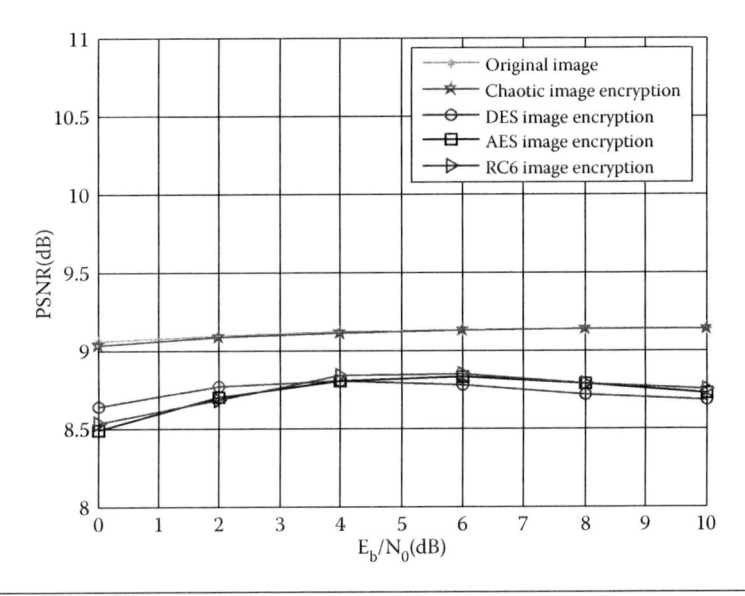

Figure 9.74 PSNR versus E_b/N_0 in the DWT-OFDM system for a Raleigh channel with $G = 16$ without PSE and equalization.

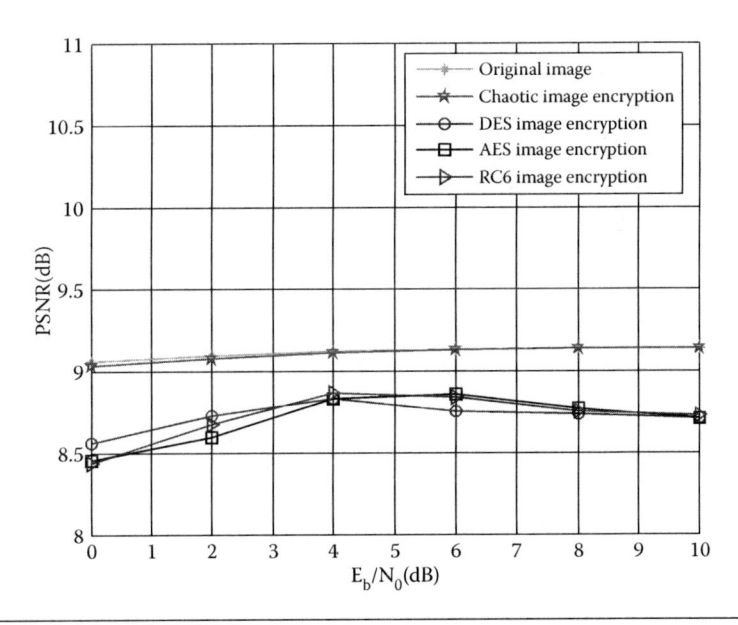

Figure 9.75 PSNR versus E_b/N_0 in the DWT-OFDM system for a Raleigh channel with $G = 32$ without PSE and equalization.

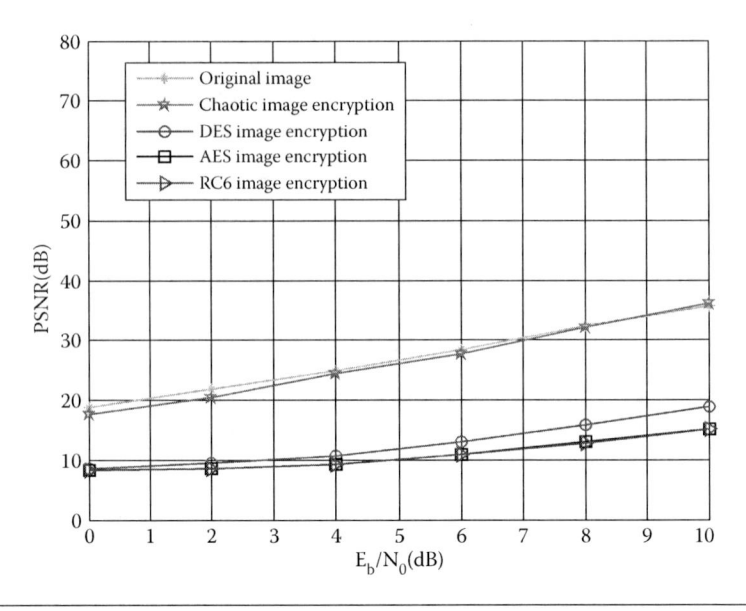

Figure 9.76   PSNR versus E_b/N_0 in the DWT-OFDM system for a Raleigh channel with $G = 0$ with PSE and equalization.

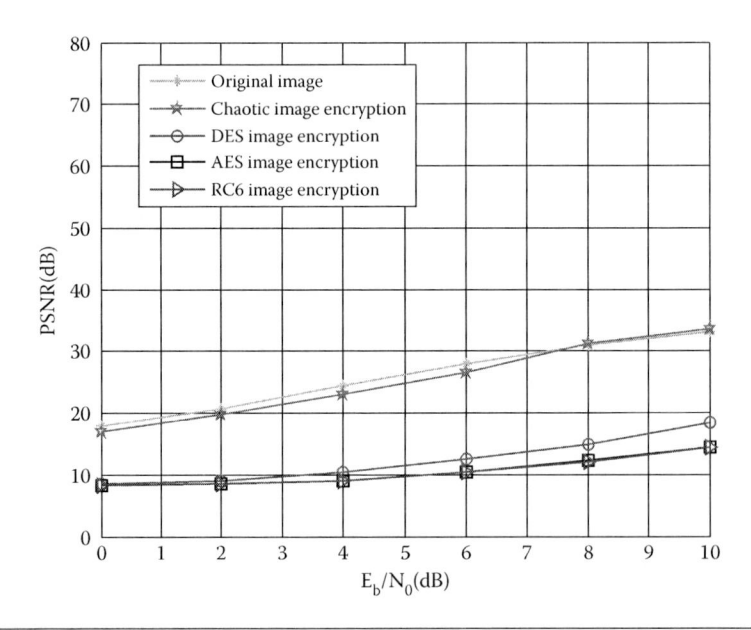

Figure 9.77   PSNR versus E_b/N_0 in the DWT-OFDM system for a Raleigh channel with $G = 16$ with PSE and equalization.

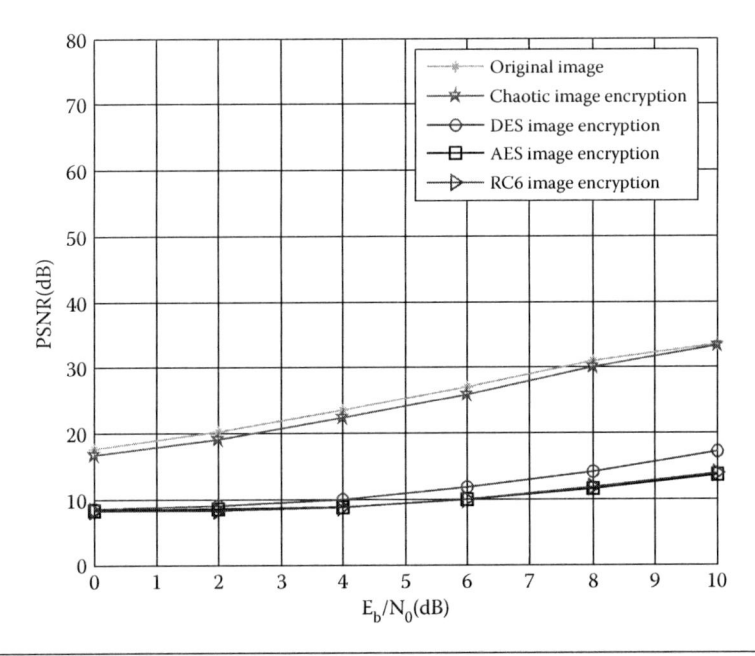

Figure 9.78 PSNR versus E_b/N_0 in the DWT-OFDM system for a Raleigh channel with $G = 32$ with PSE and equalization.

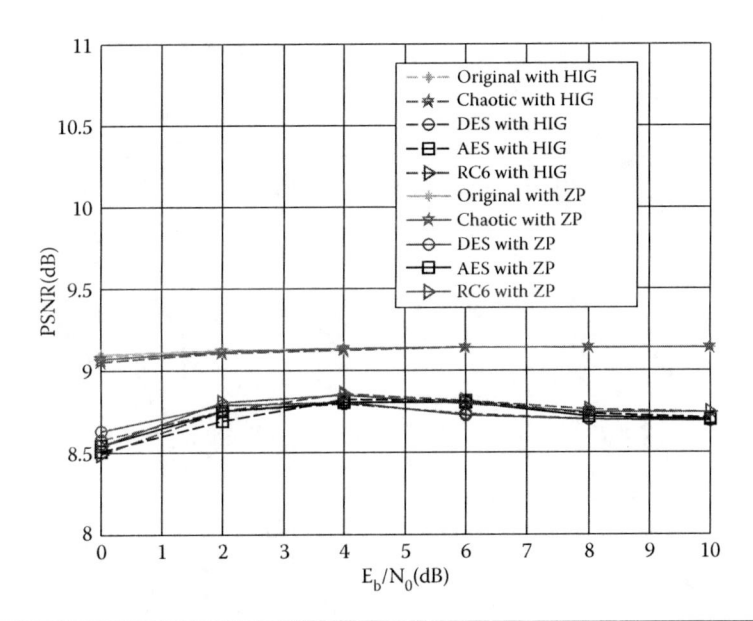

Figure 9.79 PSNR versus E_b/N_0 in the DWT-OFDM system with ZP and HGI for a Raleigh channel with $N = 128$ and $G = 32$ without PSE and equalization.

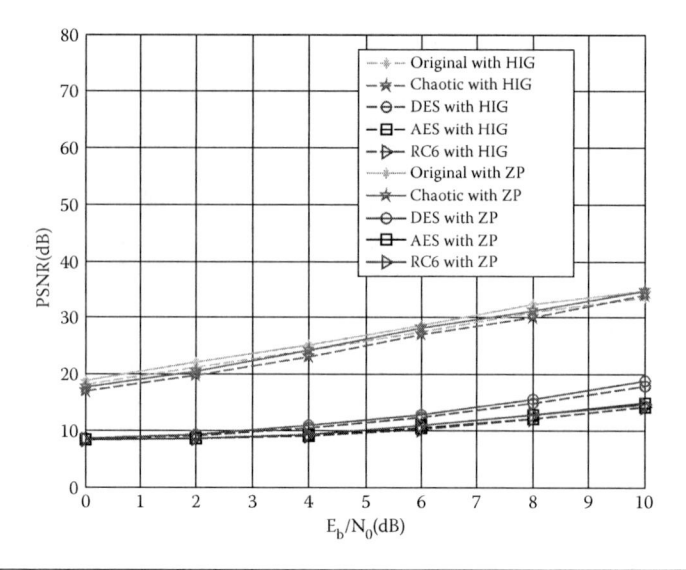

Figure 9.80 PSNR versus E_b/N_0 in the DWT-OFDM system with ZP and HGI for a Raleigh channel with $N = 128$ and $G = 32$ with PSE and equalization.

(a) $E_b/N_0 = 0$ dB
PSNR = 14.6839

(b) $E_b/N_0 = 5$ dB
PSNR = 23.7368

(c) $E_b/N_0 = 10$ dB
PSNR = 45.9748

(d) $E_b/N_0 = 15$ dB
PSNR >60

Figure 9.81 Original image with DWT-OFDM over an AWGN channel.

(a) $E_b/N_0 = 0$ dB
PSNR = 17.5731

(b) $E_b/N_0 = 5$ dB
PSNR = 24.9482

(c) $E_b/N_0 = 10$ dB
PSNR = 32.6435

(d) $E_b/N_0 = 15$ dB
PSNR = 37.7109

Figure 9.82 Original image with DWT-OFDM over a fading channel.

Figures 9.107 to 9.109. The best value of PSNR is obtained with chaotic-ECB. In Figures 9.110 to Figure 9.112, they are considered. It is shown that the PSNR performance of the ZP-OFDM systems is better than that of the CP-OFDM systems. In Figures 9.113 to 9.115, it is illustrated that over an AWGN channel the PSNR performance of the OFDM systems with CFO is better than that of OFDM systems without CFO.

Figure 9.116 compares the AES encryption algorithm with AES with preprocessing in an AWGN channel when $N = 128$ and $G = 32$ to find that FFT-OFDM is the best, then DCT-OFDM, and finally DWT-OFDM. Figure 9.117 illustrates a fading channel case without PSE and equalization; Figure 9.118 illustrates a fading channel also, but with PSE and equalization. CFO is considered without

(a) $E_b/N_0 = 0$ dB
PSNR = 14.7090

(b) $E_b/N_0 = 5$ dB
PSNR = 23.5544

(c) $E_b/N_0 = 10$ dB
PSNR = 44.6552

(d) $E_b/N_0 = 15$ dB
PSNR >60

Figure 9.83   Chaotic-encrypted image with DWT-OFDM over an AWGN channel.

compensation in Figure 9.119; Figure 9.120 illustrates CFO compensation. Figures 9.121 and 9.122 consider an AWGN channel with CP-OFDM and ZP-OFDM, respectively.

Figure 9.123 compares the RC6 encryption algorithm with RC6 with preprocessing in an AWGN channel when $N = 128$ and $G = 32$; FFT-OFDM is the best, then DCT-OFDM, and finally DWT-OFDM. Figure 9.124 illustrates a fading channel case without PSE equalization; Figure 9.125 illustrates a fading channel case also, but with PSE and equalization. CFO is considered in Figure 9.126 without compensation; Figure 9.127 illustrates CFO compensation. Figures 9.128 and 9.129 consider an AWGN channel with CP-OFDM and ZP-OFDM, respectively.

(a) $E_b/N_0 = 0$ dB
PSNR = 16.3689

(b) $E_b/N_0 = 5$ dB
PSNR = 23.8108

(c) $E_b/N_0 = 10$ dB
PSNR = 32.7138

(d) $E_b/N_0 = 15$ dB
PSNR = 37.5838

Figure 9.84 Chaotic-encrypted image with DWT-OFDM over a fading channel.

9.4 Simulation Experiments of PAPR Reduction Methods

Several experiments were carried out to test the effect of the PAPR reduction methods on the process of encrypted image transmission with OFDM. A convolutional code with rate 1/2, constraint length 7, and octal generator polynomial (133,171) are adopted to reduce the effect of PAPR reduction on the quality of received images. The compared PAPR reduction methods are clipping, companding, and hybrid clipping-and-companding methods.

The performance of the process of original image transmission with the chaotic algorithm and three different diffusion algorithms for (FFT/DCT/DWT)-OFDM systems with $N = 128$ and $G = 32$ over an AWGN is shown in Figures 9.130 to 9.132. From these figures, it

(a) $E_b/N_0 = 0$ dB
PSNR = 8.3954

(b) $E_b/N_0 = 5$ dB
PSNR = 11.2270

(c) $E_b/N_0 = 10$ dB
PSNR = 32.2373

(d) $E_b/N_0 = 15$ dB
PSNR >60

Figure 9.85   DES-encrypted image with DWT-OFDM over an AWGN channel.

is clear that the PSNR performance of the original image transmission process with FFT-OFDM is better with $\mu = 4$ than with $\mu = 0.1$; for the DCT-OFDM and the DWT-OFDM, it is better with $\mu = 0.1$ than with $\mu = 4$.

Several experiments were repeated to study the effect of PAPR reduction techniques with encrypted image transmission. From Figures 9.133 to 9.135, the PSNR performance of the chaotic-encrypted image transmission with FFT/DCT/DWT-OFDM was better with $\mu = 0.1$ than with $\mu = 4$. At high SNR values, it is possible to receive the chaotic encrypted images with a very large PSNR if error correction codes were utilized.

(a) $E_b/N_0 = 0$ dB
PSNR = 8.4729

(b) $E_b/N_0 = 5$ dB
PSNR = 10.7940

(c) $E_b/N_0 = 10$ dB
PSNR = 17.4856

(d) $E_b/N_0 = 15$ dB
PSNR = 25.9770

Figure 9.86   DES-encrypted image with DWT-OFDM over a fading channel.

Similar experiments were repeated for AES-encrypted images, and the results are shown in Figures 9.136 to 9.138. We can come to the conclusion that error-correcting codes are feasible, but the performance is much less than that for the chaotic encryption case due to the diffusion mechanism of the encryption algorithm.

Similar experiments were repeated for DES-encrypted images, and the results are shown in Figures 9.139 to 9.141. Similar experiments were repeated for RC6-encrypted images, and the results are shown in Figures 9.142 to 9.144.

(a) $E_b/N_0 = 0$ dB
PSNR = 8.4049

(b) $E_b/N_0 = 5$ dB
PSNR = 9.3407

(c) $E_b/N_0 = 10$ dB
PSNR = 28.8218

(d) $E_b/N_0 = 15$ dB
PSNR >60

Figure 9.87 AES-encrypted image with DWT-OFDM over an AWGN channel.

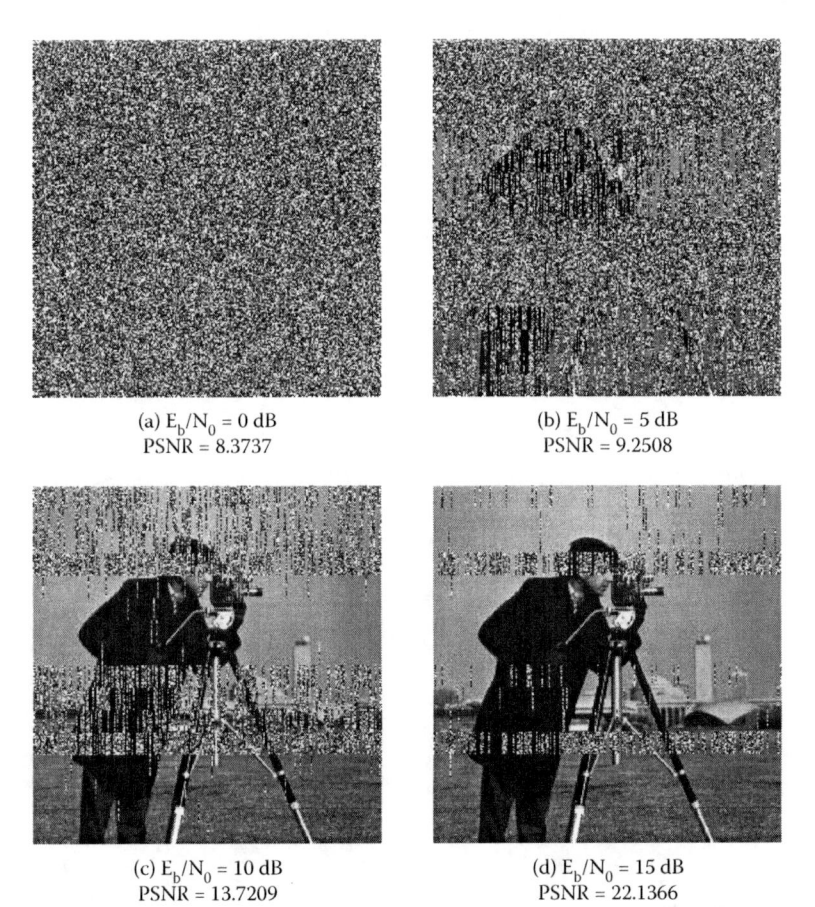

(a) $E_b/N_0 = 0$ dB
PSNR = 8.3737

(b) $E_b/N_0 = 5$ dB
PSNR = 9.2508

(c) $E_b/N_0 = 10$ dB
PSNR = 13.7209

(d) $E_b/N_0 = 15$ dB
PSNR = 22.1366

Figure 9.88   AES-encrypted image with DWT-OFDM over a fading channel.

(a) $E_b/N_0 = 0$ dB
PSNR = 8.4344

(b) $E_b/N_0 = 5$ dB
PSNR = 9.3419

(c) $E_b/N_0 = 10$ dB
PSNR = 29.1988

(d) $E_b/N_0 = 15$ dB
PSNR >60

Figure 9.89   RC6-encrypted image with DWT-OFDM over an AWGN channel.

(a) $E_b/N_0 = 0$ dB
PSNR = 8.3983

(b) $E_b/N_0 = 5$ dB
PSNR = 9.2497

(c) $E_b/N_0 = 10$ dB
PSNR = 13.9828

(d) $E_b/N_0 = 15$ dB
PSNR = 18.7859

Figure 9.90 RC6-encrypted image with DWT-OFDM over a fading channel

Table 9.2 Summarized Results

E_B/N_0	$E_B/N_0 = 0$ dB	$E_B/N_0 = 5$ dB	$E_B/N_0 = 10$ dB
ORIGINAL + AWGN CHANNEL			
FFT	14.6993	23.7838	48.0593
DCT	15.7172	26.3188	53.5993
DWT	14.6839	23.7368	45.9748
ORIGINAL + FADING CHANNEL			
FFT	17.8214	29.4547	53.8719
DCT	16.7843	23.2367	27.9452
DWT	17.5731	24.9482	32.6435
CHAOTIC + AWGN CHANNEL			
FFT	15.1527	25.1709	50.5529
DCT	15.0481	24.4613	53.4971
DWT	14.7090	23.5544	44.6552
CHAOTIC + FADING CHANNEL			
FFT	18.3597	30.0828	53.8881
DCT	16.3858	23.0375	30.3995
DWT	16.3689	23.8108	32.7138
DES + AWGN CHANNEL			
FFT	8.3766	11.1743	36.4777
DCT	8.3762	11.1133	34.4818
DWT	8.3954	11.2270	32.2373
DES + FADING CHANNEL			
FFT	9.6637	16.7520	41.4821
DCT	9.0390	11.7067	16.5200
DWT	8.4729	10.7940	17.4856
AES + AWGN CHANNEL			
FFT	8.4026	9.4551	31.5464
DCT	8.4211	9.4006	30.5205
DWT	8.4049	9.3407	28.8218
AES + FADING CHANNEL			
FFT	8.6932	14.1673	36.8018
DCT	8.5829	10.2435	14.1012
DWT	8.3737	9.2508	13.7209
RC6 + AWGN CHANNEL			
FFT	8.3916	9.4165	30.9051
DCT	8.3961	9.3528	33.1261
DWT	8.4344	9.3419	29.1988
RC6 + FADING CHANNEL			
FFT	8.7540	13.8893	40.0734
DCT	8.5691	10.2276	14.1490
DWT	8.3983	9.2497	13.9828

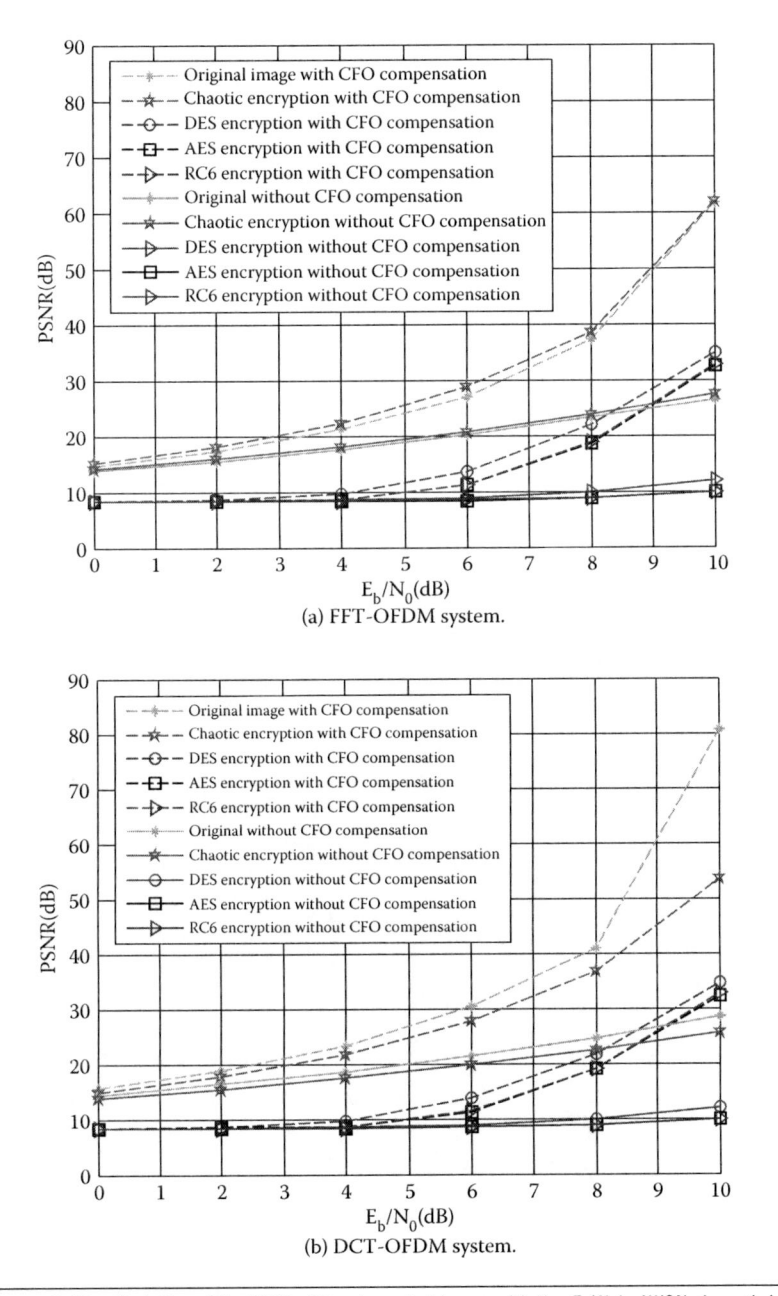

Figure 9.91 Variation of the PSNR of the decrypted image with the E_b/N_0 in AWGN channel: (a) FFT-OFDM system; (b) DCT-OFDM system.

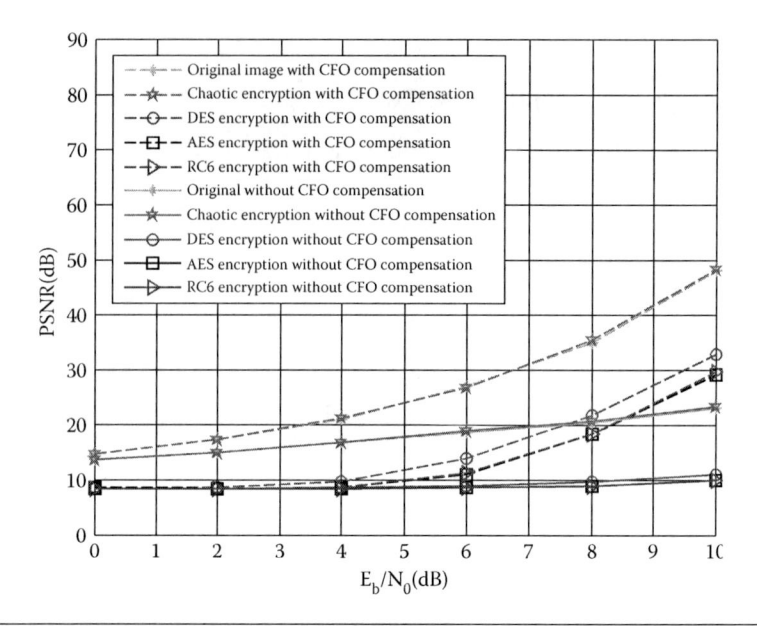

Figure 9.91 (*Continued*)   Variation of the PSNR of the decrypted image with the E_b/N_0 in AWGN channel: DWT-OFDM system.

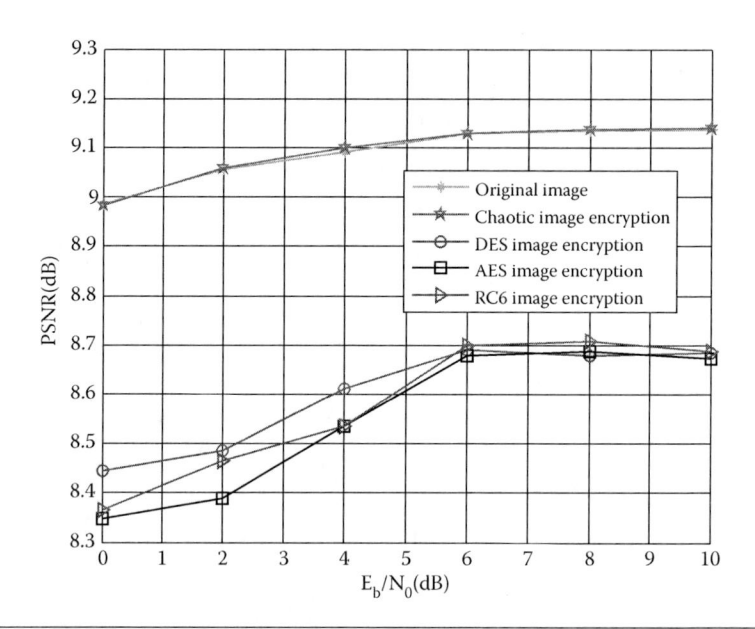

Figure 9.92   PSNR versus E_b/N_0 in the FFT-OFDM system for a Raleigh channel with $f_d = 300$ Hz without CFO compensation.

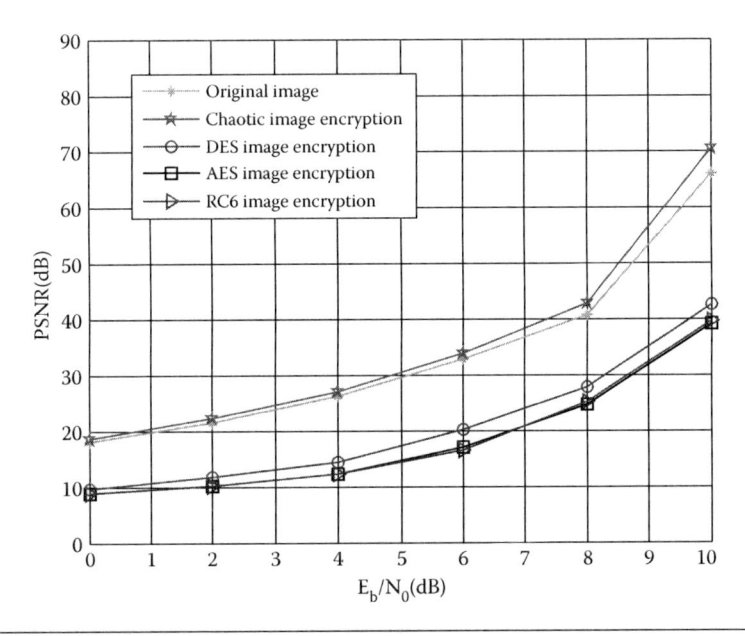

Figure 9.93 PSNR versus E_b/N_0 in the FFT-OFDM system for a Raleigh channel with $f_d = 300$ Hz with CFO compensation.

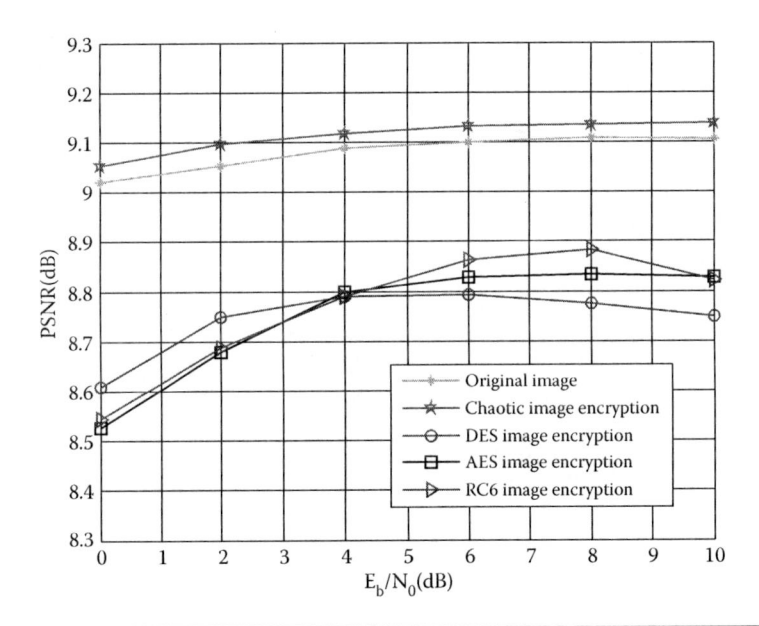

Figure 9.94 PSNR versus E_b/N_0 in the DCT-OFDM system for a Raleigh channel with $f_d = 300$ Hz without CFO compensation.

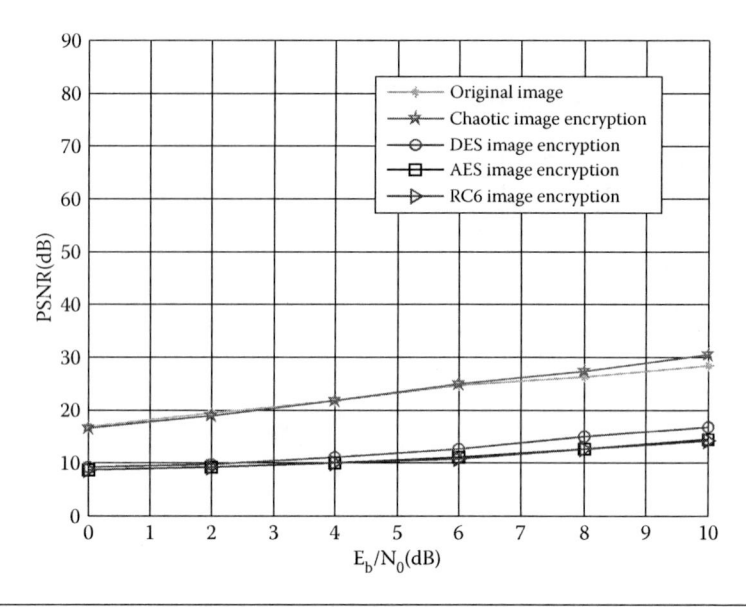

Figure 9.95   PSNR versus E_b/N_0 in the DCT-OFDM system for a Raleigh channel with $f_d = 300$ Hz with CFO compensation.

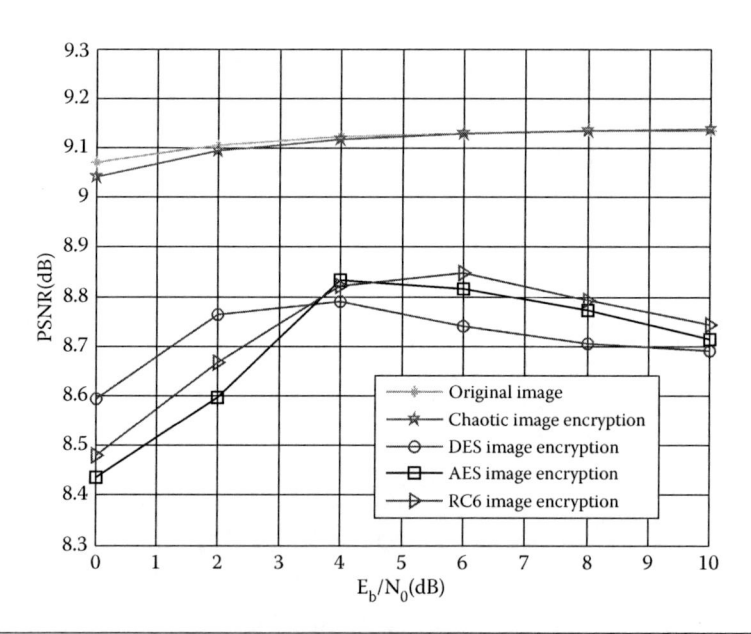

Figure 9.96   PSNR versus E_b/N_0 in the DWT-OFDM system for a Raleigh channel with $f_d = 300$ Hz without CFO compensation.

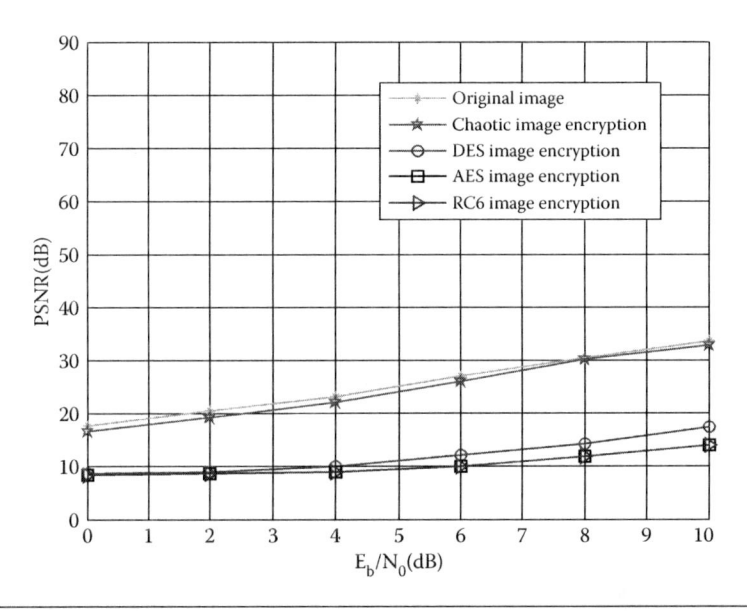

Figure 9.97 PSNR versus E_b/N_0 in the DWT-OFDM system for a Raleigh channel with $f_d = 300$ Hz with CFO compensation.

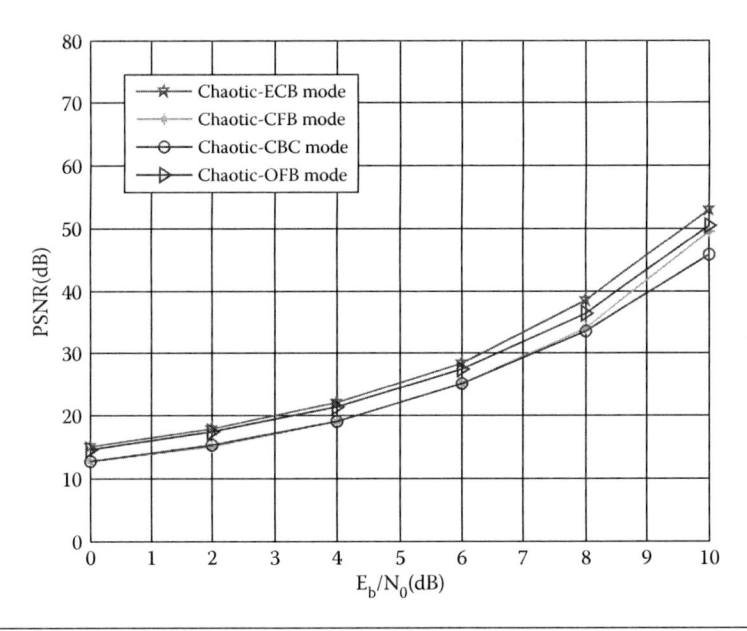

Figure 9.98 PSNR versus E_b/N_0 in the FFT-OFDM system for an AWGN channel with $N = 128$ and $G = 32$.

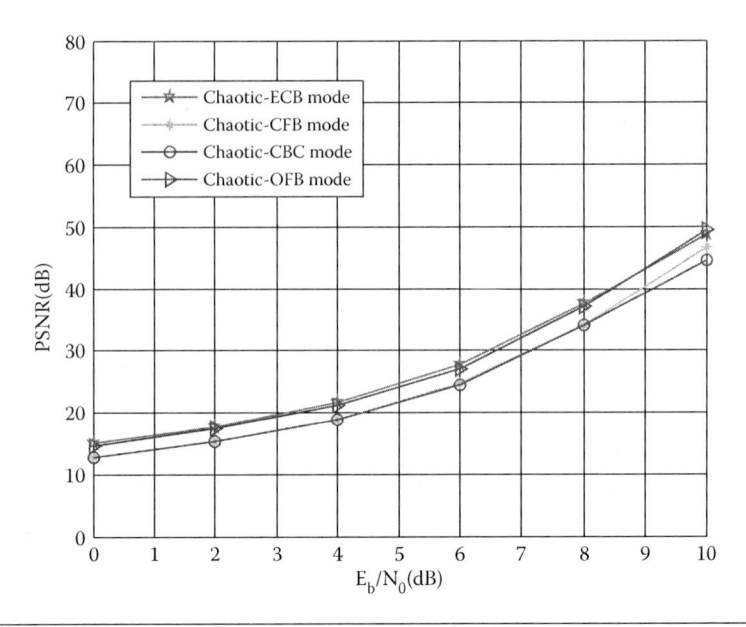

Figure 9.99   PSNR versus E_b/N_0 in the DCT-OFDM system for an AWGN channel with $N = 128$ and $G = 32$.

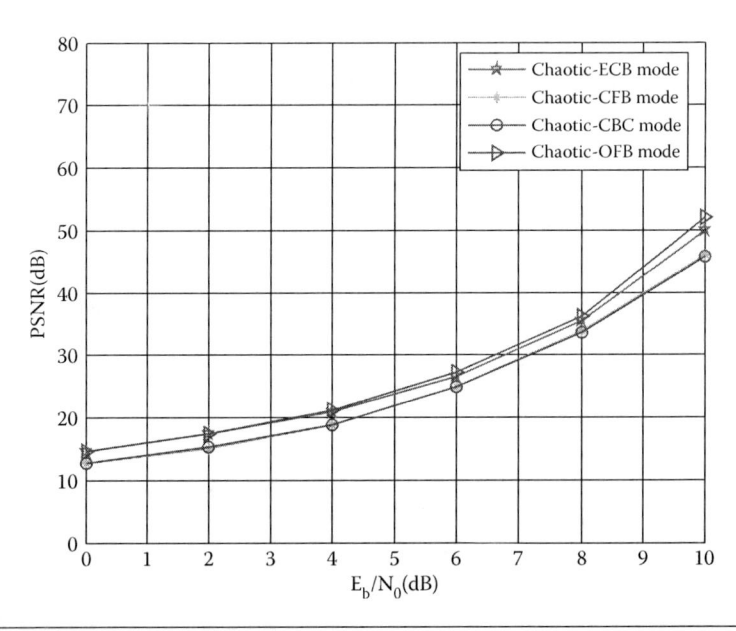

Figure 9.100   PSNR versus E_b/N_0 in the DWT-OFDM system for an AWGN channel with $N = 128$ and $G = 32$.

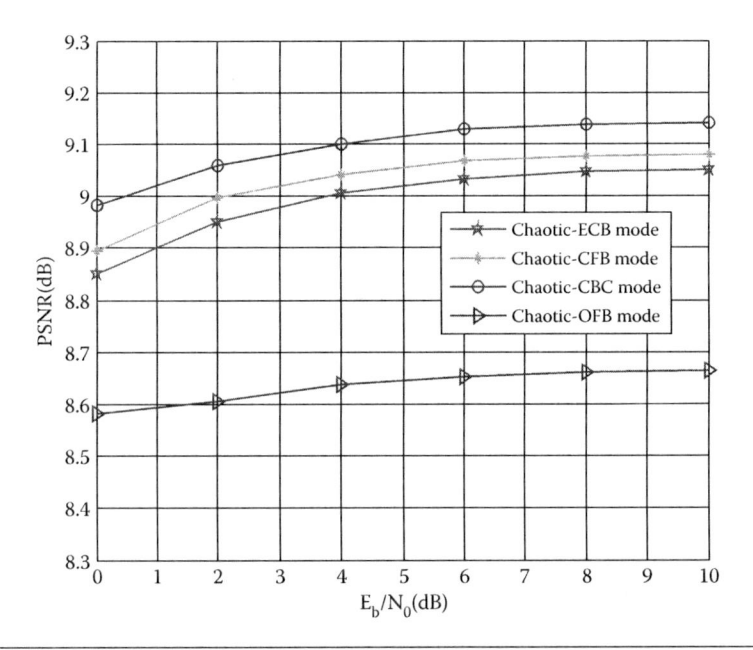

Figure 9.101 PSNR versus E_b/N_0 in the FFT-OFDM system for a Raleigh channel with $f_d = 300$ Hz without CFO compensation.

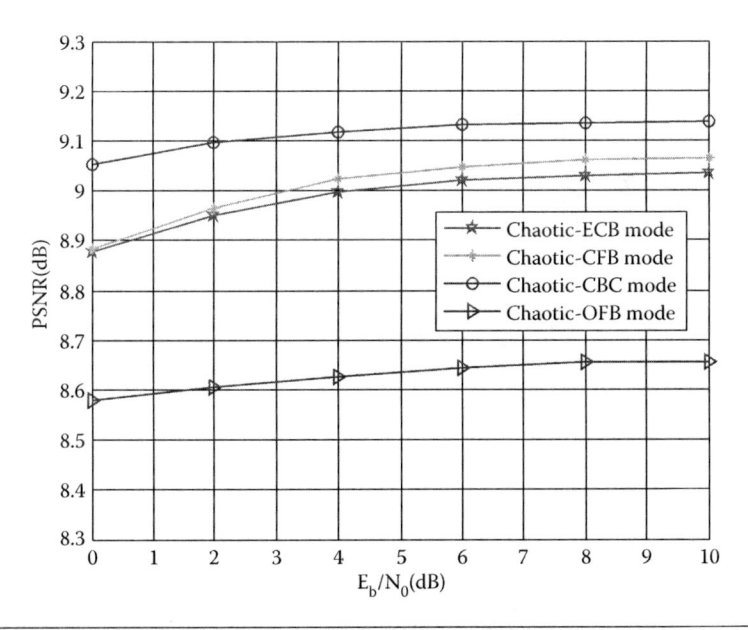

Figure 9.102 PSNR versus E_b/N_0 in the DCT-OFDM system for a Raleigh channel with $f_d = 300$ Hz without CFO compensation.

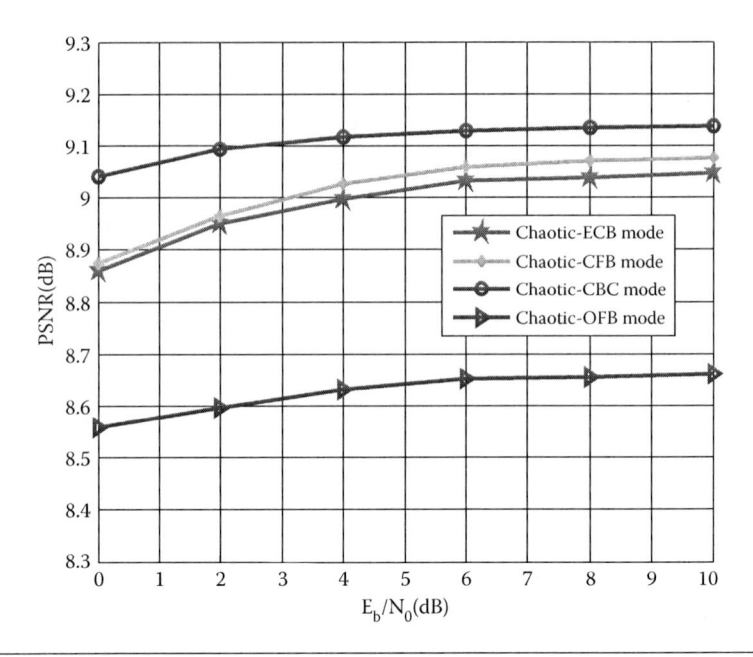

Figure 9.103   PSNR versus E_b/N_0 in the DWT-OFDM system for a Raleigh channel with $f_d = 300$ Hz without CFO compensation.

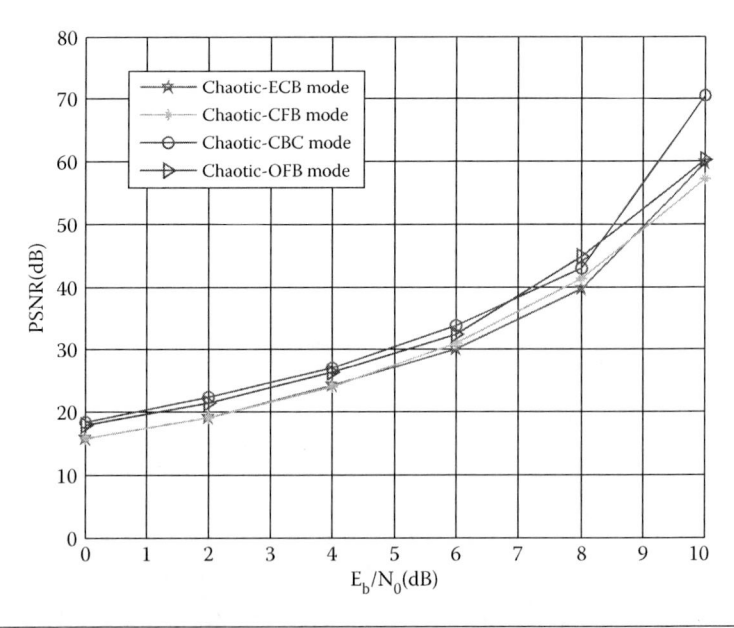

Figure 9.104   PSNR versus E_b/N_0 in the FFT-OFDM system for a Raleigh channel with $f_d = 300$ Hz with CFO compensation.

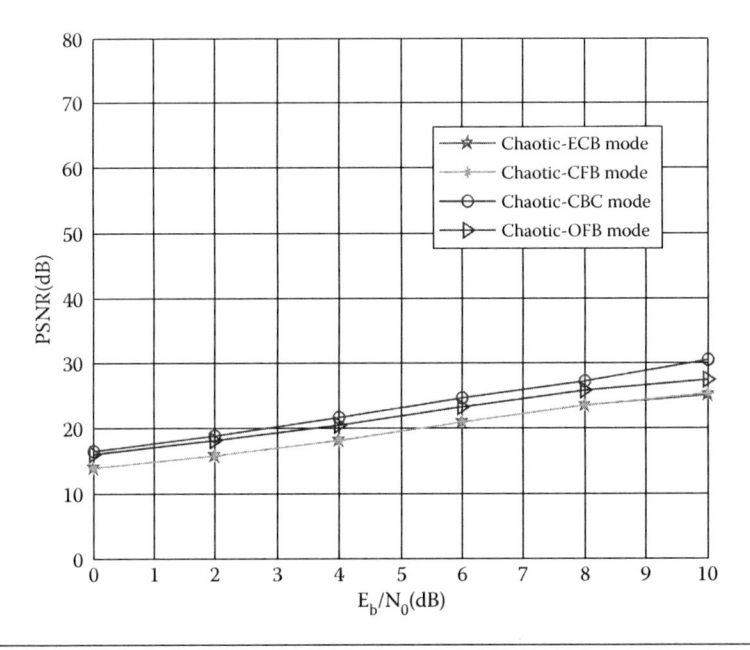

Figure 9.105   PSNR versus E_b/N_0 in the DCT-OFDM system for a Raleigh channel with $f_d = 300$ Hz with CFO compensation.

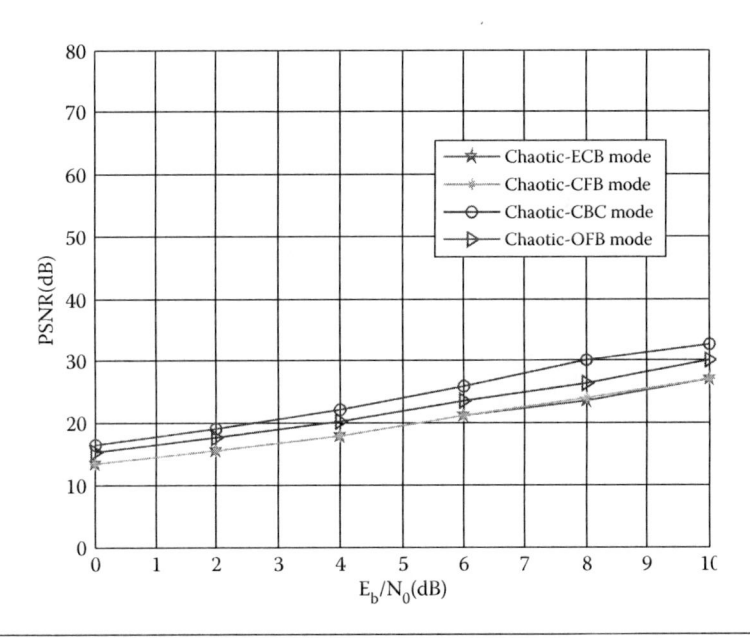

Figure 9.106   PSNR versus E_b/N_0 in the DWT-OFDM system for a Raleigh channel with $f_d = 300$ Hz with CFO compensation.

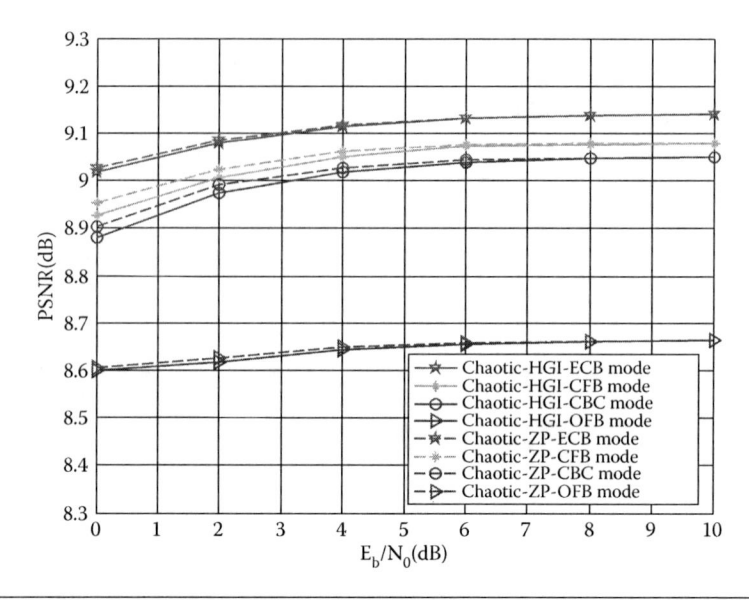

Figure 9.107   PSNR versus E_b/N_0 in the FFT-OFDM system with ZP and HGI for a Raleigh channel without PSE and equalization at $N = 128$ and $G = 32$.

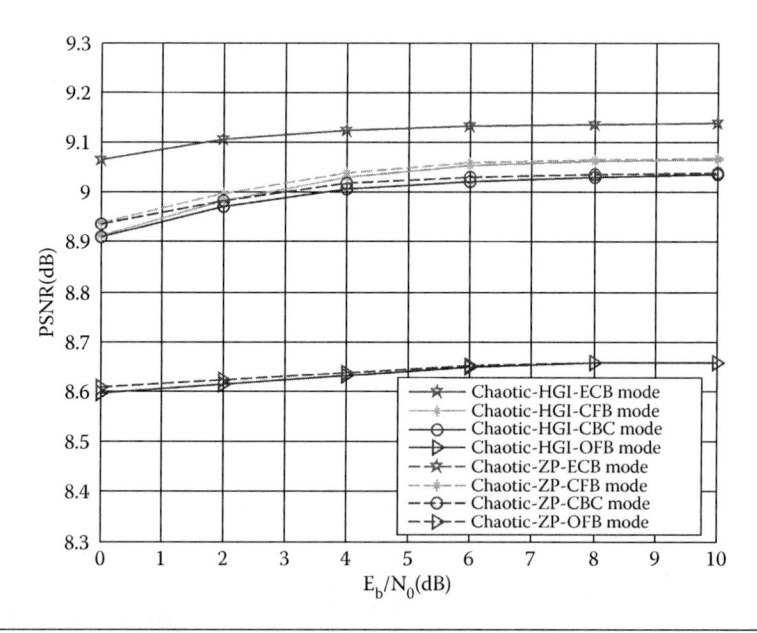

Figure 9.108   PSNR versus E_b/N_0 in the DCT-OFDM system with ZP and HGI for a Raleigh channel without PSE and equalization at $N = 128$ and $G = 32$.

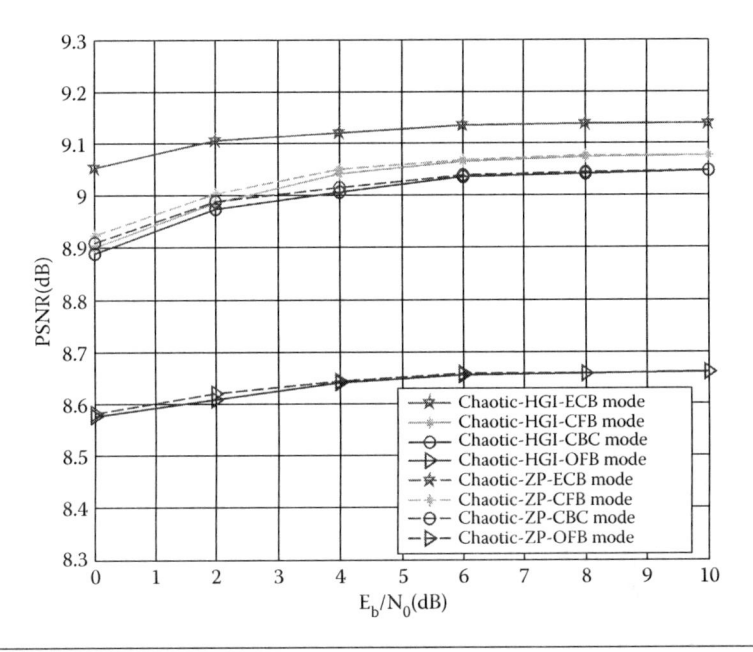

Figure 9.109 PSNR versus E_b/N_0 in the DWT-OFDM system with ZP and HGI for a Raleigh channel without PSE and equalization at $N = 128$ and $G = 32$.

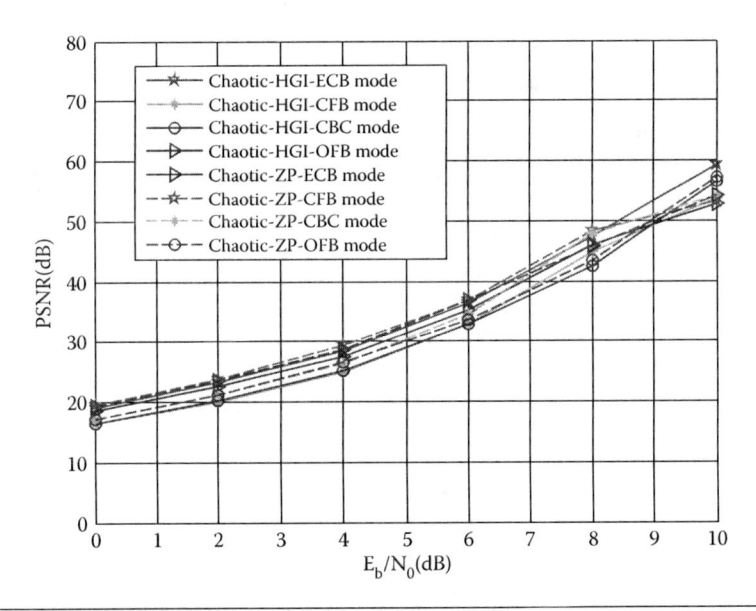

Figure 9.110 PSNR versus E_b/N_0 in the FFT-OFDM system with ZP and HGI for a Raleigh channel with PSE and equalization at $N = 128$ and $G = 32$.

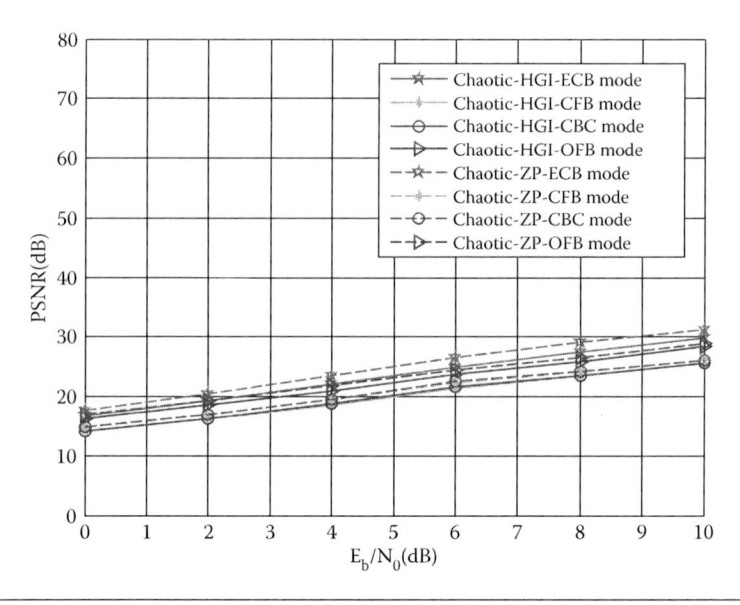

Figure 9.111 PSNR versus E_b/N_0 in the DCT-OFDM system with ZP and HGI for a Raleigh channel with PSE and equalization at $N = 128$ and $G = 32$.

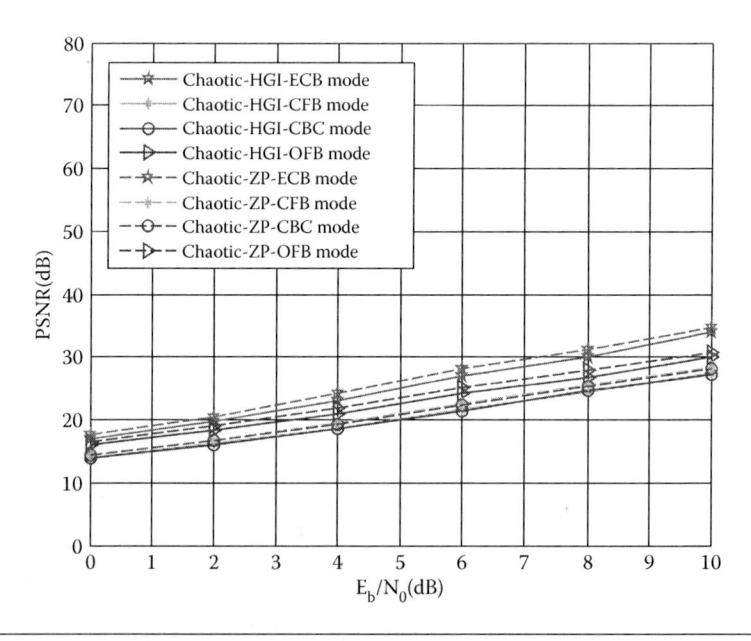

Figure 9.112 PSNR versus E_b/N_0 in the DWT-OFDM system with ZP and HGI for a Raleigh channel with PSE and equalization at $N = 128$ and $G = 32$.

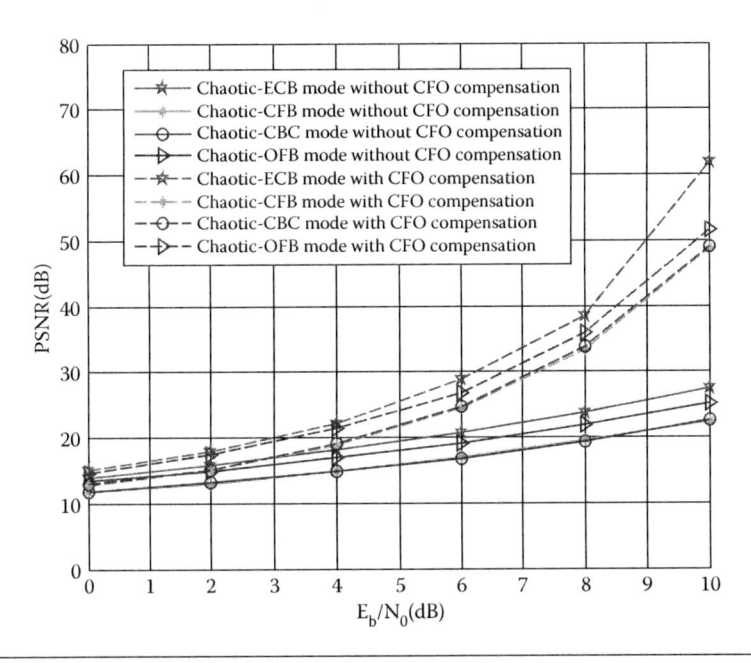

Figure 9.113   PSNR versus E_b/N_0 in the FFT-OFDM system for an AWGN channel.

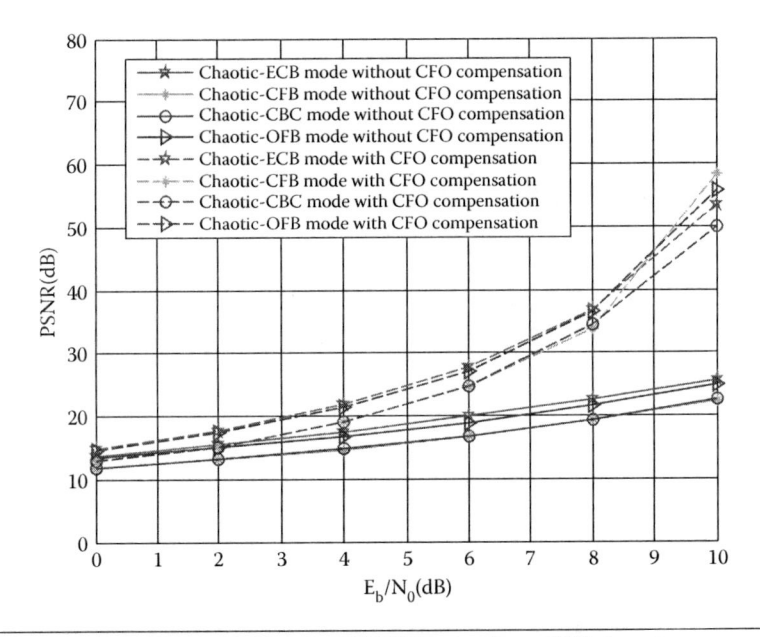

Figure 9.114   PSNR versus E_b/N_0 in the DCT-OFDM system for an AWGN channel.

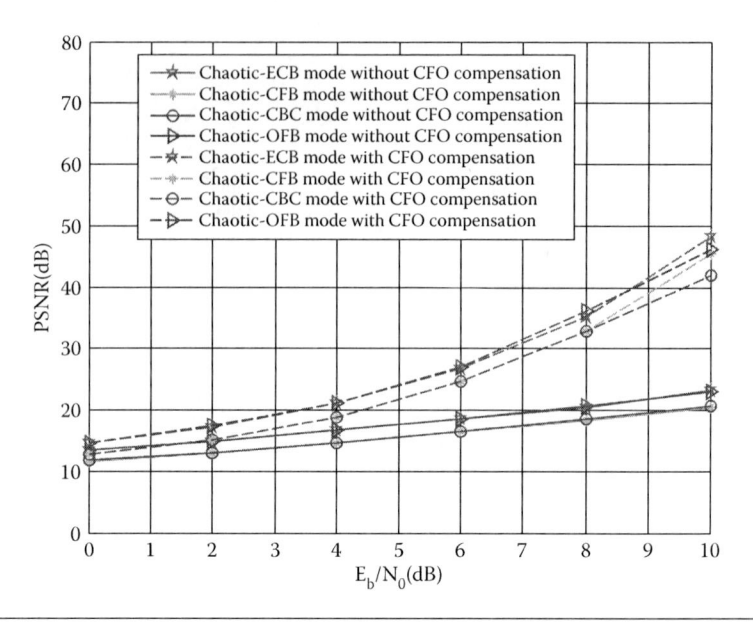

Figure 9.115   PSNR versus E_b/N_0 in the DWT-OFDM system for an AWGN channel.

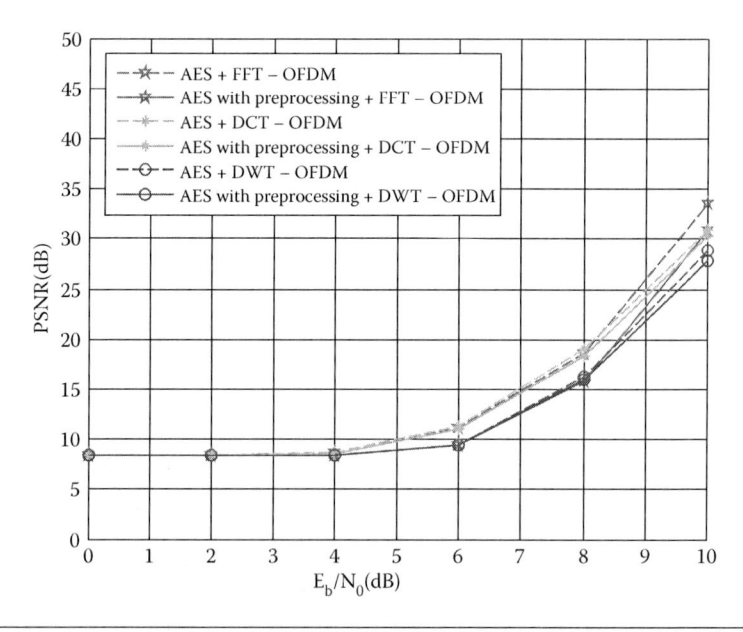

Figure 9.116   PSNR versus E_b/N_0 in the OFDM system for an AWGN channel with $G = 32$ and $N = 128$.

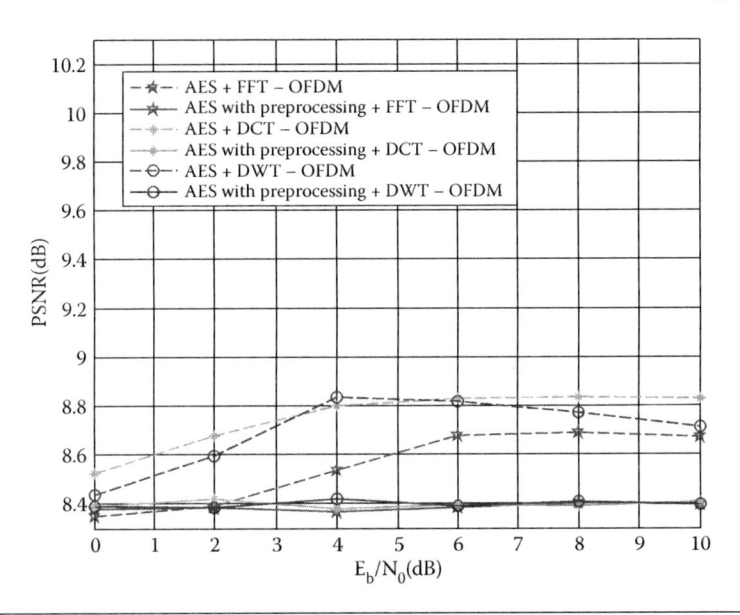

Figure 9.117   PSNR versus E_b/N_0 in the OFDM system for a Raleigh channel without PSE and equalization at $N = 128$ and $G = 32$.

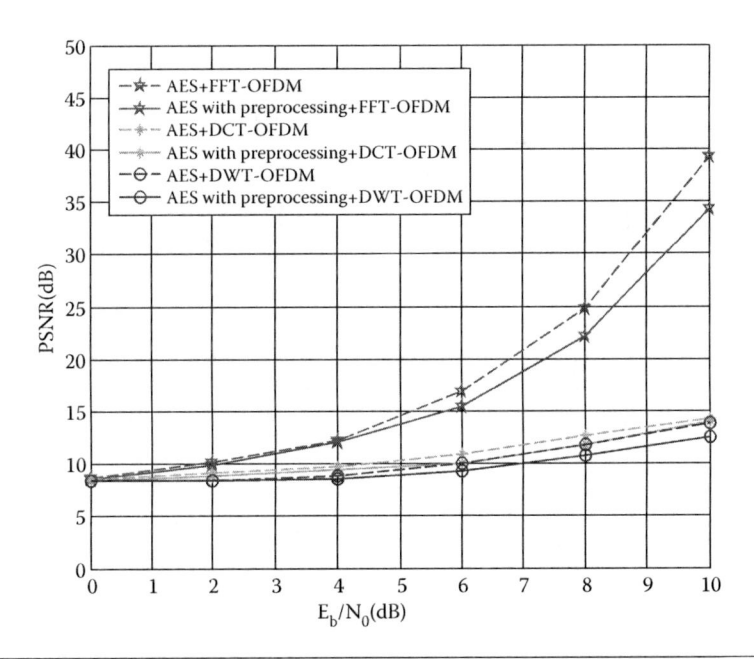

Figure 9.118   PSNR versus E_b/N_0 in the OFDM system for a Raleigh channel with PSE and equalization at $N = 128$ and $G = 32$.

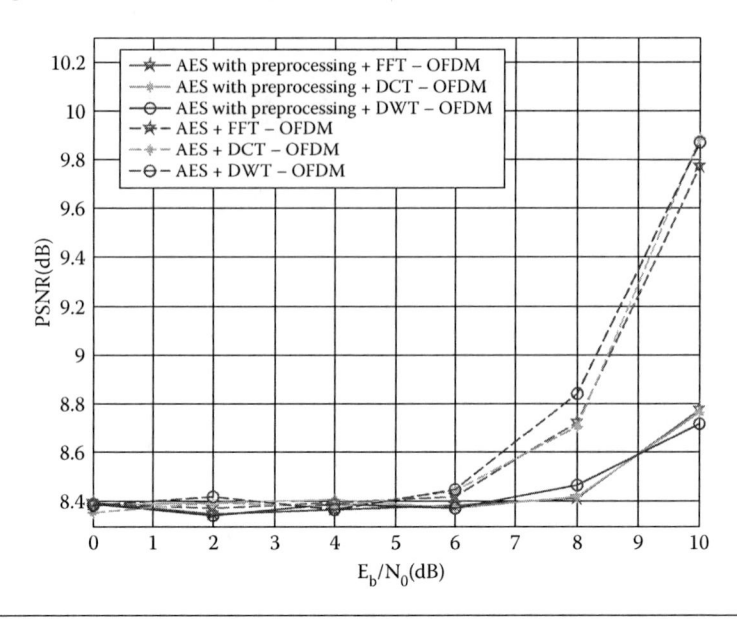

Figure 9.119   PSNR versus E_b/N_0 in the OFDM system for an AWGN channel without CFO compensation.

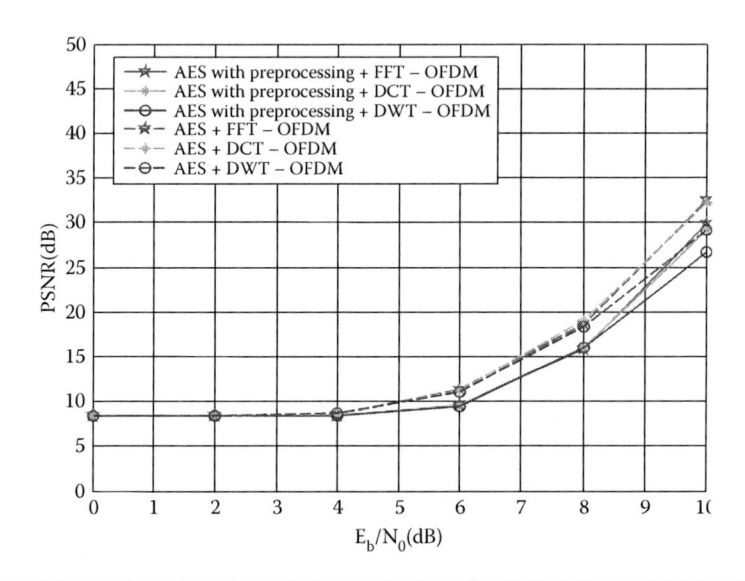

Figure 9.120   PSNR versus E_b/N_0 in the OFDM system for an AWGN channel with CFO compensation.

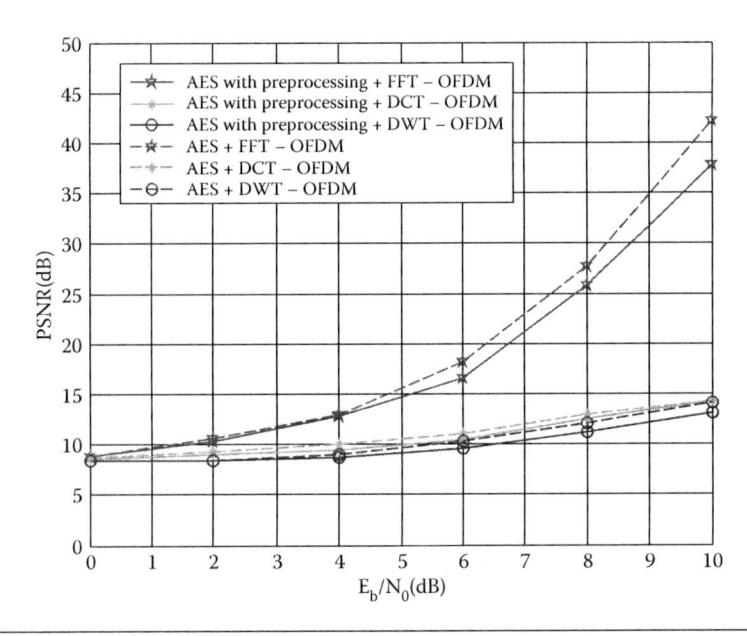

Figure 9.121   PSNR versus E_b/N_0 in the OFDM system with HGI for a Raleigh channel with $N = 128$ and $G = 32$ with PSE and equalization.

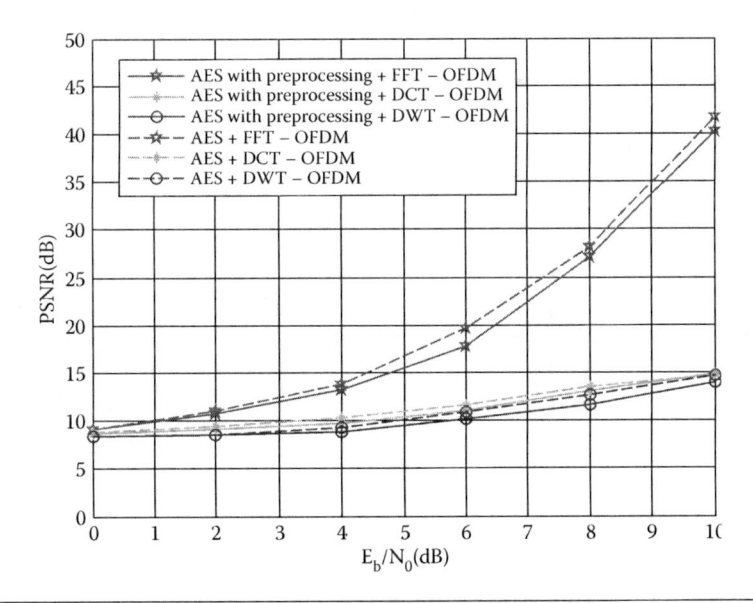

Figure 9.122   PSNR versus E_b/N_0 in the DWT-OFDM system with ZP for a Raleigh channel with $N = 128$ and $G = 32$ with PSE and equalization.

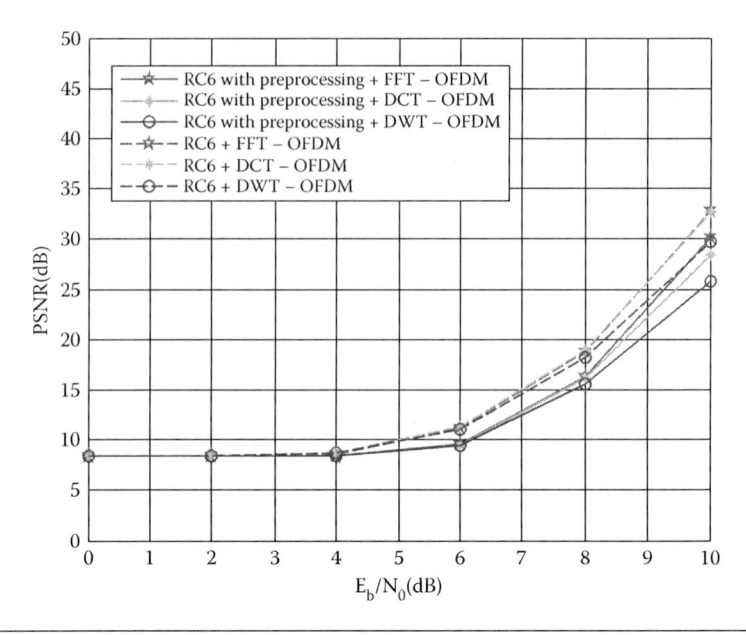

Figure 9.123 PSNR versus E_b/N_0 in the OFDM system for an AWGN channel with $N = 128$ and $G = 32$.

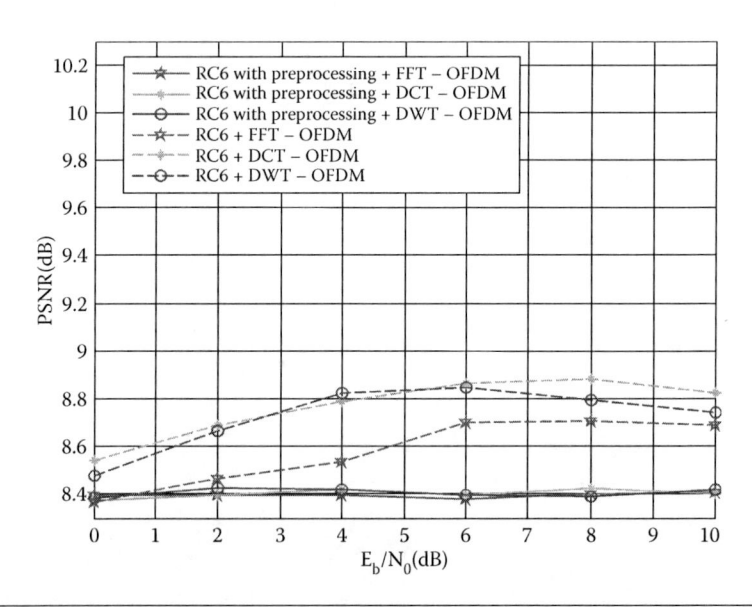

Figure 9.124 PSNR versus E_b/N_0 in the OFDM system for a Raleigh channel with $f_d = 300$ Hz without PSE and equalization.

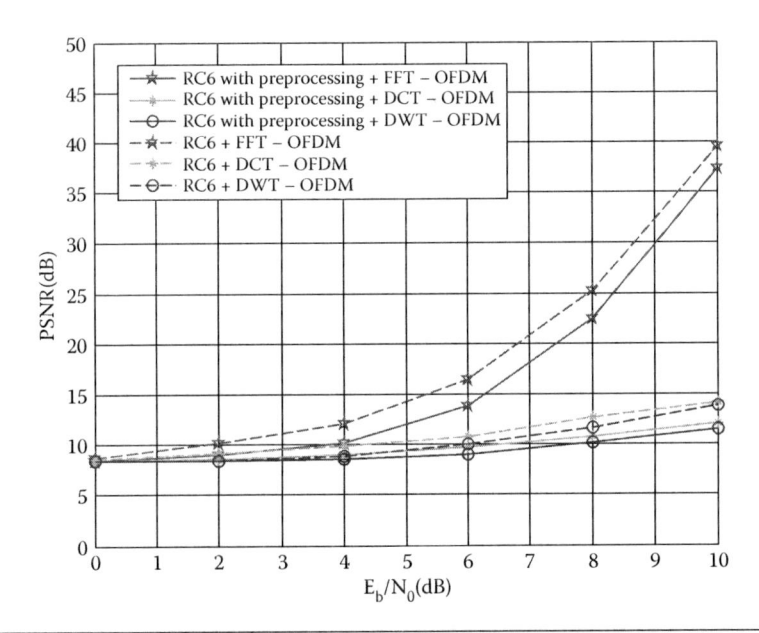

Figure 9.125 PSNR versus E_b/N_0 in the OFDM system for a Raleigh channel with $f_d = 300$ Hz with PSE and equalization.

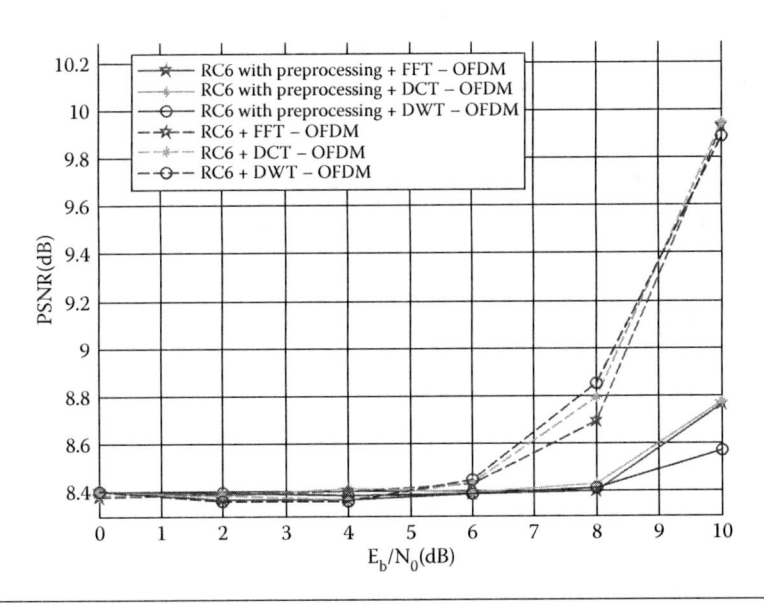

Figure 9.126 PSNR versus E_b/N_0 in the OFDM system for an AWGN channel without CFO compensation.

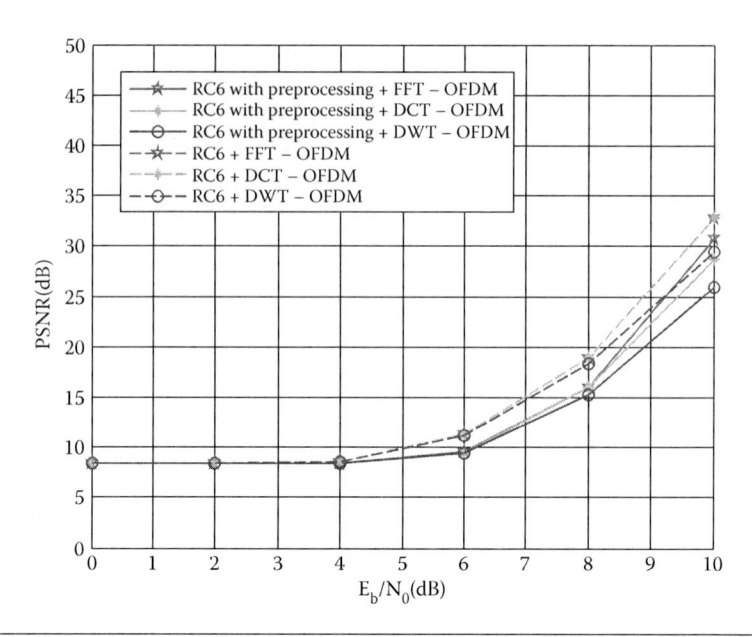

Figure 9.127 PSNR versus E_b/N_0 in the OFDM system for an AWGN channel with CFO compensation.

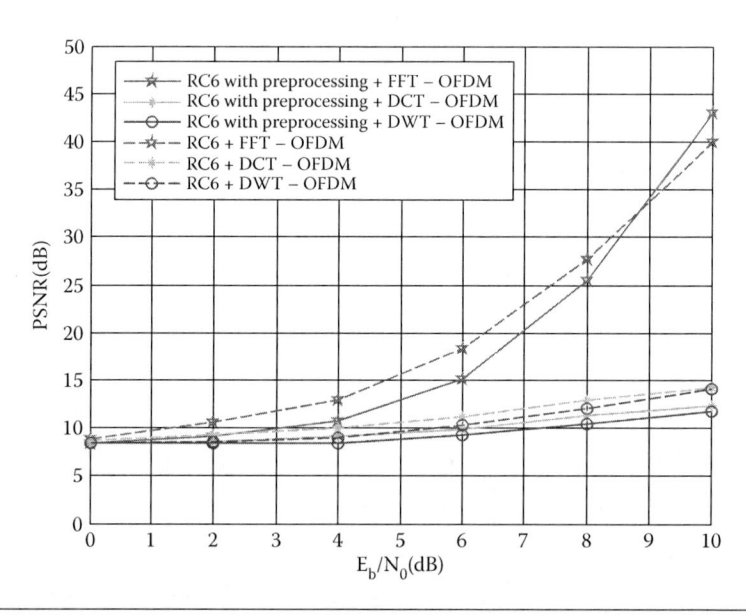

Figure 9.128 PSNR versus E_b/N_0 in the OFDM system for a Raleigh channel without CFO compensation.

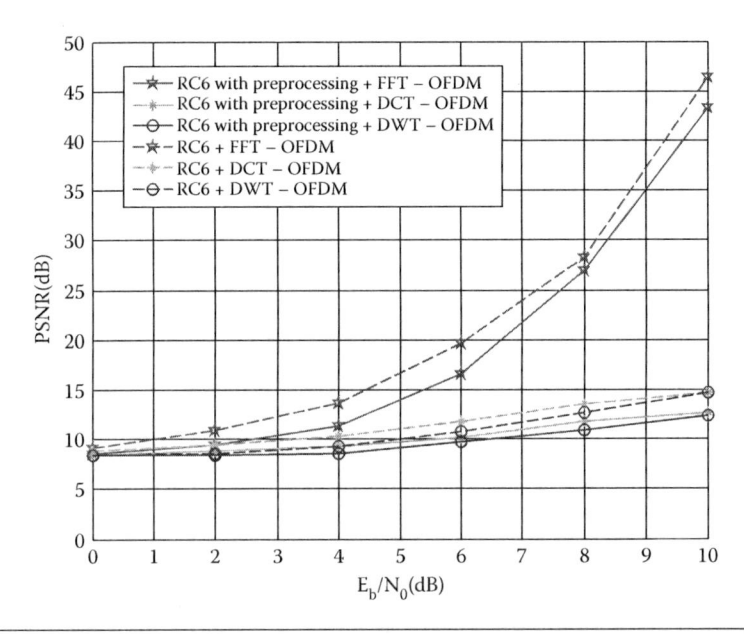

Figure 9.129 PSNR versus E_b/N_0 in the OFDM system for a Raleigh channel with CFO compensation.

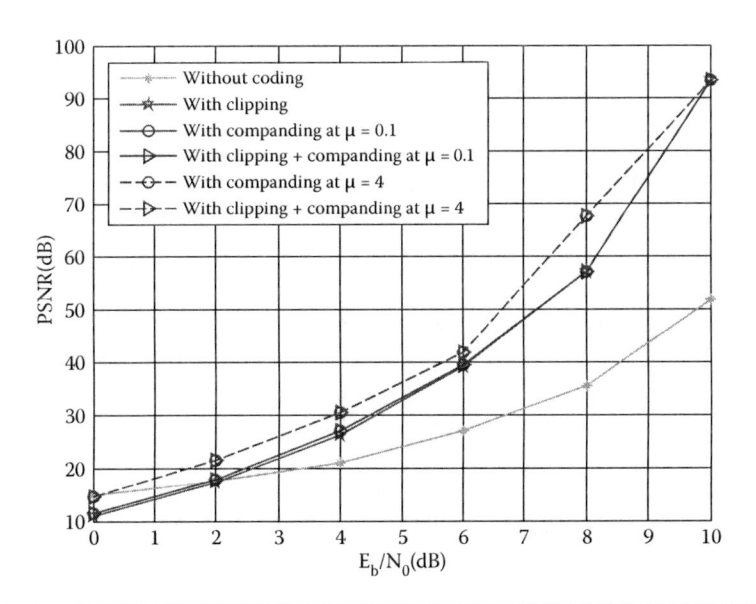

Figure 9.130 PSNR versus E_b/N_0 for original-image transmission with FFT-OFDM system over an AWGN channel. Convolutional coding is applied with all cases except the first one.

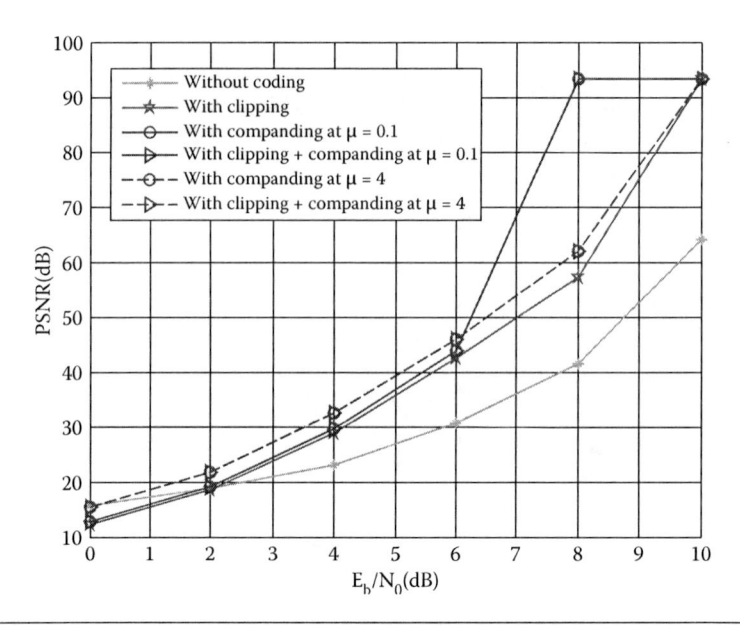

Figure 9.131 PSNR versus E_b/N_0 for original-image transmission with DCT-OFDM system over an AWGN channel. Convolutional coding is applied with all cases except the first one.

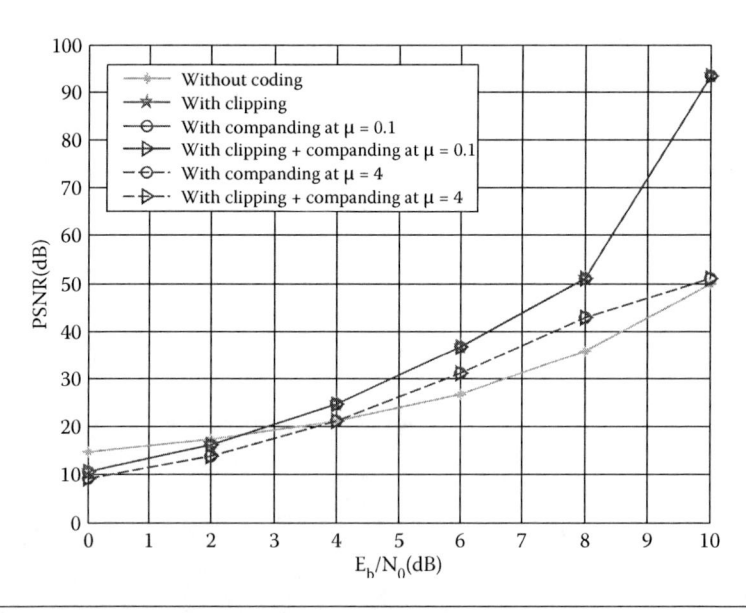

Figure 9.132 PSNR versus E_b/N_0 for original-image transmission with DWT-OFDM system over an AWGN channel. Convolutional coding is applied with all cases except the first one.

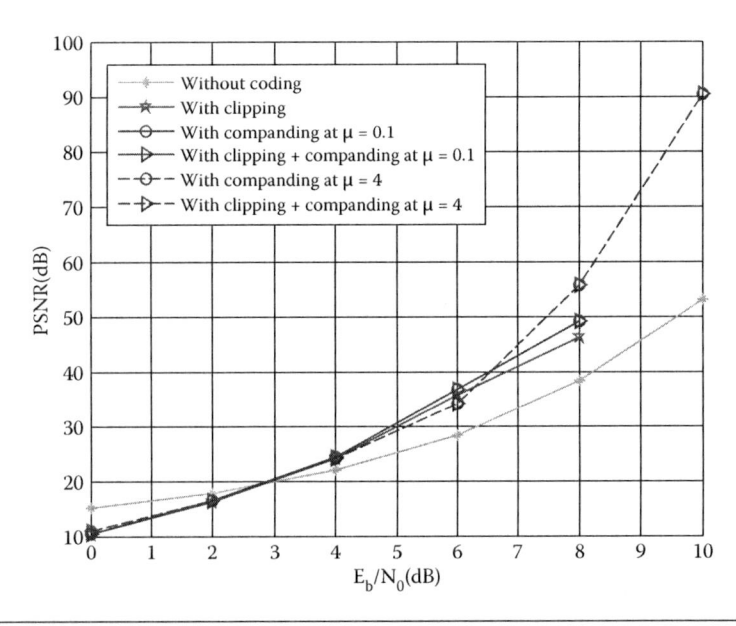

Figure 9.133 PSNR versus E_b/N_0 for chaotic-encrypted image transmission with FFT-OFDM system over an AWGN channel. Convolutional coding is applied with all cases except the first one.

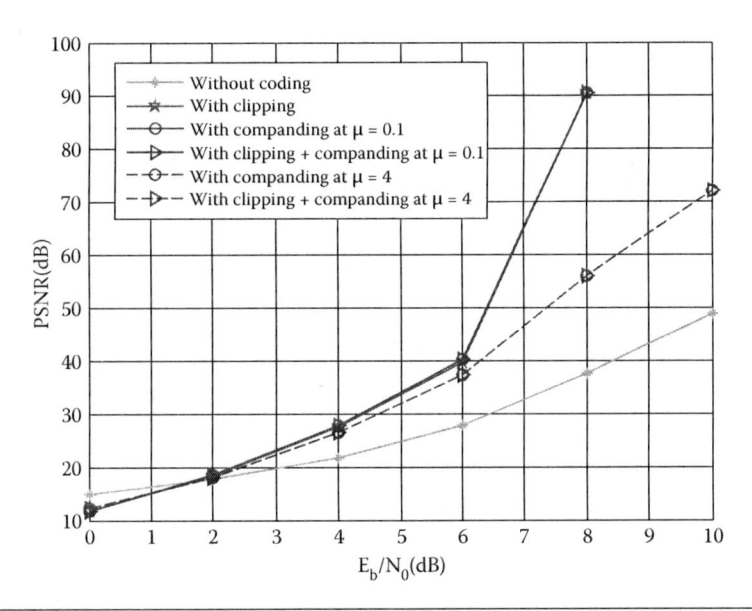

Figure 9.134 PSNR versus E_b/N_0 for chaotic-encrypted image transmission with DCT-OFDM system over an AWGN channel. Convolutional coding is applied with all cases except the first one.

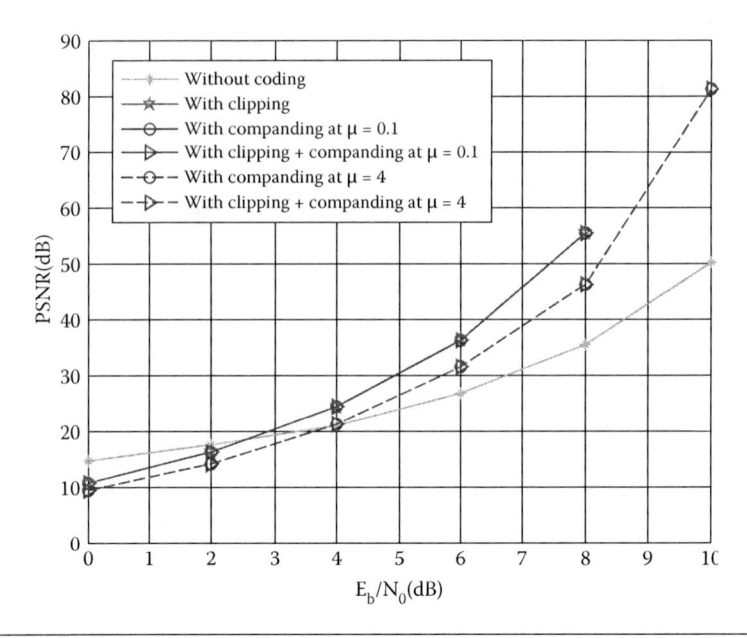

Figure 9.135 PSNR versus E_b/N_0 for chaotic-encrypted image transmission with DWT-OFDM system over an AWGN channel. Convolutional coding is applied with all cases except the first one.

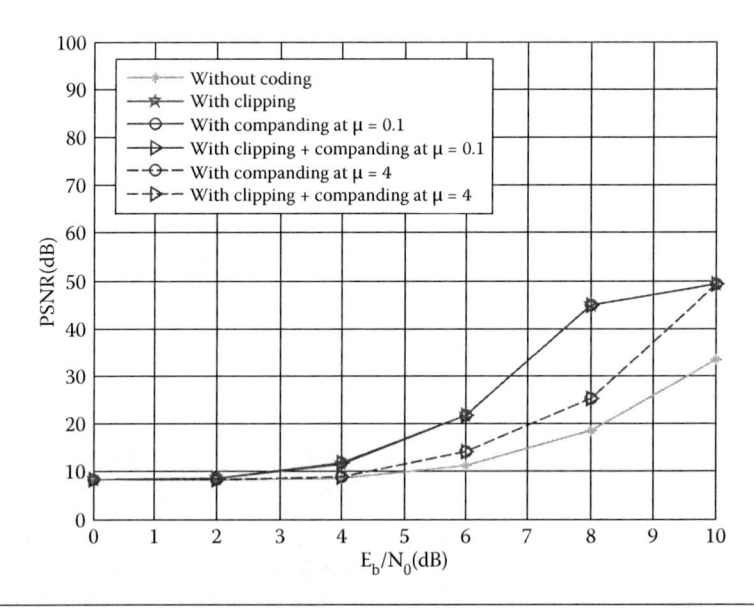

Figure 9.136 PSNR versus E_b/N_0 for AES-encrypted image transmission with FFT-OFDM system over an AWGN channel. Convolutional coding is applied with all cases except the first one.

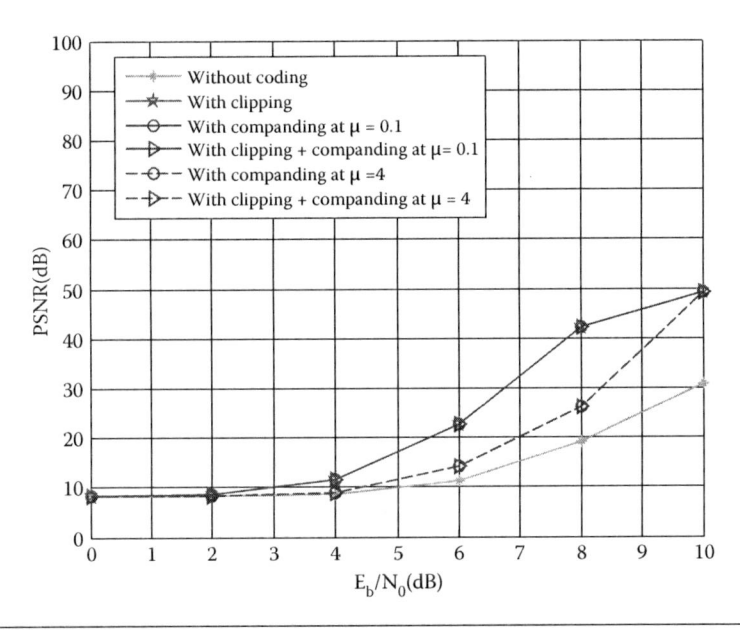

Figure 9.137 PSNR versus E_b/N_0 for AES-encrypted image transmission with DCT-OFDM system over an AWGN channel. Convolutional coding is applied with all cases except the first one.

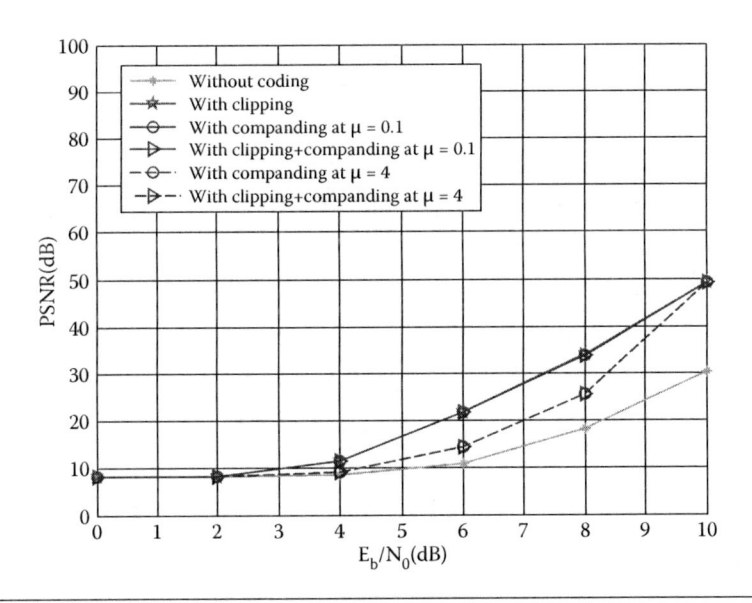

Figure 9.138 PSNR versus E_b/N_0 for AES-encrypted image transmission with DWT-OFDM system over an AWGN channel. Convolutional coding is applied with all cases except the first one.

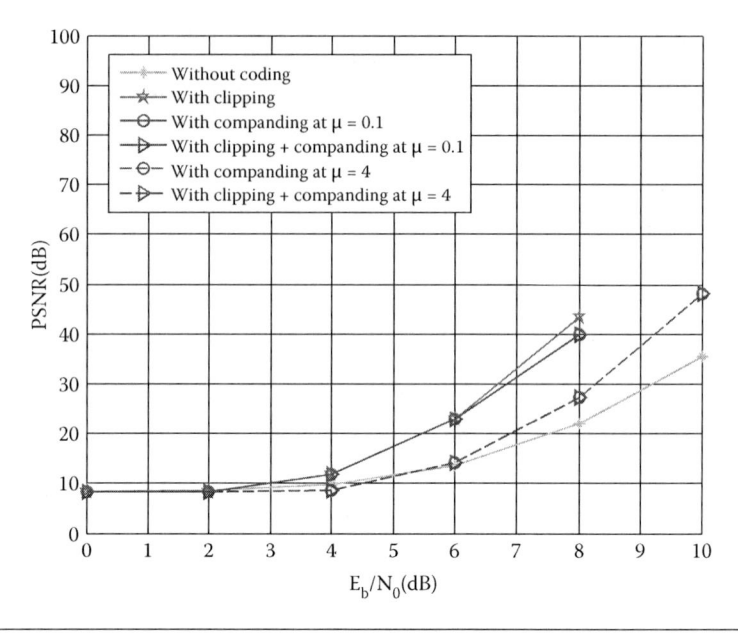

Figure 9.139   PSNR versus E_b/N_0 for DES-encrypted image transmission with FFT-OFDM system over an AWGN channel. Convolutional coding is applied with all cases except the first one.

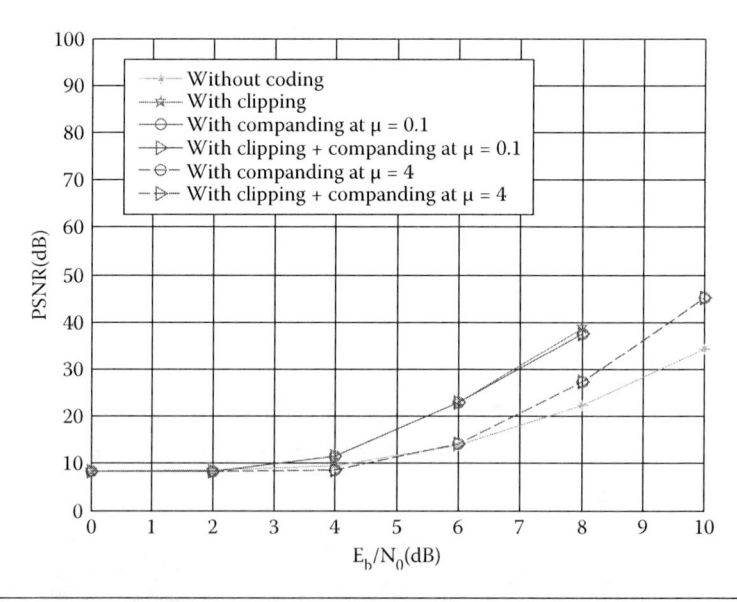

Figure 9.140   PSNR versus E_b/N_0 for DES-encrypted image transmission with DCT-OFDM system over an AWGN channel. Convolutional coding is applied with all cases except the first one.

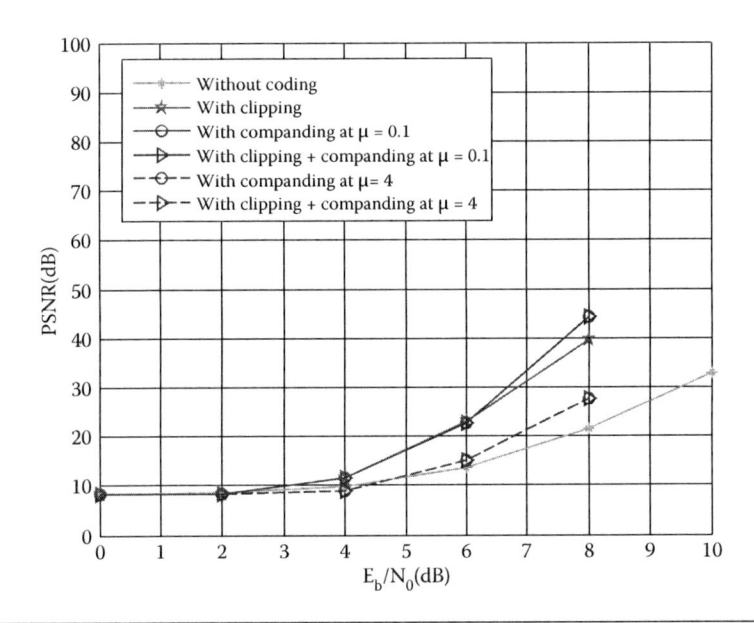

Figure 9.141 PSNR versus E_b/N_0 for DES-encrypted image transmission with DWT-OFDM system over an AWGN channel. Convolutional coding is applied with all cases except the first one.

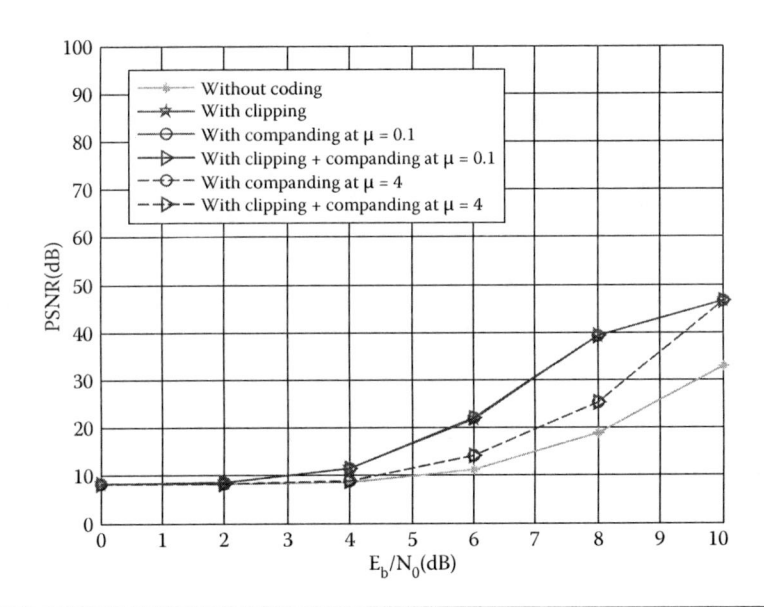

Figure 9.142 PSNR versus E_b/N_0 for RC6-encrypted image transmission with FFT-OFDM system over an AWGN channel. Convolutional coding is applied with all cases except the first one.

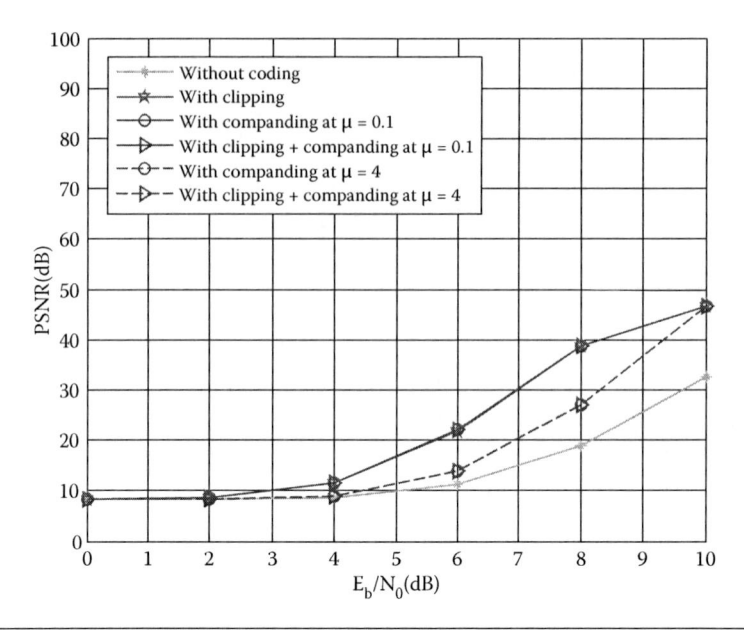

Figure 9.143   PSNR versus E_b/N_0 for RC6-encrypted image transmission with DCT-OFDM system over an AWGN channel. Convolutional coding is applied with all cases except the first one.

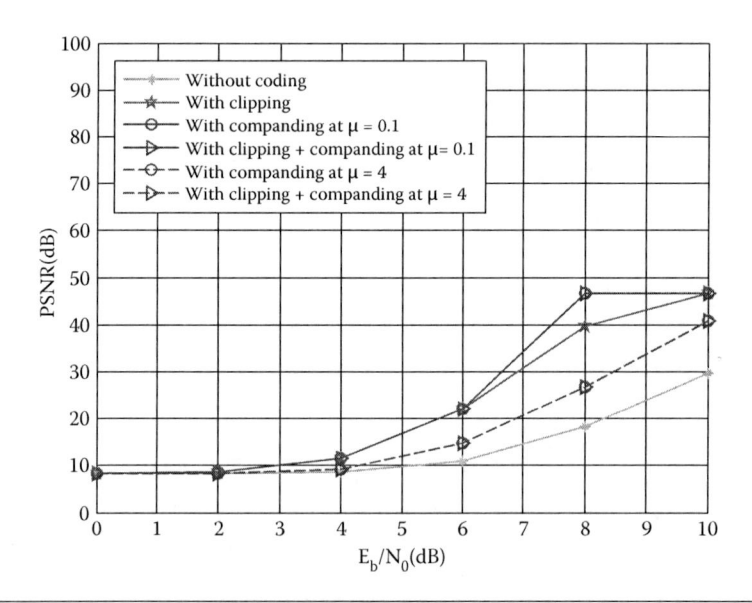

Figure 9.144   PSNR versus E_b/N_0 for RC6-encrypted image transmission with DWT-OFDM system over an AWGN channel. Convolutional coding is applied with all cases except the first one.

9.5 Summary

From all obtained results, we can see that PAPR reduction technique effects can be mitigated to a great extent with error-correcting codes such as the convolutional code. Another conclusion is that chaotic encryption preserves its rank as the best encryption algorithm from the immunity to noise perspective even in the presence of PAPR reduction. A simple clipping, companding, or hybrid clipping-and-companding method may be feasible for PAPR reduction in the case of encrypted image transmission.

References

1. O. S. Faragallah, Utilization of Security Techniques for Multimedia Applications, PhD thesis, Department of Computer Science and Engineering, Faculty of Electronic Engineering, Menoufiya University, Egypt, 2007.
2. A. J. Menezes, P. C. V. Oorschot, and S. Vanstone, *Handbook of Applied Cryptography*, CRC Press, Boca Raton, FL, 1996.
3. L. Qiao, Multimedia Security and Copyright Protection, PhD thesis, Department of Computer Science, University of Illinois at Urbana-Champaign, 1998.
4. W. Stallings, *Cryptography and Network Security Principles and Practice*, Prentice-Hall, Upper Saddle River, NJ, 1999.
5. C. E. Shannon, Communication Theory of Secrecy Systems, *Bell System Technical Journal*, Vol. 28, No. 4, pp. 656–715, October 1949.
6. S. Li, G. Chen, and X. Zheng, Chaos-Based Encryption for Digital Images and Videos, Chapter 4 in *Multimedia Security Handbook*, CRC Press, Boca Raton, FL, February 2004.
7. Y. Mao and M. Wu, A Joint Signal Processing and Cryptographic Approach to Multimedia Encryption, *IEEE Transactions on Image Processing*, Vol. 15, No. 7, pp. 2061–2075, July 2006.
8. Y. Mao, Research on Chaos-Based Image Encryption and Watermarking Technology, PhD thesis, Department of Automation, Nanjing University of Science and Technology, China, August 2003.
9. J. Daemen and V. Rijmen, AES Proposal: Rijndael, AES Algorithm Submission to the National Institute of Standards and Technology, 1999.
10. R. L. Rivest, M. J. B. Robshaw, R. Sidney, and Y. L. Yin, *The RC6^{TM} Block Cipher*, MIT Laboratory for Computer Science, Cambridge, MA, 1998.
11. R. F. Sewell, Bulk Encryption Algorithm for Use with RSA, *Electronics Letters*, Vol. 29, No. 25, pp. 2183–2185, December 9, 1993.

12. H. E. H. Ahmed, H. M. Kalash, and O. S. Faragallah, Encryption Efficiency Analysis and Security Evaluation of RC6 Block Cipher for Digital Images, *International Conference on Electrical Engineering (ICEE '07)*, pp. 1–7, 11–12, April 2007.

13. C. E. Shannon, A Mathematical Theory of Communication, *Bell System Technical Journal*, Vol. 27, pp. 379–423, July 1948.

14. M. Asim and V. Jeoti, On Image Encryption: Comparison between AES and a Novel Chaotic Encryption Scheme, *International Conference on Signal Processing, Communications and Networking (ICSCN '07)*, pp. 65–69, 22–24, February 2007.

15. R. Matthews, On the Derivation of a Chaotic Encryption Algorithm, *Cryptologia*, Vol. 8, No. 1, pp. 29–41, 1989.

16. B. Schneier, *Applied Cryptography*, 2nd edition, Wiley, New York, 1996.

17. L. Kocarev, Chaos-Based Cryptography: A Brief Overview, *IEEE Circuits and Systems Magazine*, Vol. 1, No. 3, pp. 6–21, 2001.

18. D. Stinson, *Cryptography: Theory and Practice*, 2nd edition, Chapman & Hall/CRC Press, Boca Raton, FL, 2002.

19. Y. Mao, G. Chen, and S. Lian, A Novel Fast Image Encryption Scheme Based on 3D Chaotic Baker Maps, *International Journal of Bifurcation and Chaos*, Vol. 14, No. 10, pp. 3613–3624, 2004.

20. S. Li, Analyses and New Designs of Digital Chaotic Ciphers, PhD thesis, School of Electronics and Information Engineering, Xi'an Jiaotong University, Xi'an, China, June 2003.

21. National Bureau of Standards, *Data Encryption Standard Modes of Operation*, Federal Information Processing Standards Publication 81, U.S. Government Printing Office, Washington, DC, 1980.

22. S. Li, X. Mou, and Y. Cai, Pseudo-Random Bit Generator Based on Couple Chaotic Systems and Its Applications in Stream Cipher Cryptography, *Lecture Notes in Computer Science*, Vol. 2247, pp. 316–329, 2001.

23. B. A. Forouzan, *Cryptography and Network Security*, McGraw-Hill, New Delhi, India, 2007.

24. R. Kusters and M. Tuengerthal, Universally Composable Symmetric Encryption, *2nd IEEE Computer Security Foundations Symposium (CSF '09)*, pp. 293–307, July 2009.

25. H. Jin, Z. Liao, D. Zou, and C. Li, Asymmetrical Encryption Based Automated Trust Negotiation Model, *2nd IEEE International Conference on Digital Ecosystems and Technologies (DEST 2008)*, pp. 363–368, February 2008.

26. S. G. Lian, J. Sun, and Z. Wang, A Novel Image Encryption Scheme Based on JPEG Encoding, *Proceedings of 8th International Conference on Information Visualization*, pp. 217–220, July 14–16, 2004.

27. M. V. Droogenbroeck and R. Benedett, Techniques for a Selective Encryption of Uncompressed and Compressed Images, *Proceedings of Advanced Concepts for Intelligent Vision Systems (ACIVS)*, Ghent, Belgium, pp. 90–97, September 9–11, 2002.

28. F. Dachselt, K. Kelber, and W. Schwarz, Chaotic Coding and Cryptoanalysis, *Proceedings of IEEE International Symposium on Circuits and Systems,* Hong Kong, pp. 1061–1064, June 9–12, 1997.

29. S. Li and X. Zheng, Cryptanalysis of a Chaotic Image Encryption Method, *Proceedings of IEEE International Symposium on Circuits and Systems (ISCAS),* Vol. 2, pp. 708–711, 2002.

30. J. Wei, X. Liao, K. W. Wong, and T. Zhou, Cryptanalysis of Cryptosystem Using Multiple One-Dimensional Chaotic Maps, *Communications in Nonlinear Science and Numerical Simulation,* Vol. 12, pp. 814–822, 2007.

31. H. Gao, Y. Zhang, S. Liang, and D. Li, A New Chaotic Algorithm for Image Encryption, *Chaos, Solitons & Fractals,* Vol. 29, No. 2, pp. 393–399, July 2006.

32. X. Li, J. Knipe, and H. Cheng, Image Compression and Encryption Using Tree Structures, *Pattern Recognition Letters,* Vol. 18, No. 8, pp. 2439–2451, 1997.

33. J. I. Guo, J. C. Yen, and J. C. Yeh, The Design and Realization of a New Hierarchical Chaotic Image Encryption Algorithm, *Proceedings of the International Symposium on Communications (ISCOM'99),* pp. 210–214, 1999.

34. L. Zhang, X. Liao, and X. Wang, An Image Encryption Approach Based on Chaotic Maps, *Chaos, Solitons & Fractals,* Vol. 24, No. 3, pp. 759–765, May 2005.

35. L. Kocarev and G. Jakimoski, Logistic Map as a Block Encryption Algorithm, *Physics Letters A,* Vol. 289, No. 4–5, pp. 199–206, October 22, 2001.

36. T. Xiang, X. Liao, G. Tang, Y. Chen, and K.W. Wong, A Novel Block Cryptosystem Based on Iterating a Chaotic Map, *Physics Letters A,* Vol. 349, No. 1–4, pp. 109–115, January 9, 2006.

37. S. Contini, R. L. Rivest, M. J. B. Robshaw, and Y. L. Yin, *The Security of the RC6™ Block Cipher,* Version 1.0, RSA Laboratories, MIT Laboratory for Computer Science, Cambridge, MA, August 20, 1998.

38. M. Salleh, S. Ibrahim, and I. F. Isnin, Enhanced Chaotic Image Encryption Algorithm Based on Baker›s Map, *Proceedings of the 2003 International Symposium on Circuits and Systems (ISCAS ‹03),* Vol. 2, pp. 508–511, May 2003.

39. D. Chen, A Feasible Chaotic Encryption Scheme for Image, *International Workshop on Chaos-Fractals Theories and Applications (IWCFTA·09),* pp. 172–176, November 6–85, 2009.

40. A. Palacios and H. Juarez, Cryptography with Cycling Chaos, *Physics Letters A,* Vol. 303, No. 5–6, pp. 345–351, October 28, 2002.

41. H. Cheng and X. Li, Partial Encryption of Compressed Images and Videos, *IEEE Transactions on Signal Processing,* Vol. 48, No. 8, pp. 2439–2451, August 2000.

42. A. Servetti and J. C. De Martin, Perception-Based Partial Encryption of Compressed Speech, *IEEE Transactions on Speech and Audio Processing,* Vol. 10, No. 8, pp. 637–643, November 2002.

43. X. Wu, H. Hu, and B. Zhang, Analyzing and Improving a Chaotic Encryption Method, *Chaos, Solitons & Fractals*, Vol. 22, No. 2, pp. 367–373, October 2004.

44. R. C. Merkle and M. Hellman, On the Security of Multiple Encryption, *Communications of the ACM*, Vol. 24, No. 7, pp. 465–467, July 1981.

45. P. Kitsos, S. Goudevenos, and O. Koufopavlou, VLSI Implementations of the Triple-DES Block Cipher, *Proceedings of the 10th IEEE International Conference on Electronics, Circuits and Systems (ICECS)*, Vol. 1, pp. 76–79, December 14–17, 2003.

46. National Bureau of Standards, *Data Encryption Standard*, Federal Information Processing Standards Publication 46, U.S. Government Printing Office, Washington, DC, 1977.

47. W. Jian, L. Xu, and J. Xiaoyong, A Secure Communication System with Multiple Encryption Algorithms, *Proceedings of the International Conference on E-Business and E-Government*, pp. 3574–3577, May 7–9, 2010.

48. R. L. Rivest, The RC5 Encryption Algorithm, *Proceedings of the 2nd Workshop on Fast Software Encryption*, pages 86–96, 1995.

49. R. Baldwin and R. Rivest, RFC 2040: The RC5, RC5-CBC, RC5-CBC-Pad, and RC5-CTS Algorithms, October 30, 1996.

50. R. Ratchkov, *Advanced Encryption Standard (AES) Algorithm and Modes of Operation*, LSI Logic, San Jose, CA, August 2002.

51. J. C. Yen and J. I. Guo, A New Hierarchical Chaotic Image Encryption Algorithm and Its Hardware Architecture, *Proceedings 9th VLSI Design/CAD Symposium*, Taiwan, pp. 358–362, 1998.

52. J. Fridrich, Symmetric Ciphers Based on Two-Dimensional Chaotic Maps, *International Journal of Bifurcation and Chaos*, Vol. 8, No. 6, pp. 1259–1284, 1998.

53. W. Xiao, J. Zhang, and W. Wu, A Watermarking Algorithm Based on Chaotic Encryption, *Proceedings of IEEE TENCON*, pp. 545–548, 2002.

54. S. Li and X. Zheng, On the Security of an Image Encryption Method, *Proceedings IEEE International Conference on Image Processing (ICIP)*, Vol. 2, pp. 925–928, 2002.

55. M. Ashtiyani, P. M. Birgani, and H. M. Hosseini, Chaos-Based Medical Image Encryption Using Symmetric Cryptography, *3rd International Conference on Information and Communication Technologies: From Theory to Applications (ICTTA)*, pp. 1–5, 7–11 April 2008.

56. N. Bourbakis and C. Alexopoulos, Picture Data Encryption Using SCAN Patterns, *Pattern Recognition*, Vol. 25, No. 6, pp. 567–581, 1992.

57. J. Scharinger, Fast Encryption of Image Data Using Chaotic Kolmogrov Flow, *Journal of Electronic Engeering*, Vol. 7, No. 2, pp. 318–325, 1998.

58. J. C. Yen and J. I. Guo, A New Image Encryption Algorithm and Its VLSI Architecture, *Proceedings of the IEEE Workshop Signal Processing Systems*, pp. 430–437, 1999.

59. J. C. Yen and J. I. Guo, A New Chaotic Key Based Design for Image Encryption and Decryption, *Proceedings of the IEEE International Symposium Circuits and Systems*, Vol. 4, pp. 49–52, 2000.

60. G. Chen, Y. Mao, and C. K. Chui, A Symmetric Image Encryption Scheme Based on 3D Chaotic Cat Maps, *Chaos, Solitons and Fractals*, Vol. 21, No. 3, pp. 749–761, 2004.

61. K. Kelber and W. Schwarz, General Design Rules for Chaos-Based Encryption Systems, *International Symposium on Nonlinear Theory and Its Applications (NOLTA2005)*, pp. 465–468, Bruges, Belgium, October 18–21, 2005.

62. M. Yang, N. Bourbakis, and S. Li, Data-Image-Video Encryption, *Potentials IEEE,* Vol. 23, No. 3, pp. 28–34, 2004.

63. Y. Feng and X. Yu, A Novel Symmetric Image Encryption Approach Based on an Invertible Two-Dimensional Map, *35th Annual Conference of IEEE Industrial Electronics (IECON '09)*, pp. 1973–1978, November 3–5, 2009.

64. C. J. Kuo, Novel Image Encryption Technique and Its Application in Progressive Transmission, *Journal of Electronic Imaging*, Vol. 2, No. 4, pp. 345–351, 1993.

65. H. K. Chang and J. L. Liou, An Image Encryption Scheme Based on Quadtree Compression Scheme, *Proceedings of the International Computer Symposium*, Taiwan, pp. 230–237, 1994.

66. J. Scharinger, Secure and Fast Encryption Using Chaotic Kolmogorov Flows, *Proceedings of the IEEE Industrial Electronics and Applications Conference*, pp. 3662–3666, 1998.

67. H. M. Elkamchouchi and M. A. Makar, Measuring Encryption Quality for Bitmap Images Encrypted with Rijndael and KAMKAR Block Ciphers, *Proceedings of the National Radio Science Conference of Egypt*, pp. 277–284, 2005.

68. S. A. Yeung, S. Zhu, and B. Zeng, Quality Assessment for a Perceptual Video Encryption System, *Proceedings of the IEEE Wireless Communications, Networking and Information Security (WCNIS) Conference,* pp. 102–106, 2010.

69. A. Sinha and K. Singh, A Technique for Image Encryption Using Digital Signature, *Optics Commununication*, Vol. 218, No. 4–6, pp. 229–234, 2003.

70. D. E. Newton, *Encyclopedia of Cryptology*, ABC-CLIO, Santa Barbara, CA, 1997.

71. Z. Yun-Peng, L. Wei, C. Shui-Ping, Z. Zheng-Jun, N. Xuan, and D. Wei-Di, Digital Image Encryption Algorithm Based on Chaos and Improved DES, *IEEE International Conference on Systems, Man and Cybernetics,* pp. 474–479, 2009.

72. J. Cheng, F. Zhang, K. Yu, and J. Ma, The Dynamic and Double Encryption System Based on Two-Dimensional Image, *International Conference on Computational Intelligence and Security (CIS '09)*, pp. 458–462, 2009.

73. A. H. M. Ragab, N. A. Ismail, and O. S. FaragAllah, Enhancements and Implementation of RC6 Block Cipher for Data Security, *Proceedings of International Conference on Electrical and Electronic Technology*, Vol. 1, pp. 133–137, 2001.

74. H. Feistel, Cryptography and Computer Privacy, *Scientific American*, Vol. 228, No. 5, pp. 15–23, May 1973.

75. H. H. Nien, S. K. Changchien, S. Y. Wu, and C. K. Huang, A New Pixel-Chaotic-Shuffle Method for Image Encryption, *10th International Conference on Control, Automation, Robotics and Vision (ICARCV)*, pp. 883–887, December 17–20, 2008.

76. H. M. Heys Analysis of the Statistical Cipher Feedback Mode of Block Ciphers, *IEEE Transactions on Computers*, Vol. 52, No. 1, pp. 77–92, January 2003.

77. C. Lu and S. Tseng, Integrated Design of AES (Advanced Encryption Standard) Encrypter and Decrypter, *Proceedings of The IEEE International Conference on Application-Specific Systems, Architectures and Processors*, pp. 277–285, July 2002.

78. M. Borsc and H. Shinde, Wireless Security & Privacy, *IEEE International Conference on Personal Wireless Communications (ICPWC)*, pp. 424–428, January 2005.

79. A. G. Chefranov and T. A. Mazurova, Pseudo-Random Number Generator (RC4) Period Improvement, *International Conference on Automation, Quality and Testing, Robotics*, Vol. 2, pp. 38–41, May 2006.

80. I. F. Elashry, O. S. Farag Allah, A. M. Abbas, S. El-Rabaie, and F. E. Abd El-Samie, Homomorphic Image Encryption, *Electronic Imaging*, Vol. 18, No. 3, 033002, 2009.

81. W. K. Wong, L. P. Lee, and K. W. Wong, A Modified Chaotic Cryptographic Method, *Computer Physics Communication*, Vol. 13, No. 8, pp. 234–236, 2001.

82. M. Ahmad, C. Gupta, and A. Varshney, Digital Image Encryption Based on Chaotic Map for Secure Transmission, *International Multimedia, Signal Processing and Communication Technologies (IMPACT '09)*, pp. 292–295, March 2009.

83. D. Garg and S. Verma, Improvement Over Public Key Cryptographic Algorithm, *IEEE International Advance Computing Conference (IACC)*, pp. 734–739, March 2009.

84. A. Ramzi, A. N. El-Kassar, and B. M. Shebaro, A Comparative Study of Elgamal Based Digital Signature Algorithms, *World Automation Congress (WAC '06)*, pp. 1–6, July 2006.

85. G. Kim, J. Kim, and G. Cho, An Improved RC6 Algorithm with the Same Structure of Encryption and Decryption, *11th International Conference on Advanced Communication Technology (ICACT)*, Vol. 2, pp. 1211–1215, February 2009.

86. Q. Zhu, L. Li, J. Liu, and N. Xu, The Analysis and Design of Accounting Information Security System Based on AES Algorithm, *International Conference on Machine Learning and Cybernetics*, Vol. 5, pp. 2713–2718, July 2009.

87. S. C. Koduru and V. Chandrasekaran, Integrated Confusion-Diffusion Mechanisms for Chaos-Based Image Encryption, *IEEE 8th International Conference on Computer and Information Technology Workshops*, pp. 260–263, July 2008.

88. S. A. N. Gilani and M. A. Bangash, Enhanced Block Based Color Image Encryption Technique with Confusion, *IEEE International Multitopic Conference (INMIC)*, pp. 200–206, December 2008.

89. D. Coppersmith, The Data Encryption Standard (DES) and Its Strength Against Attacks, *IBM Journal of Research and Development*, Vol. 38, No. 3, pp. 243–250, 1994.

90. J. Daemen and V. Rijmen, *Advanced Encryption Standard (AES)*, FIPS 197, Technical report, Catholic University, ESAT, Leuven, Belgium, November 2001.

91. W. Meier and L. R. Knudsen, Correlations in RC6 with a Reduced Number of Rounds Source, *Proceedings of the 7th International Workshop on Fast Software Encryption*, pp. 94–108, 2000.

92. K. L. Chung and L. C. Chang, Large Encrypting Binary Images with Higher Security, *Pattern Recognition Letters*, Vol. 19, No. 5–6, pp. 461–468, April 1998.

93. A. M. Fiskiran and R. B. Lee, Performance Impact of Addressing Modes on Encryption Algorithms, *Proceedings of the International Conference on Computer Design (ICCD)*, pp. 542–545, September 2001.

94. M. Fouad, D. H. Salem, and I. Ziedan, Application of Data Encryption Standard to Bitmap and JPEG Images, *Proceedings of the Twentieth National Radio Science Conference (NRSC)*, pp. 1–8, March 2003.

95. W. Ying, Z. W. Zhao, and Z. Lelin, A Fault-Tolerable Encryption Algorithm for Two-Dimensional Digital Image, *2nd IEEE Conference on Industrial Electronics and Applications (ICIEA)*, pp. 2737–2741, May 23–25, 2007.

96. S. J. Shyu, Image Encryption by Random Grids, *Pattern Recognition*, Vol. 40, No. 3, pp. 1014–1031, 2007.

97. H. El-Din, H. Ahmed, H. M. Kalash, and O. S. Faragallah, An Efficient Chaos Based Feedback Stream Cipher (ECBFSC) for Image Encryption and Decryption, *Informatica*, Vol. 31, No.1, pp. 121–129, 2007.

98. S. Li, X. Zheng, X. Mou, and Y. Cai, Chaotic Encryption Scheme for Real-Time Digital Video, *Proceedings of SPIE*, Vol. 4666, pp. 149–160, 2002.

99. S. Li, X. Mou, and Y. Cai, Improving Security of a Chaotic Encryption Approach, *Physics Letters A*, Vol. 290, No. 3–4, pp. 127–133, November 12, 2001.

100. F. Han, X. Yu and S. Han, Improved Baker Map for Image Encryption, *1st International Symposium on Systems and Control in Aerospace and Astronautics (ISSCAA)*, pp. 1273–1276, January 19–21, 2006.

101. S. Lian, J. Sun, and Z. Wang, Security Analysis of a Chaos-Based Image Encryption Algorithm, *Physica A: Statistical and Theoretical Physics*, Vol. 351, No. 2–4, pp. 645–661, June 15, 2005.

102. N. K. Pareek, V. Patidar, and K. K. Sud, Cryptography Using Multiple One-Dimensional Chaotic Maps, *Communications in Nonlinear Science and Numerical Simulation*, Vol. 10, No. 7, pp. 715–723, October 2005.

103. I. F. Elashry, O. S. Faragallah, A. M. Abbas, S. El-Rabaie, and F. E. Abd El-Samie, A New Method for Encrypting Images with Few Details Using Rijndael and RC6 Block Ciphers in the Electronic Code Book Mode, *Information Security Journal: A Global Perspective*, Vol. 21, No, 4, pp. 193–205, 2012.

104. C. A. Henk and V. Tilborg, *Encyclopedia of Cryptography and Security*, Springer Science and Business Media, New York, 2005.

105. N. El-Fishawy and O. M. Abu Zaid, Quality of Encryption Measurement of Bitmap Images with RC6, MRC6, and Rijndael Block Cipher Algorithms, *International Journal of Network Security*, Vol. 5, No. 3, pp. 241–251, 2007.

106. X. Wang and D. Zhao, Image Encryption Based on Anamorphic Fractional Fourier Transform and Three-Step Phase-Shifting Interferometry, *Optics Communications*, Vol. 268, No. 2, pp. 240–244, December 15, 2006.

107. S. Behnia, A. Akhshani, H. Mahmodi, and A. Akhavan, A Novel Algorithm for Image Encryption Based on Mixture of Chaotic Maps, *Chaos, Solitons & Fractals*, Vol. 35, No. 2, pp. 408–419, January 2008.

108. Y. B. Mao and G. Chen, Chaos-Based Image Encryption, in *Handbook of Computational Geometry for Pattern Recognition, Computer Vision, Neuralcomputing and Robotics*, Springer-Verlag, Berlin, pp. 231–265, 2005.

109. C. C. Chang, M. S. Hwang, and T. S. Chen, A New Encryption Algorithm for Image Cryptosystems, *Journal of Systems and Software*, Vol. 58, No. 2, pp. 83–91, 2001.

110. D. C. Key and J. R. Levine, *Graphics File Format*, Winderest Books/McGraw-Hill, New York, 1992.

111. D. Kim and G. L. Stüber, Residual ISI Cancellation for OFDM with Applications to HDTV Broadcasting, *IEEE Journal on Selected Areas in Communications*, Vol. 16, No. 8, 1590–1599, October 1998.

112. G. H. Yang, D. Shen, and V. O. K. Li, UEP for Video Transmission in Space-Time Coded OFDM Systems, *IEEE Infocom*, 2004.

113. M. I. Rahman, S. S. Das, and Frank H.P. Fitzek, *OFDM Based WLAN Systems*, Technical Report R-04-1002, Center for TeleInFrastruktur (CTiF), Copenhagen, Denmark, Vol. 1.2, February 18, 2005.

114. E. P. Lawrey, Adaptive Techniques for Multiuser OFDM, PhD thesis, James Cook University, Townsville, QLD, Australia, December 2001.

115. E. Lawrey, The Suitability of OFDM as a Modulation Technique for Wireless Telecommunications, with a CDMA Comparison, Bachelor thesis, James Cook University, Townsville, QLD, Australia, October 1997.

116. K. Abdullah and Z. M. Hussain, Performance of Fourier-Based and Wavelet-Based OFDM for DVB-T Systems, *Proceeding of the 2007 Australasian Telecommunication Networks and Applications Conference*, Christchurch, New Zealand, December 2nd–5th 2007.

117. J. J. van de Beek, O. Edfors, M. Sandell, S. K. Wilson, and P. O. Borjesson, On Channel Estimation in OFDM Systems, *IEEE 45th Vehicular Technology Conference*, 1995.

118. P. Tan and N. C. Beaulieu, A Comparison of DCT-Based OFDM and DFT-Based OFDM in Frequency Offset and Fading Channels, *IEEE Transactions on Communications*, Vol. 54, No. 11, 2113–2125, November 2006.

119. J. G. Andrews, A. Ghosh, and R. Muhamed, *Fundamentals of WiMAX Understanding Broadband Wireless Networking*, Prentice Hall Communications Engineering and Emerging Technologies Series, Prentice-Hall, Englewood Cliffs, NJ, pp. 113–145, February 2007.

120. A. K. Lee, Ooi, M. Drieberg, and V. Jeoti, DWT based FFT in Practical OFDM Systems, *IEEE* 2006.

121. P. Liu, B. B. Li, Z. Y. Lu, and F. K. Gong, An OFDM Bandwidth Estimation Scheme for Spectrum Monitoring, *IEEE* 2005.

122. X. F. Wang, Y. R. Shayan, and M. Zeng, On the Code and Interleaver Design of Broadband OFDM Systems, *IEEE Communications Letters*, Vol. 8, No. 11, 653–655, November 2004.

123. F. Gao, T. Cui, A. Nallanathan, and C. Tellambura, Maximum Likelihood Based Estimation of Frequency and Phase Offset in DCT OFDM Systems under Non Circular Transmissions: Algorithms, Analysis and Comparisons, *IEEE Transactions on Communications*, Vol. 56, No. 9, 1425–1429, September 2008.

124. R. Merched, On OFDM and Signal-Carrier Frequency-Domain Systems Based on Trigonometric Transforms, *IEEE Signal Processing Letters*, Vol. 13, No. 8, 473–476, August 2006.

125. E. Lawrey and C. J. Kikkert, Peak to Average Power Ratio Reduction of OFDM Signals Using Peak Reduction Carriers, *Fifth International Symposium on Signal Processing and its Applications, ISSPA '99*, Brisbane, Australia, August 22–25, 1999.

126. M. Shen, G. Li, and H. Liu, Effect of Traffic Channel Configuration on the Orthogonal Frequency Division Multiple Access Downlink Performance, *IEEE Transactions on Wireless Communications*, Vol. 4, No. 4, 1901–1913, July 2005.

127. H. Schulze and C. Luders, *Theory and Application of OFDM and CDMA Wideband Wireless Communications*, Wiley New York, pp. 145–264, 2005.

128. N. Al-Dhahir and H. Minn, A New Multicarrier Transceiver Based on the Discrete Cosine Transform, *Proceedings of the IEEE Wireless Communications and Networking Conference*, Vol. 1, pp. 45–50, March 13–17, 2005.

129. P. Tan and N. C. Beaulieu, An Improved DCT-Based OFDM Data Transmission Scheme, *Proceedings of the IEEE 16th PIMRC'05*, 2005.

130. H. Harada and R. Prasad, *Simulation and Software Radio for Mobile Communications*, House Universal Personal Communications Library, 2002.

131. M. S. El-Tanany, Y. Wu, and L. Házy, OFDM Uplink for Interactive Broadband Wireless: Analysis and Simulation in the Presence of Carrier, Clock and Timing Errors, IEEE, 2001.

132. A. Langowski, Time and Frequency Synchronisation in 4G OFDM Systems, *EURASIP Journal on Wireless Communications and Networking*, doi:10.1155/2009/641292, 2009.

133. Mary Ann Ingram, *OFDM Simulation Using Matlab*, Smart Antenna Research Laboratory, Guillermo Acosta, Georgia Tech, Atlanta, August 2000.

134. J. Zhang and B. Li, New Modulation Identification Scheme for OFDM in Multipath Rayleigh Fading Channel, *International Symposium on Computer Science and Computational Technology*, 2008.

135. H. S. Chu, B. S. Park, C. K. An, J. S. Kang, and H. G. Son, Wireless Image Transmission based on Adaptive OFDM System, *IEEE*, 2007.

136. C. S. Avilaf and R. S. Reillot, The Rijndael Block Cipher (AES Proposal): A Comparison with DES, IEEE, 2001.

137. M. S. Liu, Y. Zhang, and J. Huali, Research on Improving Security of DES by Chaotic Mapping, *Proceedings of the Eighth International Conference on Machine Learning and Cybernetics*, Baoding, July 12–15.

138. V. B. Vats, K. K. Garg, and A. Abad, Performance Analysis of DFT-OFDM, DCT-OFDM, and DWT-OFDM Systems in AWGN, *Proceedings of the IEEE Fourth International Conference on Wireless and Mobile Communications*, 2008.

139. P. Tan and N. C. Beaulieu, Precise Bit Error Probability Analysis of DCT OFDM in the Presence of Carrier Frequency Offset on AWGN Channels, *Proceedings of the IEEE Globcom 2005*, pp. 1429–1434, 2005.

140. Giridhar D. Mandyam, *On the Discrete Cosine Transform and OFDM Systems*, Nokia Research Center, Irving, TX, 2003.

141. M. Misiti, Y. Misiti, and G. Oppenheim, *Wavelets and Their Applications*, Jean-Michel Poggi, ISTE, 2007.

142. A. Prochazka, J. Uhlir, P. J. W. Rayner, and N. J. Kingsbury, *Signal Analysis and Prediction*, Birkhauser, New York, 1998.

143. B. Muquet, Z. Wang, G. B. Giannakis, M. d. Courville, and P. Duhamel, Cyclic Prefixing or Zero Padding for Wireless Multicarrier Transmissions, *IEEE Transactions on Communications*, Vol. 50, No. 12, December 2002.

144. B. Li, S. Zhou, M. Stojanovic, L. Freitag, and P. Willett, Non-Uniform Doppler Compensation for Zero-Padded OFDM over Fast-Varying Under water Acoustic Channels, *OCEANS2007-Europe*, IEEE, 2007.

145. C. R. N. Athaudage and R. R. V. Angiras, Sensitivity of FFT-Equalised Zero-Padded OFDM Systems to Time and Frequency Synchronization Errors, *IEE Proc.-Commun.*, Vol. 152, No. 6, December 2005.

146. B. Muquet, M. D. Courville, P. Duhamel, G. B. Giannakis, and P. Magniez, Turbo Demodulation of Zero-Padded OFDM Transmissions, *IEEE Transactions on Communications*, Vol. 50, No. 11, November 2002.

147. D. Huang and K. B. Letaief, An Interference-Cancellation Scheme for Carrier Frequency Offsets Correction in OFDMA Systems, *IEEE Transactions on Communication*, Vol. 53, No. 7, pp. 1155–1165, July 2005.

148. J. Paul and M. G. Linnartz, Performance Analysis of Synchronous MC-CDMA in Mobile Rayleigh Channel with Both Delay and Doppler Spreads, *IEEE Transactions on Vehicular Technology*, Vol. 50, No. 6, November 2001.

149. T. Cui, F. Gao, A. Nallanathan, and C. Tellambura, ML CFO and PO Estimation in DCT OFDM Systems under Non-Circular Transmissions, *IEEE Communications Society*, 2007.

150. P. Bansal and A. Brzezinski, Adaptive Loading in MIMO/OFDM Systems, December 13, 2001.

151. H. Hwang and H. Park, *Doppler Frequency Offset Estimation in OFDM Systems*, Mobile Telecommunication Research Division, ETRI, IEEE, 2009.

152. J. Lee, D. Toumpakaris, H. L. Lou, and J. M. Cioffi, Effect of Carrier Frequency Offset on Time-Domain Differential Demodulation in OFDM Systems, *IEEE*, 2004.

153. M. Anandpara, E. Erwa, J. Golab, R. Samanta, and H. Wang, Inter-carrier Interference Cancellation for OFDM Systems, *EE 381K-11: Wireless Communications*, May 6, 2003.

154. A. S. Baiha, M. Singh, A. J. Goldsmith, and B. R. Saltzberg, A New Approach for Evaluating Clipping Distortion in Multicarrier Systems, *IEEE Journal on Select Areas in Communications,* Vol. 20, No. 5, pp. 1037–1046, June 2002.

155. R. V. Nee and R. Prasad, *OFDM for Wireless Multimedia Communications*, Artech House, Boston, 2000.

156. R. V. Nee and A, D. Wild, Reducing the Peak-to-Average Power Ratio of OFDM, *Proceedings of the IEEE Vehicular Technology Conference (VTC'98)*, pp. 2072–2076, May 1998.

157. S. Wei, D. L. Goeckel, and P. E. Kelly, A Modern Extreme Value Theory Scheme to Calculating the Distribution of the PAPR in OFDM Systems, *Proceedings of IEEE ICC 2002*, New York, pp. 1686–90, May 2002.

158. M. Sharif, M. Gharavi-Alkhansari, and B. H. Khalaj, On the Peak-to-Average Power of OFDM Signals Based on Oversampling, *IEEE Transactions on Communications*, Vol. 51, pp. 72–78, January 2003.

159. G. Wunder and H. Boche, Upper Bounds on the Statistical Distribution of the Crestfactor in OFDM Transmission, *IEEE Transactions on Information Theory*, Vol. 49, pp. 488–494, February 2003.

160. R. Prasad, *OFDM for Wireless Communications Systems*, Artech House, Boston, 2004.

161. S. H. Han and J. H. Lee, Modified Selected Mapping Scheme for PAPR Reduction of Coded OFDM Signal, *IEEE Transactions on Broadcasting*, Vol. 50, pp. 335–341, September 2004.

162. H. Dai and H. V. Poor, Advanced Signal Processing for Power Line Communications, *IEEE Communications Magazine*, Vol. 41, No. 5, pp. 100–107, May 2003.

163. S. B. Weinstein, The History of Orthogonal Frequency Division Multiplexing, *IEEE Communications Magazine*, pp. 26–35, November 2009.

164. S. Celebi, Interblock Interference (IBI) Minimizing Time-Domain Equalizer (TEQ) for OFDM, *IEEE Signal Processing Letters*, Vol. 10, No. 8, pp. 232–234, August 2003.

165. Y. Wu and W. Y. Zou, Orthogonal Frequency Division Multiplexing: A Multicarrier Modulation Scheme, *IEEE Transactions on Consumer Electronics*, Vol. 41, No. 3, pp. 392–399, August 1995.

166. T. Pollet, P. Spruyt, and M. Moeneclaey, The BER Performance of OFDM Systems Using Non-Synchronized Sampling, *Proceedings of GLOBECOM*, 1994, pp. 253–257.

167. ETSI, *Transmission and Multiplexing (TM); Access Transmission Systems on Metallic Access Cables; Very High Speed Digital Subscriber Line (VDSL); Part 2: Transceiver Specification*, 2002, TS 101 270-2.

168. F. J. Cañete, J. A. Cortées, L. Díez, and J. T. Entrambasaguas, Modeling and Evaluation of the Indoor Power Line Transmission Medium, *IEEE Communications Magazine*, Vol. 41, No. 4, pp. 41–47, April 2003.

Appendix A

```
function y = multi(x,w)
%z = a*a1 mod 2^32
a = frombase256(x); a1 = frombase256(w);
z = mod(a*a1,2^32);
y = tobase256(z);

function permutation = p(substitution)
% p function permutes an input "substitution" based on x
x = [16 7 20 21 29 12 28 17 01 15 23 26 05 18 31 10 2
8 24 14 32 27 3 9 19 13 30 6 22 11 4 25];
permutation = substitution(x);

function y = padding(x)
%Padding x with zeros if x is shorter than 16
L = length(x);
if mod(L,16)==0
    y = x;
else
    pad = 16-mod(L,16);
    y = zeros(1,L+pad);
    y(1:L) = x;
end;
function RC5enctextRC5 = RC5CBC(plaintext1,m,n,CO,r,
keyi)
% Encrypt data using RC5 in CBC mode.
RC5enctextRC5 = zeros(m*n/8,8);
```

```
for i = 1:m*n/8
    plaintext = plaintext1(i,:);
    if i==1
        plaintexts = bitxor(plaintext,CO);
    else
        plaintexts = bitxor(plaintext,RC5enctext);
    end;
    RC5enctext = RC5enc(plaintexts,r,keyi);
    RC5enctextRC5(i,:) = RC5enctext;

end;

function plaintextRC5 = RC5CBCDec(RC5enctext1,m,n,CO,r,
keyi)
% Decrypt data using RC5 in CBC mode.
plaintextRC5 = zeros(m*n/8,8);
for i = 1:m*n/8
    RC5enctextss = RC5enctext1(i,:);
    re_plaintext = RC5decry(RC5enctextss,r,keyi);
    if i==1
        plaintext = bitxor(re_plaintext,CO);
    else
        plaintext = bitxor(re_plaintext,RC5enctext1
        ((i-1),:));
    end;
    plaintextRC5(i,:) = plaintext;
end;

function RC5enctextRC5 = RC5CFB(plaintext1,m,n,CO,r,
keyi)
% Encrypt data using RC5 in CFB mode.
RC5enctextRC5 = zeros(m*n/8,8);
for i = 1:m*n/8
    plaintext = plaintext1(i,:);
    if i==1
        cr1 = RC5enc(CO,r,keyi);

        RC5enctext = bitxor(plaintext,cr1);
    else
        cr1 = RC5enc(RC5enctext,r,keyi);

        RC5enctext = bitxor(plaintext,cr1);
    end;
    RC5enctextRC5(i,:) = RC5enctext;

end;
```

```
function plaintextRC5 = RC5CFBDec(RC5enctext1,m,n,CO,r,
keyi)
% Decrypt data using RC5 in CFB mode.
plaintextRC5 = zeros(m*n/8,8);
for i = 1:m*n/8
    RC5enctextss = RC5enctext1(i,:);
    if i==1
        cr1 = RC5enc(CO,r,keyi);

        re_plaintext = bitxor(RC5enctextss,cr1);
    else
        cr1 = RC5enc(RC5enctext1(i-1,:),r,keyi);

        re_plaintext = bitxor(RC5enctextss,cr1);
    end;

    plaintextRC5(i,:) = re_plaintext;
end;

function [xa1, Fs, nbits] = RC5DecCBCAudio(x1, Fs,
nbits,key,r,CO)
% Decrypt Audio using RC5 in CBC mode.
keyi = RC5keygen(key,r);
%CBC L = length(x1); y1 = double(x1);
plaintextRC5 = zeros(L/8,8);
RC5enctext1 = reshape(y1,L/8,8);
for i = 1:L/8
    RC5enctextss = RC5enctext1(i,:);
    re_plaintext = RC5decry(RC5enctextss,r,keyi);
    if i==1

        plaintext = bitxor(re_plaintext,CO);
    else
        plaintext = bitxor(re_plaintext,RC5enctext1
        ((i-1),:));
    end;
    plaintextRC5(i,:) = plaintext;

end;
xa1 = uint8(plaintextRC5(:));

function xa1 = RC5DecCBCImage(x1,key,r,CO)
% Decrypt Image using RC5 in CBC mode.
keyi = RC5keygen(key,r);
%CBC
```

```
[m,n] = size(x1);
y1 = double(x1);
plaintextRC5 = zeros(m*n/8,8);
RC5enctext1 = reshape(y1',8,m*n/8)';
for i = 1:m*n/8
    RC5enctextss = RC5enctext1(i,:);
    re_plaintext = RC5decry(RC5enctextss,r,keyi);
    if i = =1

        plaintext = bitxor(re_plaintext,CO);
    else
        plaintext = bitxor(re_plaintext,RC5enctext1
        ((i-1),:));
    end;
    plaintextRC5(i,:) = plaintext;

end;
ya1 = reshape(plaintextRC5',n,m)';

xa1 = uint8(ya1);


function xa1 = RC5DecCBCimageC(x,key,r,CO)
% Decrypt Colored Image using RC5 in CBC mode.
keyi = RC5keygen(key,r);
%CBC
x1 = x(:,:,1);
x2 = x(:,:,2);
x3 = x(:,:,3);
y1 = double(x1);
y2 = double(x2);
y3 = double(x3);
[m,n] = size(x1);
plaintext(:,:,1) = reshape(y1',8,m*n/8)';
plaintext(:,:,2) = reshape(y2',8,m*n/8)';
plaintext(:,:,3) = reshape(y3',8,m*n/8)';
ciphertext = x;
for i = 1:3 plaintextRC5 = plaintext(:,:,i);
    RC5enctext = RC5CBCDec(plaintextRC5,m,n,CO,
    r,keyi);
    RC5enctext = reshape(RC5enctext',n,m)';
    ciphertext(:,:,i) = RC5enctext;
end

xa1 = uint8(ciphertext);
```

```
function [xa1, Fs, nbits] = RC5DecCFBAudio(x1, Fs,
nbits,key,r,CO)
%Decrypt Audio using RC5 in CFB mode.
keyi = RC5keygen(key,r);
L = length(x1); y1 = double(x1);
plaintextRC5 = zeros(L/8,8);
RC5enctext1 = reshape(y1,L/8,8);
for i = 1:L/8
    RC5enctextss = RC5enctext1(i,:);
    if i = =1
        cr1 = RC5enc(CO,r,keyi);

        re_plaintext = bitxor(RC5enctextss,cr1);
    else
        cr1 = RC5enc(RC5enctext1(i-1,:),r,keyi);

        re_plaintext = bitxor(RC5enctextss,cr1);
    end;

    plaintextRC5(i,:) = re_plaintext;
end;
xa1 = uint8(plaintextRC5(:));

function xa1 = RC5DecCFBImage(x1,key,r,CO)
%Decrypt Image using RC5 in CFB mode.
keyi = RC5keygen(key,r);
[m,n] = size(x1); y1 = double(x1);
plaintextRC5 = zeros(m*n/8,8);
RC5enctext1 = reshape(y1',8,m*n/8)';
for i = 1:m*n/8
    RC5enctextss = RC5enctext1(i,:);
    if i = =1
        cr1 = RC5enc(CO,r,keyi);

        re_plaintext = bitxor(RC5enctextss,cr1);
    else
        cr1 = RC5enc(RC5enctext1(i-1,:),r,keyi);
        re_plaintext = bitxor(RC5enctextss,cr1);
    end;

    plaintextRC5(i,:) = re_plaintext;
end;
ya1 = reshape(plaintextRC5',n,m)';

xa1 = uint8(ya1);
```

```
function xa1 = RC5DecCFBImageC(x,key,r,CO)
%Decrypt Colored Image using RC5 in CFB mode.
keyi = RC5keygen(key,r);
%CFB

x1 = x(:,:,1);
x2 = x(:,:,2);
x3 = x(:,:,3);
y1 = double(x1);
y2 = double(x2);
y3 = double(x3);
[m,n] = size(x1);
plaintext(:,:,1) = reshape(y1',8,m*n/8)';
plaintext(:,:,2) = reshape(y2',8,m*n/8)';
plaintext(:,:,3) = reshape(y3',8,m*n/8)';
ciphertext = x;
for i = 1:3 plaintextRC5 = plaintext(:,:,i);
    RC5enctext = RC5CFBDec(plaintextRC5,m,n,CO,r,
    keyi);
    RC5enctext = reshape(RC5enctext',n,m)';
    ciphertext(:,:,i) = RC5enctext;
end

xa1 = uint8(ciphertext);

function [xa1, Fs, nbits] = RC5DecECBAudio(x1, Fs,
nbits,key,r)
%Decrypt Audio using RC5 in ECB mode.
keyi = RC5keygen(key,r);
%ECB L = length(x1);
y1 = double(x1);
plaintextRC5 = zeros(L/8,8);
RC5enctext1 = reshape(y1,L/8,8);
for i = 1:L/8
    RC5enctextss = RC5enctext1(i,:);
    re_plaintext = RC5decry(RC5enctextss,r,keyi);
    plaintextRC5(i,:) = re_plaintext;

end;
xa1 = uint8(plaintextRC5(:));

function xa1 = RC5DecECBImage(x1,key,r)
%Decrypt Image using RC5 in ECB mode.
keyi = RC5keygen(key,r);
```

```
[m,n] = size(x1); y1 = double(x1);
plaintextRC5 = zeros(m*n/8,8);
RC5enctext1 = reshape(y1',8,m*n/8)';
for i = 1:m*n/8
    RC5enctextss = RC5enctext1(i,:);
    re_plaintext = RC5decry(RC5enctextss,r,keyi);
    plaintextRC5(i,:) = re_plaintext;

end;
ya1 = reshape(plaintextRC5',n,m)';

xa1 = uint8(ya1);

function xa1 = RC5DecECBImageC(x,key,r)
%Decrypt Colored Image using RC5 in ECB mode.
keyi = RC5keygen(key,r);
%ECB
x1 = x(:,:,1);
x2 = x(:,:,2);
x3 = x(:,:,3);
y1 = double(x1);
y2 = double(x2);
y3 = double(x3);
[m,n] = size(x1);
plaintext(:,:,1) = reshape(y1',8,m*n/8)';
plaintext(:,:,2) = reshape(y2',8,m*n/8)';
plaintext(:,:,3) = reshape(y3',8,m*n/8)';
ciphertext = x;
for i = 1:3 plaintextRC5 = plaintext(:,:,i);
    RC5enctext = RC5ECBDec(plaintextRC5,m,n,r,keyi);
    RC5enctext = reshape(RC5enctext',n,m)';
    ciphertext(:,:,i) = RC5enctext;
end
xa1 = uint8(ciphertext);

function [xa1, Fs, nbits] = RC5DecOFBAudio(x1, Fs,
nbits,key,r,CO)
%Decrypt Audio using RC5 in OFB mode.
keyi = RC5keygen(key,r);
%OFB

L = length(x1);
y1 = double(x1);
plaintextRC5 = zeros(L/8,8);
RC5enctext1 = reshape(y1,L/8,8);
```

```
for i = 1:L/8
    RC5enctextss = RC5enctext1(i,:);
    if i = =1
        cr = RC5enc (CO,r,keyi);
        re_plaintext = bitxor(RC5enctextss,cr);
    else cr = RC5enc(cr,r,keyi);
        re_plaintext = bitxor(RC5enctextss,cr);
    end;

    plaintextRC5(i,:) = re_plaintext;

end;
xa1 = uint8(plaintextRC5(:));

function xa1 = RC5DecOFBImage(x1,key,r,CO)
%Decrypt Image using RC5 in OFB mode.
keyi = RC5keygen(key,r);

[m,n] = size(x1); y1 = double(x1);
plaintextRC5 = zeros(m*n/8,8);
RC5enctext1 = reshape(y1',8,m*n/8)';
for i = 1:m*n/8
    RC5enctextss = RC5enctext1(i,:);
    if i = =1
        cr = RC5enc (CO,r,keyi);
        re_plaintext = bitxor(RC5enctextss,cr);
    else cr = RC5enc(cr,r,keyi);
        re_plaintext = bitxor(RC5enctextss,cr);
    end;

    plaintextRC5(i,:) = re_plaintext;

end;
ya1 = reshape(plaintextRC5',n,m)';

xa1 = uint8(ya1);

function xa1 = RC5DecOFBImagec(x,key,r,CO)
%Decrypt Colored Image using RC5 in OFB mode.
keyi = RC5keygen(key,r);
%OFB

x1 = x(:,:,1);
x2 = x(:,:,2);
x3 = x(:,:,3);
```

```
y1 = double(x1);
y2 = double(x2);
y3 = double(x3);
[m,n] = size(x1);
plaintext(:,:,1) = reshape(y1',8,m*n/8)';
plaintext(:,:,2) = reshape(y2',8,m*n/8)';
plaintext(:,:,3) = reshape(y3',8,m*n/8)';
ciphertext = x;
for i = 1:3 plaintextRC5 = plaintext(:,:,i);
    RC5enctext = RC5OFBDec(plaintextRC5,m,n,CO,r,k
    eyi);
    RC5enctext = reshape(RC5enctext',n,m)';
    ciphertext(:,:,i) = RC5enctext;
end

xa1 = uint8(ciphertext);

function y = RC5decry(plaintext,round,s)
  %RC5 Decryption
  a = plaintext(1:4); b = plaintext(5:8); for
  i = round:-1:1
      b = RC5sub(b,s(2*i+2,:)); b = shifting
      (b,-LSB5(a)); b = bitxor(b',a);
      a = RC5sub(a,s(2*i+1,:)); a = shifting
      (a,-LSB5(b)); a = bitxor(a',b);
  end b = RC5sub(b,s(2,:));
  a = RC5sub(a,s(1,:)); y(1:4) = a;
  y(5:8) = b;

function RC5enctextRC5 = RC5ECB(plaintext1,m,n,r,keyi)
%Encrypt data using RC5 in ECB mode
RC5enctextRC5 = zeros(m*n/8,8);
for i = 1:m*n/8 plaintext = plaintext1(i,:);
    RC5enctext = RC5enc(plaintext,r,keyi);
    RC5enctextRC5(i,:) = RC5enctext;

end;

function plaintextRC5 = RC5ECBDec(RC5enctext1,m,n,r,
keyi)
%Decrypt data using RC5 in ECB mode
plaintextRC5 = zeros(m*n/8,8);
for i = 1:m*n/8
    RC5enctextss = RC5enctext1(i,:);
    re_plaintext = RC5decry(RC5enctextss,r,keyi);
```

```
    plaintextRC5(i,:) = re_plaintext;

end;

function y = RC5enc(plaintext,round,s)
%RC5 Encryption
a = plaintext(1:4); b = plaintext(5:8);
a = add(a,s(1,:)); b = add(b,s(2,:)); for
i = 1:round
    a = bitxor(a,b); a = shifting(a,LSB5(b));
    a = add(a,s(2*i+1,:)); b = bitxor(b,a);
    b = shifting(b,LSB5(a)); b = add(b,s(2*i+2,:));
end y(1:4) = a; y(5:8) = b;

function [xa, Fs, nbits] = RC5EncCBCAudio(x, Fs,
nbits,key,r,CO)
%Encrypt Audio using RC5 in CBC mode
L = length(x);
y = double(x); keyi = RC5keygen(key,r);
plaintext1 = reshape(y,L/8,8);
RC5enctextRC5 = zeros(L/8,8);
  %CBC
for i = 1:L/8 plaintext = plaintext1(i,:); if i = =1
        plaintexts = bitxor(plaintext,CO);
    else plaintexts = bitxor(plaintext,RC5enctext);
    end;
    RC5enctext = RC5enc(plaintexts,r,keyi);
    RC5enctextRC5(i,:) = RC5enctext;

end;
xa = uint8(RC5enctextRC5(:));

function xa = RC5EncCBCImage(x,key,r,CO)
%Encrypt Image using RC5 in CBC mode
[m,n] = size(x);
y = double(x); keyi = RC5keygen(key,r);
plaintext1 = reshape(y',8,m*n/8)';

RC5enctextRC5 = zeros(m*n/8,8);
%CBC
for i = 1:m*n/8 plaintext = plaintext1(i,:); if i = =1
        plaintexts = bitxor(plaintext,CO);
    else plaintexts = bitxor(plaintext,RC5enctext);
    end;
    RC5enctext = RC5enc(plaintexts,r,keyi);
```

```
        RC5enctextRC5(i,:) = RC5enctext;

end; ya = reshape(RC5enctextRC5',n,m)';
xa =  uint8(ya);

function xa = RC5EncCBCImageC(x,key,r,CO)
%Encrypt Colored Image using RC5 in CBC mode

x1 = x(:,:,1); x2 = x(:,:,2); x3 = x(:,:,3);
y1 = double(x1); y2 = double(x2);
y3 = double(x3); [m,n] = size(x1);

keyi = RC5keygen(key,r);
plaintext(:,:,1) = reshape(y1',8,m*n/8)';
plaintext(:,:,2) = reshape(y2',8,m*n/8)';
plaintext(:,:,3) = reshape(y3',8,m*n/8)';


RC5enctext = x;
%CBC

for i = 1:3 plaintext1 = plaintext(:,:,i);
    RC5enctextRC5 = RC5CBC(plaintext1,m,n,CO,r,keyi);
    RC5enctextRC5 = reshape(RC5enctextRC5',n,m)';
    RC5enctext(:,:,i) = RC5enctextRC5;
end;
xa = uint8(RC5enctext);


function [xa, Fs, nbits] = RC5EncCFBAudio(x, Fs,
nbits,key,r,CO)
%Encrypt Audio using RC5 in CBC mode
L = length(x);
y = double(x); keyi = RC5keygen(key,r);
plaintext1 = reshape(y,L/8,8);
RC5enctextRC5 = zeros(L/8,8);
 % CFB
for i = 1:L/8 plaintext = plaintext1(i,:); if i = =1
        cr1 = RC5enc(CO,r,keyi);

        RC5enctext = bitxor(plaintext,cr1);
    else cr1 = RC5enc(RC5enctext,r,keyi);

        RC5enctext = bitxor(plaintext,cr1);

    end;
```

```
      RC5enctextRC5(i,:) = RC5enctext;

end;
xa = uint8(RC5enctextRC5(:));

function xa = RC5EncCFBImage(x,key,r,CO)
%Encrypt Image using RC5 in CFB mode
[m,n] = size(x);
y = double(x); keyi = RC5keygen(key,r);
plaintext1 = reshape(y',8,m*n/8)';

RC5enctextRC5 = zeros(m*n/8,8);
for i = 1:m*n/8 plaintext = plaintext1(i,:); if i = =1
        cr1 = RC5enc(CO,r,keyi);

        RC5enctext = bitxor(plaintext,cr1);
    else cr1 = RC5enc(RC5enctext,r,keyi);

        RC5enctext = bitxor(plaintext,cr1);
    end;

    RC5enctextRC5(i,:) = RC5enctext;

end; ya = reshape(RC5enctextRC5',n,m)';
xa = uint8(ya);

function xa = RC5EncCFBImageC(x,key,r,CO)
%Encrypt Colored Image using RC5 in CFB mode
x1 = x(:,:,1); x2 = x(:,:,2); x3 = x(:,:,3);
y1 = double(x1); y2 = double(x2);
y3 = double(x3); [m,n] = size(x1);

keyi = RC5keygen(key,r);
plaintext(:,:,1) = reshape(y1',8,m*n/8)';
plaintext(:,:,2) = reshape(y2',8,m*n/8)';
plaintext(:,:,3) = reshape(y3',8,m*n/8)';

RC5enctext = x;
% CFB
for i = 1:3 plaintext1 = plaintext(:,:,i);
    RC5enctextRC5 = RC5CFB(plaintext1,m,n,CO,r,keyi);
    RC5enctextRC5 = reshape(RC5enctextRC5',n,m)';
    RC5enctext(:,:,i) = RC5enctextRC5;
end; xa = uint8(RC5enctext);
```

```
imwrite(xa,'onion1RC5encCFB.tif','tif')
imshow('onion1RC5encCFB.tif')

function [xa, Fs, nbits] = RC5EncECBAudio(x, Fs,
nbits,key,r)
%Encrypt Audio using RC5 in ECB mode
L = length(x);
y = double(x); keyi = RC5keygen(key,r);
plaintext1 = reshape(y,L/8,8);
RC5enctextRC5 = zeros(L/8,8);
% ECB
for i = 1:L/8 plaintext = plaintext1(i,:);
    RC5enctext = RC5enc(plaintext,r,keyi);
    RC5enctextRC5(i,:) = RC5enctext;

end;
xa = uint8(RC5enctextRC5(:));

function xa = RC5EncECBImage(x,key,r)
%Encrypt Image using RC5 in ECB mode
[m,n] = size(x);
y = double(x); keyi = RC5keygen(key,r);
plaintext1 = reshape(y',8,m*n/8)';

RC5enctextRC5 = zeros(m*n/8,8);
 % ECB
for i = 1:m*n/8
plaintext = plaintext1(i,:);
    RC5enctext = RC5enc(plaintext,r,keyi);
    RC5enctextRC5(i,:) = RC5enctext;

end; ya = reshape(RC5enctextRC5',n,m)';
xa = uint8(ya);

function xa = RC5EncECBImageC(x,key,r)
%Encrypt Colored Image using RC5 in ECB mode
x1 = x(:,:,1); x2 = x(:,:,2); x3 = x(:,:,3);
y1 = double(x1); y2 = double(x2);
y3 = double(x3); [m,n] = size(x1);

keyi = RC5keygen(key,r);

plaintext(:,:,1) = reshape(y1',8,m*n/8)';
plaintext(:,:,2) = reshape(y2',8,m*n/8)';
plaintext(:,:,3) = reshape(y3',8,m*n/8)';
```

```
RC5enctext = x;% ECB
for i = 1:3 plaintext1 = plaintext(:,:,i);
    RC5enctextRC5 = RC5ECB(plaintext1,m,n,r,keyi);
    RC5enctextRC5 = reshape(RC5enctextRC5',n,m)';
    RC5enctext(:,:,i) = RC5enctextRC5;
end;
xa = uint8(RC5enctext);

function [xa, Fs, nbits] = RC5EncOFBAudio(x, Fs,
nbits,key,r,CO)
%Encrypt Audio using RC5 in OFB mode
L = length(x);
y = double(x); keyi = RC5keygen(key,r);
plaintext1 = reshape(y,L/8,8);
RC5enctextRC5 = zeros(L/8,8);
% OFB
for i = 1:L/8 plaintext = plaintext1(i,:);

    if i = =1 cr = RC5enc(CO,r,keyi);

    else cr = RC5enc(cr,r,keyi);

    end; RC5enctext = bitxor(plaintext,cr);
    RC5enctextRC5(i,:) = RC5enctext;

end;
xa = uint8(RC5enctextRC5(:));

function xa = RC5EncOFBImage(x,key,r,CO)
%Encrypt Image using RC5 in OFB mode
[m,n] = size(x);
y = double(x); keyi = RC5keygen(key,r);
plaintext1 = reshape(y',8,m*n/8)';

RC5enctextRC5 = zeros(m*n/8,8);
 % OFB
for i = 1:m*n/8 plaintext = plaintext1(i,:);

    if i = =1 cr = RC5enc(CO,r,keyi);

    else cr = RC5enc(cr,r,keyi);

    end; RC5enctext = bitxor(plaintext,cr);
    RC5enctextRC5(i,:) = RC5enctext;
```

```
end; ya = reshape(RC5enctextRC5',n,m)';
xa =  uint8(ya);


function xa = RC5EncOFBImageC(x,key,r,CO)
%Encrypt Colored Image using RC5 in OFB mode


x1 = x(:,:,1); x2 = x(:,:,2); x3 = x(:,:,3);
y1 = double(x1); y2 = double(x2);
y3 = double(x3); [m,n] = size(x1);
keyi = RC5keygen(key,r);
plaintext(:,:,1) = reshape(y1',8,m*n/8)';
plaintext(:,:,2) = reshape(y2',8,m*n/8)';
plaintext(:,:,3) = reshape(y3',8,m*n/8)';
RC5enctext = x;
% OFB
for i = 1:3 plaintext1 = plaintext(:,:,i);
    RC5enctextRC5 = RC5OFB(plaintext1,m,n,CO,r,keyi);
    RC5enctextRC5 = reshape(RC5enctextRC5',n,m)';
    RC5enctext(:,:,i) = RC5enctextRC5;
end;
xa = uint8(RC5enctext);


function s = keygen(key,r)
% RC5/RC6 Key Generation
p = [99 81 225 183];
q = [185 121 55 158];
s(1,:) = p;
for i = 2:2*r+4 s(i,:) = add(s(i-1,:),q);
end; i = 1; a = zeros(1,4);
b = zeros(1,4); v = 3*(2*r+4); for
h = 1:v
    s(i,:) = add(s(i,:),a); s(i,:) = add(s(i,:),b);
    s(i,:) = shifting(s(i,:),3); a = s(i,:); key =
    add(key,a); key = add(key,b);
    key = shifting(key,LSB5(add(a,b)));
    b = key;
    if i = =2*r+4 i = 1;
    end end


function RC5enctextRC5 = RC5OFB(plaintext1,m,n,CO,r,k
eyi)
%Encrypt Data using RC5 in OFB mode
RC5enctextRC5 = zeros(m*n/8,8);
for i = 1:m*n/8 plaintext = plaintext1(i,:);
```

```
    if i = =1 cr = RC5enc(CO,r,keyi);

    else cr = RC5enc(cr,r,keyi);

    end; RC5enctext = bitxor(plaintext,cr);
    RC5enctextRC5(i,:) = RC5enctext;

end;

function plaintextRC5 = RC5OFBDec(RC5enctext1,m,n,CO,r,
keyi)
%Decrypt data using RC5 in OFB mode

plaintextRC5 = zeros(m*n/8,8);
for i = 1:m*n/8
    RC5enctextss = RC5enctext1(i,:);
    if i = =1
        cr = RC5enc (CO,r,keyi);
        re_plaintext = bitxor(RC5enctextss,cr);
    else cr = RC5enc(cr,r,keyi);
        re_plaintext = bitxor(RC5enctextss,cr);
    end;

    plaintextRC5(i,:) = re_plaintext;

end;


function y = RC5sub(x,w)
%z = a-a1 mod 2^32
a = frombase256(x); a1 = frombase256(w);
z = mod(a-a1,2^32); y = tobase256(z);

function RC6enctextRC6 = CBC(plaintext1,m,n,CO,r,keyi)
%Encrypt Data using RC6 in CBC mode
RC6enctextRC6 = zeros(m*n/16,16);
for i = 1:m*n/16 plaintext = plaintext1(i,:); if i = =1
        plaintexts = bitxor(plaintext,CO);
    else plaintexts = bitxor(plaintext,RC6enctext);
    end;
    RC6enctext = RC6enc(plaintexts,r,keyi);
    RC6enctextRC6(i,:) = RC6enctext;

end;
```

```
function plaintextRC6 = CBCDec(RC6enctext1,m,n,CO,
r,keyi)
%Decrypt data using RC6 in CBC mode

plaintextRC6 = zeros(m*n/16,16);
for i = 1:m*n/16
    RC6enctextss = RC6enctext1(i,:);
    re_plaintext = RC6dec(RC6enctextss,r,keyi);
    if i = =1 plaintext = bitxor(re_plaintext,CO);
    else
                    plaintext = bitxor(re_plaintext,R
                    C6enctext1((i-1),:));
    end;
    plaintextRC6(i,:) = plaintext;
end;

function RC6enctextRC6 = CFB(plaintext1,m,n,CO,r,keyi)
%Encrypt data using RC6 in CFB mode

RC6enctextRC6 = zeros(m*n/16,16);
for i = 1:m*n/16 plaintext = plaintext1(i,:); if i = =1
        cr1 = RC6enc(CO,r,keyi);

        RC6enctext = bitxor(plaintext,cr1);
    else cr1 = RC6enc(RC6enctext,r,keyi);

        RC6enctext = bitxor(plaintext,cr1); end;
    RC6enctextRC6(i,:) = RC6enctext;

end;

function plaintextRC6 = CFBDec(RC6enctext1,m,n,CO,r,
keyi)
%Decrypt data using RC6 in CFB mode
plaintextRC6 = zeros(m*n/16,16);
for i = 1:m*n/16
    RC6enctextss = RC6enctext1(i,:);
    if i = =1 cr1 = RC6enc(CO,r,keyi);

        re_plaintext = bitxor(RC6enctextss,cr1);
    else
        cr1 = RC6enc(RC6enctext1(i-1,:),r,keyi);

        re_plaintext = bitxor(RC6enctextss,cr1);
    end;
```

```
        plaintextRC6(i,:) = re_plaintext;

end;

function y = RC6dec(plaintext,round,s)
% RC6 Decryption
a = plaintext(1:4); b = plaintext(5:8);
c = plaintext(9:12); d = plaintext(13:16);
c = RC6sub(c,s(2*round+4,:));
a = RC6sub(a,s(2*round+3,:)); for i = round:-1:1
    temp = d; d = c; c = b; b = a;
    a = temp;
    u = multi(d,(2*d+1)); u = shifting(u,5);
    t = multi(b,(2*b+1)); t = shifting(t,5);
    c = RC6sub(c,s(2*i+2,:)); c = shifting
    (c,-LSB5(t)); c = bitxor(c,u);
    a = RC6sub(a,s(2*i+1,:)); a = shifting
    (a,-LSB5(u)); a = bitxor(a,t);
end d = RC6sub(d,s(2,:));
b = RC6sub(b,s(1,:)); y(1:4) = a;
y(5:8) = b; y(9:12) = c; y(13:16) = d;

function [xa1,Fs,nbits] = RC6DecCBCAudio(x1, Fs,
nbits,key,r,CO)

%Decrypt Audio using RC6 in CBC mode
%CBC keyi = RC6keygen(key,r); L = length(x1);
y1 = double(x1); y1 = reshape(y1,L/16,16);
plaintextRC6 = zeros(L/16,16);
for i = 1:L/16
    RC6enctextss = y1(i,:);
    re_plaintext = RC6dec(RC6enctextss,r,keyi);
    if i = =1

        plaintext = bitxor(re_plaintext,CO);
    else
        plaintext = bitxor(re_plaintext,y1((i-1),:));
    end;
    plaintextRC6(i,:) = plaintext; end;
xa1 = uint8(plaintextRC6(:));

function xa1 = RC6DecCBCImage(x1,key,r,CO)
%Decrypt Image using RC6 in CBC mode

keyi = RC6keygen(key,r);
```

```
%CBC [m,n] = size(x1);
y1 = double(x1);
plaintextRC6 = zeros(m*n/16,16);
RC6enctext1 = reshape(y1',16,m*n/16)'; for i = 1:m*n/16

   RC6enctextss = RC6enctext1(i,:);
   re_plaintext = RC6dec(RC6enctextss,r,keyi);
   if i = =1

       plaintext = bitxor(re_plaintext,CO);
   else
       plaintext = bitxor(re_plaintext,RC6enctext1
       ((i-1),:));
   end;
   plaintextRC6(i,:) = plaintext;

end;
ya1 = reshape(plaintextRC6',n,m)';

xa1 = uint8(ya1);

function xa1 = RC6DecCBCImageC(x,key,r,CO)
%Decrypt Colored Image using RC6 in CBC mode

x1 = x(:,:,1); x2 = x(:,:,2); x3 = x(:,:,3);
y1 = double(x1); y2 = double(x2);
y3 = double(x3); [m,n] = size(x1);
plaintext(:,:,1) = reshape(y1',16,m*n/16)';
plaintext(:,:,2) = reshape(y2',16,m*n/16)';
plaintext(:,:,3) = reshape(y3',16,m*n/16)';
ciphertext = x;
keyi = RC6keygen(key,r);
%CBC
for i = 1:3 plaintextRC6 = plaintext(:,:,i);
    RC6enctext = RC6CBCDec(plaintextRC6,m,n,CO,r,keyi);
    RC6enctext = reshape(RC6enctext',n,m)';
    ciphertext(:,:,i) = RC6enctext;
end

xa1 = uint8(ciphertext);

function [xa1, Fs, nbits] = RC6DecCFBAudio(x1, Fs,
nbits,key,r,CO)

%Decrypt Audio using RC6 in CFB mode
```

```
%CFB keyi = RC6keygen(key,r); L = length(x1);
y1 = double(x1); y1 = reshape(y1,L/16,16);
plaintextRC6 = zeros(L/16,16);
for i = 1:L/16
    RC6enctextss = y1(i,:);
    if i = =1 cr1 = RC6enc(CO,r,keyi);
        re_plaintext = bitxor(RC6enctextss,cr1);
    else
        cr1 = RC6enc(y1(i-1,:),r,keyi);
        re_plaintext = bitxor(RC6enctextss,cr1);
    end;
    plaintextRC6(i,:) = re_plaintext;

end;
xa1 = uint8(plaintextRC6(:));

function xa1 = RC6DecCFBImage(x1,key,r,CO)
%Decrypt Image using RC6 in CFB mode

keyi = RC6keygen(key,r);
%CFB [m,n] = size(x1);
y1 = double(x1);
plaintextRC6 = zeros(m*n/16,16);
RC6enctext1 = reshape(y1',16,m*n/16)'; for i = 1:m*n/16
  RC6enctextss = RC6enctext1(i,:);
  if i = =1 cr1 = RC6enc(CO,r,keyi);

                  re_plaintext =
                  bitxor(RC6enctextss,cr1);
  else
      cr1 = RC6enc(RC6enctext1(i-1,:),r,keyi);

                  re_plaintext =
                  bitxor(RC6enctextss,cr1);
  end;

  plaintextRC6(i,:) = re_plaintext;

end;
ya1 = reshape(plaintextRC6',n,m)';

xa1 = uint8(ya1);

function xa1 = RC6DecCFBImageC(x,key,r,CO)
%Decrypt Colored Image using RC6 in CFB mode
```

```
x1 = x(:,:,1); x2 = x(:,:,2); x3 = x(:,:,3);
y1 = double(x1); y2 = double(x2);
y3 = double(x3); [m,n] = size(x1);
plaintext(:,:,1) = reshape(y1',16,m*n/16)';
plaintext(:,:,2) = reshape(y2',16,m*n/16)';
plaintext(:,:,3) = reshape(y3',16,m*n/16)';
ciphertext = x;
keyi = RC6keygen(key,r);
%CFB
for i = 1:3 plaintextRC6 = plaintext(:,:,i);
    RC6enctext = RC6CFBDec(plaintextRC6,m,n,CO,r,keyi);
    RC6enctext = reshape(RC6enctext',n,m)';
    ciphertext(:,:,i) = RC6enctext;
end

xa1 = uint8(ciphertext);

function [xa1, Fs, nbits] = RC6DecECBAudio(x1, Fs,
nbits,key,r)
%Decrypt Audio using RC6 in ECB mode

%ECB keyi = RC6keygen(key,r); L = length(x1);
y1 = double(x1); y1 = reshape(y1,L/16,16);
plaintextRC6 = zeros(L/16,16);
for i = 1:L/16
    RC6enctextss = y1(i,:);
    re_plaintext = RC6dec(RC6enctextss,r,keyi);
    plaintextRC6(i,:) = re_plaintext;

end;

xa1 = uint8(plaintextRC6(:));

function xa1 = RC6DecECBImage(x1,key,r)
%Decrypt Image using RC6 in ECB mode
keyi = RC6keygen(key,r);
%ECB [m,n] = size(x1);
y1 = double(x1);
plaintextRC6 = zeros(m*n/16,16);
RC6enctext1 = reshape(y1',16,m*n/16)'; for i = 1:m*n/16
  RC6enctextss = RC6enctext1(i,:);
  re_plaintext = RC6dec(RC6enctextss,r,keyi);
  plaintextRC6(i,:) = re_plaintext;

end;
```

```
ya1 = reshape(plaintextRC6',n,m)';

xa1 = uint8(ya1);

function xa1 = RC6DecECBImageC(x,key,r)
%Decrypt Colored Image using RC6 in ECB mode

x1 = x(:,:,1); x2 = x(:,:,2); x3 = x(:,:,3);
y1 = double(x1); y2 = double(x2);
y3 = double(x3); [m,n] = size(x1);
plaintext(:,:,1) = reshape(y1',16,m*n/16)';
plaintext(:,:,2) = reshape(y2',16,m*n/16)';
plaintext(:,:,3) = reshape(y3',16,m*n/16)';
ciphertext = x;
keyi = RC6keygen(key,r);
%ECB
for i = 1:3 plaintextRC6 = plaintext(:,:,i);
    RC6enctext = RC6ECBDec(plaintextRC6,m,n,r,keyi);
    RC6enctext = reshape(RC6enctext',n,m)';
    ciphertext(:,:,i) = RC6enctext;
end xa1 = uint8(ciphertext);

function [xa1, Fs, nbits] = RC6DecOFBAudio(x1, Fs,
nbits,key,r,CO)
%Decrypt Audio using RC6 in OFB mode

%OFB keyi = RC6keygen(key,r); L = length(x1);
y1 = double(x1); y1 = reshape(y1,L/16,16);
plaintextRC6 = zeros(L/16,16);
for i = 1:L/16
    RC6enctextss = y1(i,:);
    if i = =1
        cr = RC6enc (CO,r,keyi);
        re_plaintext = bitxor(RC6enctextss,cr);
    else cr = RC6enc(cr,r,keyi);
        re_plaintext = bitxor(RC6enctextss,cr);
    end;
    plaintextRC6(i,:) = re_plaintext;
end;
xa1 = uint8(plaintextRC6(:));

function xa1 = RC6DecOFBImage(x1,key,r,CO)
%Decrypt Image using RC6 in OFB mode

keyi = RC6keygen(key,r);
```

```
%OFB [m,n] = size(x1);
y1 = double(x1);
plaintextRC6 = zeros(m*n/16,16);
RC6enctext1 = reshape(y1',16,m*n/16)'; for i = 1:m*n/16
  RC6enctextss = RC6enctext1(i,:);
  if i = =1
               cr = RC6enc (CO,r,keyi);
       re_plaintext = bitxor(RC6enctextss,cr);
  else cr = RC6enc(cr,r,keyi);
       re_plaintext = bitxor(RC6enctextss,cr);
  end;

  plaintextRC6(i,:) = re_plaintext;

end;
ya1 = reshape(plaintextRC6',n,m)';

xa1 = uint8(ya1);

function xa1 = RC6DecOFBImageC(x,key,r,CO)
%Decrypt Colored Image using RC6 in OFB mode

x1 = x(:,:,1); x2 = x(:,:,2); x3 = x(:,:,3);
y1 = double(x1); y2 = double(x2);
y3 = double(x3); [m,n] = size(x1);
plaintext(:,:,1) = reshape(y1',16,m*n/16)';
plaintext(:,:,2) = reshape(y2',16,m*n/16)';
plaintext(:,:,3) = reshape(y3',16,m*n/16)';
ciphertext = x;
keyi = RC6keygen(key,r);
%OFB
for i = 1:3 plaintextRC6 = plaintext(:,:,i);
    RC6enctext = RC6OFBDec(plaintextRC6,m,n,CO,r,keyi);
    RC6enctext = reshape(RC6enctext',n,m)';
    ciphertext(:,:,i) = RC6enctext;
end

xa1 = uint8(ciphertext);

function RC6enctextRC6 = RC6ECB(plaintext1,m,n,r,keyi)
%Encrypt data using RC6 in ECB mode

RC6enctextRC6 = zeros(m*n/16,16);
for i = 1:m*n/16 plaintext = plaintext1(i,:);
    RC6enctext = RC6enc(plaintext,r,keyi);
```

```
    RC6enctextRC6(i,:) = RC6enctext;

end;

function plaintextRC6 =
RC6ECBDec(RC6enctext1,m,n,r,keyi)
%Decrypt Data using RC6 in ECB mode

plaintextRC6 = zeros(m*n/16,16);
for i = 1:m*n/16
    RC6enctextss = RC6enctext1(i,:);
    re_plaintext = RC6dec(RC6enctextss,r,keyi);
    plaintextRC6(i,:) = re_plaintext;

end;

function y = RC6enc(plaintext,round,s)
%RC6 Encryption
a = plaintext(1:4); b = plaintext(5:8);
c = plaintext(9:12); d = plaintext(13:16);
b = add(b,s(1,:)); d = add(d,s(2,:));
for i = 1:round t = multi(b,(2*b+1)); t = shifting(t,5);
    u = multi(d,(2*d+1)); u = shifting(u,5);
    a = bitxor(a,t'); a = shifting(a,LSB5(u));
    a = add(a,s(2*i+1,:)); c = bitxor(c,u');
    c = shifting(c,LSB5(t)); c = add(c,s(2*i+2,:));
    temp = a;
    a = b; b = c; c = d; d = temp;
end a = add(a,s(2*round+3,:)); c =
add(c,s(2*round+4,:)); y(1:4) = a;
y(5:8) = b; y(9:12) = c; y(13:16) = d;

function [xa, Fs, nbits] = RC6EncCBCAudio(x, Fs,
nbits,key,r,CO)
%Encrypt Audio using RC6 in CBC mode
L = length(x);
y = double(x); keyi = RC6keygen(key,r);
plaintext1 = reshape(y,L/16,16);
RC6enctextRC6 = zeros(L/16,16);
            %CBC
for i = 1:L/16
plaintext = plaintext1(i,:);
if i = =1
plaintexts = bitxor(plaintext,CO);
else plaintexts = bitxor(plaintext,RC6enctext);
```

```
end;
RC6enctext = RC6enc(plaintexts,r,keyi);
RC6enctextRC6(i,:) = RC6enctext;
end;
xa = uint8(RC6enctextRC6(:));

function xa = RC6EncCBCImage(x,key,r,CO)
%Encrypt Image using RC6 in CBC mode
[m,n] = size(x);
y = double(x);
keyi = RC6keygen(key,r);
plaintext1 = reshape(y',16,m*n/16)';
RC6enctextRC6 = zeros(m*n/16,16);
  %CBC
for i = 1:m*n/16
    plaintext = plaintext1(i,:);
    if i = =1
        plaintexts = bitxor(plaintext,CO);
    else
        plaintexts = bitxor(plaintext,RC6enctext);
    end;
    RC6enctext = RC6enc(plaintexts,r,keyi);
    RC6enctextRC6(i,:) = RC6enctext;

end;
ya = reshape(RC6enctextRC6',n,m)';
xa = uint8(ya);

function xa = RC6EncCBCImageC(x,key,r,CO)
%Encrypt Colored Image using RC6 in CBC mode
x1 = x(:,:,1);
x2 = x(:,:,2);
x3 = x(:,:,3);
[m,n] = size(x1);
y1 = double(x1);
y2 = double(x2);
y3 = double(x3);
keyi = RC6keygen(key,r);
plaintext(:,:,1) = reshape(y1',16,m*n/16)';
plaintext(:,:,2) = reshape(y2',16,m*n/16)';
plaintext(:,:,3) = reshape(y3',16,m*n/16)';
RC6enctext = x;
  %CBC
for i = 1:3 plaintext1 = plaintext(:,:,i);
    RC6enctextRC6 = RC6CBC(plaintext1,m,n,CO,r,keyi);
```

```
        RC6enctextRC6 = reshape(RC6enctextRC6',n,m)';
        RC6enctext(:,:,i) = RC6enctextRC6;
end;
xa = uint8(RC6enctext);

function [xa, Fs, nbits] = RC6EncCFBAudio(x, Fs,
nbits,key,r,CO)
%Encrypt Audio using RC6 in CFB mode

L = length(x);
y = double(x);
keyi = RC6keygen(key,r);
plaintext1 = reshape(y,L/16,16);
RC6enctextRC6 = zeros(L/16,16);
  % CFB
for i = 1:L/16
    plaintext = plaintext1(i,:);
    if i = =1
        cr1 = RC6enc(CO,r,keyi);

        RC6enctext = bitxor(plaintext,cr1);
    else
        cr1 = RC6enc(RC6enctext,r,keyi);

        RC6enctext = bitxor(plaintext,cr1);

    end;

    RC6enctextRC6(i,:) = RC6enctext;

end;

xa = uint8(RC6enctextRC6(:));

function xa = RC6EncCFBImage(x,key,r,CO)
%Encrypt Image using RC6 in CFB mode
[m,n] = size(x);
y = double(x);
keyi = RC6keygen(key,r);
plaintext1 = reshape(y',16,m*n/16)';
RC6enctextRC6 = zeros(m*n/16,16);
% CFB
for i = 1:m*n/16
    plaintext = plaintext1(i,:);
    if i = =1
```

```
        cr1 = RC6enc(CO,r,keyi);

        RC6enctext = bitxor(plaintext,cr1);
    else
        cr1 = RC6enc(RC6enctext,r,keyi);
        RC6enctext = bitxor(plaintext,cr1);

    end;

    RC6enctextRC6(i,:) = RC6enctext;
    end;

ya = reshape(RC6enctextRC6',n,m)';
xa = uint8(ya);

function xa = RC6EncCFBImageC(x,key,r,CO)
%Encrypt Colored Image using RC6 in CFB mode
x1 = x(:,:,1);
x2 = x(:,:,2);
x3 = x(:,:,3);
[m,n] = size(x1);
y1 = double(x1);
y2 = double(x2);
y3 = double(x3);
keyi = RC6keygen(key,r);
plaintext(:,:,1) = reshape(y1',16,m*n/16)';
plaintext(:,:,2) = reshape(y2',16,m*n/16)';
plaintext(:,:,3) = reshape(y3',16,m*n/16)';
RC6enctext = x;
  % CFB
for i = 1:3 plaintext1 = plaintext(:,:,i);
    RC6enctextRC6 = RC6CFB(plaintext1,m,n,CO,r,keyi);
    RC6enctextRC6 = reshape(RC6enctextRC6',n,m)';
    RC6enctext(:,:,i) = RC6enctextRC6;
end;
xa = uint8(RC6enctext);

function [xa, Fs, nbits] = RC6EncECBAudio(x, Fs,
nbits,key,r)
%Encrypt Audio using RC6 in ECB mode
L = length(x);
y = double(x);
keyi = RC6keygen(key,r);
plaintext1 = reshape(y,L/16,16);
RC6enctextRC6 = zeros(L/16,16);
```

```
    % ECB
for i = 1:L/16
    plaintext = plaintext1(i,:);
    RC6enctext = RC6enc(plaintext,r,keyi);
    RC6enctextRC6(i,:) = RC6enctext;
end;

xa = uint8(RC6enctextRC6(:));

function xa = RC6EncECBImage(x,key,r)
%Encrypt Image using RC6 in ECB mode
[m,n] = size(x);
y = double(x);
keyi = RC6keygen(key,r);
plaintext1 = reshape(y',16,m*n/16)';
RC6enctextRC6 = zeros(m*n/16,16);
% ECB
for i = 1:m*n/16
    plaintext = plaintext1(i,:);
    RC6enctext = RC6enc(plaintext,r,keyi);
    RC6enctextRC6(i,:) = RC6enctext;

end;
ya = reshape(RC6enctextRC6',n,m)';
xa = uint8(ya);

function xa = RC6EncECBImageC(x,key,r)
%Encrypt Colored Image using RC6 in ECB mode
x1 = x(:,:,1);
x2 = x(:,:,2);
x3 = x(:,:,3);
[m,n] = size(x1);
y1 = double(x1);
y2 = double(x2);
y3 = double(x3);
keyi = RC6keygen(key,r);
plaintext(:,:,1) = reshape(y1',16,m*n/16)';
plaintext(:,:,2) = reshape(y2',16,m*n/16)';
plaintext(:,:,3) = reshape(y3',16,m*n/16)';
RC6enctext = x;
% ECB
for i = 1:3 plaintext1 = plaintext(:,:,i);
    RC6enctextRC6 = RC6ECB(plaintext1,m,n,r,keyi);
    RC6enctextRC6 = reshape(RC6enctextRC6',n,m)';
    RC6enctext(:,:,i) = RC6enctextRC6;
```

```
end;
xa = uint8(RC6enctext);

function [xa, Fs, nbits] = RC6EncOFBAudio(x, Fs,
  nbits,key,r,CO)
%Encrypt Audio using RC6 in OFB mode
L = length(x);
y = double(x);
keyi = RC6keygen(key,r);
plaintext1 = reshape(y,L/16,16);
RC6enctextRC6 = zeros(L/16,16);
 % OFB
 for i = 1:L/16

 plaintext = plaintext1(i,:);
  if i == 1
       cr = RC6enc(CO,r,keyi);
  else
       cr = RC6enc(cr,r,keyi);
  end;
  RC6enctext = bitxor(plaintext,cr);
  RC6enctextRC6(i,:) = RC6enctext;
end;
xa = uint8(RC6enctextRC6(:));

function xa = RC6EncOFBImage(x,key,r,CO)
    %Encrypt Image using RC6 in OFB mode
    [m,n] = size(x);
    y = double(x);
    keyi = RC6keygen(key,r);
    plaintext1 = reshape(y',16,m*n/16)';
    RC6enctextRC6 = zeros(m*n/16,16);
    % OFB
    for i = 1:m*n/16
        plaintext = plaintext1(i,:);

        if i == 1
            cr = RC6enc(CO,r,keyi);
        else
            cr = RC6enc(cr,r,keyi);

        end;
        RC6enctext = bitxor(plaintext,cr);
        RC6enctextRC6(i,:) = RC6enctext;

end;
```

```
ya = reshape(RC6enctextRC6',n,m)';
xa = uint8(ya);

function xa = RC6EncOFBImageC(x,key,r,CO)
%Encrypt Colored Image using RC6 in OFB mode
x1 = x(:,:,1);
x2 = x(:,:,2);
x3 = x(:,:,3);
[m,n] = size(x1);
y1 = double(x1);
y2 = double(x2);
y3 = double(x3);
keyi = RC6keygen(key,r);
plaintext(:,:,1) = reshape(y1',16,m*n/16)';
plaintext(:,:,2) = reshape(y2',16,m*n/16)';
plaintext(:,:,3) = reshape(y3',16,m*n/16)';
RC6enctext = x;
% OFB
for i = 1:3 plaintext1 = plaintext(:,:,i);
    RC6enctextRC6 = RC6OFB(plaintext1,m,n,CO,r,keyi);
    RC6enctextRC6 = reshape(RC6enctextRC6',n,m)';
    RC6enctext(:,:,i) = RC6enctextRC6;
end;
xa = uint8(RC6enctext);

function s = keygen(key,r)
%RC6 Key Generation
p = [99 81 225 91];
q = [185 121 55 158];
s(1,:) = p;
for i = 2:2*r+4
    s(i,:) = add(s(i-1,:),q);
end; i = 1;
a = zeros(1,4);
b = zeros(1,4);
v = 3*(2*r+4);
for h = 1:v
    s(i,:) = add(s(i,:),a);
    s(i,:) = add(s(i,:),b);
    s(i,:) = shifting(s(i,:),3);
    a = s(i,:);
    key = add(key,a);
    key = add(key,b);
    key = shifting(key,LSB5(add(a,b)));
    b = key;
```

```
    if i == 2*r+4
        i = 1;
    end
end

function RC6enctextRC6 = OFB(plaintext1,m,n,CO,r,keyi)
%Encrypt data using RC6 in OFB mode
RC6enctextRC6 = zeros(m*n/16,16);
for i = 1:m*n/16
    plaintext = plaintext1(i,:);

    if i == 1
        cr = RC6enc(CO,r,keyi);

    else
        cr = RC6enc(cr,r,keyi);

    end;

RC6enctext = bitxor(plaintext,cr); RC6enctextRC6(i,:)
  = RC6enctext;

end;

function plaintextRC6 = RC6OFBDec(RC6enctext1,m,n,CO,r,
  keyi)
%Decrypt data using RC6 in OFB mode
plaintextRC6 = zeros(m*n/16,16);
for i = 1:m*n/16
    RC6enctextss = RC6enctext1(i,:);
    if i = =1
    cr = RC6enc (CO,r,keyi);
        re_plaintext = bitxor(RC6enctextss,cr);
    else

    cr = RC6enc(cr,r,keyi);
    re_plaintext = bitxor(RC6enctextss,cr);
    end;

    plaintextRC6(i,:) = re_plaintext;

end;

function y = RC6sub(x,w)
%z = a-a1mod 2^32
```

```
a = frombase256(x);
a1 = frombase256(w);
z = mod(a-a1,2^32);
y = tobase256(z);

function out = shifting (w,n)
% shifting w by n
y = dec2bin(w,8);
y = rot90 (y); y = y(:);
y = circshift(y,n);
y = reshape(y,8,4)';
out = bin2dec(y(:,end:-1:1));

function y = shiftleft(key,round)
% RC6 Shifting
if (round = =1||round = =2||round = =9||round = =16)
  y = circshift(key,1); else
  y = circshift(key,2); end

function y = tobase256(x)
%convert Hex to 256 bits base
for i = 4:-1:1
    y(i) = fix(x/256^(i-1)); x = x-(y(i)
    *256^(i-1));
end

function c = add(a,b)

%c = a+b mode 2^8
c = zeros(1,4);
for i = 1:4
    c(i) = mod(a(i)+
    b(i),2^8); if
    (i+1)~ = 5
        a(i+1) = a(i+1)+fix((a(i)+b(i))/256);
        end;
        end;

function y = binvec2decA(x)
%convert binary to decimal
y = 0;
for i = 1:length(x)
    y = y+x(i)*2^(i-1);
end;
```

```
function cipherdes1 = cipher(plaindes1,r,keyi)
%%%%%%%%%%%%%%%%%%%%%%%%%%%%%%%%%%%%%%%%%%%%%%%%%%%%%%%%%%
%%%%%%%%%%%%%%%%%%%
%this file for the encryption of the plaintext to a
ciphertext using des
% clc
% clear all bin = zeros(8);
for i = 1:8
        bin(i,:) = dec2binvecA(plaindes1(i),8);
end;
plaindes = rot90(bin);
plaindes = plaindes(:);
ip = InitialPermutation(plaindes);
left = ip(1:32);
right = ip(33:64);
for round = 1:r
    expansion = expl(right);
    xor_one = bitxor(expansion',keyi(round,:));
    substitution = DESsub(xor_one);
    permutation = p(substitution);
    xor_two = bitxor(left,permutation);
if round ~ = r left = right;
    right = xor_two;
else

        left = xor_two;
end;
end;
y(1:32) = left; y(33:64) = right;

cipherdesBIN1 = finalpermutation(y);

cipherdesBIN1 = reshape(cipherdesBIN1,8,8)';
cipherdes1 = zeros(1,8);
for i = 1:8
    cipherdes1(i) = binvec2decA(cipherdesBIN1
    (i,end:-1:1));
end;

function y = Convert256toHex(x)
%convert 256 base numbers to Hex.
a = length(x);
j = 1;
y = zeros(1,2*a);
for i = 1:a
```

```
y(j) = fix(x(i)/16); y(j+1) = x(i)-y(j)*16; j = j+2;
end

function y = dec2binvecA(x,size)
%Convert decimal to binary
y = zeros(1,size);
z = zeros(1,size+1);
z(1) = x;
for i = 1:size
    y(i) = mod(z(i),2^i)/2^(i-1);
    z(i+1) = z(i)-y(i)*2^(i-1);
end;
% y = y(end:-1:1);

function plaindes1 = decry(cipherdes1,r,keyi)
%DES decryption
bin = zeros(8);
for i = 1:8
    bin(i,:) = dec2binvecA(cipherdes1(i),8);
end;
cipherdes = rot90(bin);
cipherdes = cipherdes(:);

ip = InitialPermutation(cipherdes);

left = ip(1:32);
right = ip(33:64);

for round = r:-1:1 expansion = exp1(right);
    xor_one = bitxor(expansion',keyi(round,:));
    substitution = DESsub(xor_one);
    permutation = p(substitution);
    xor_two = bitxor(left,permutation);
if round ~ = 1
    left = right;
    right = xor_two;
else
    left = xor_two;
end;
end;

y(1:32) = left;
y(33:64) = right;

plaindesBIN1 = finalpermutation(y);
```

```
plaindesBIN1 = reshape(plaindesBIN1,8,8)';
plaindes1 = zeros(1,8);
for i = 1:8
    plaindes1(i) = binvec2decA(plaindesBIN1(i,
end:-1:1));
end;

function DESenctextDES = DESCBC(plaintext1,m,n,CO,r,
    keyi)
%Encrypt data using DES in CBC mode
DESenctextDES = zeros(m*n/8,8);
for i = 1:m*n/8
    plaintext = plaintext1(i,:);
    if i == 1
        plaintexts = bitxor(plaintext,CO);
    else
        plaintexts = bitxor(plaintext,DESenctext);
    end;
    DESenctext = cipher(plaintexts,r,keyi);
    DESenctextDES(i,:) = DESenctext;

end;

function plaintextDES = DESCBCDec(DESenctext1,m,n,CO,
    r,keyi)
%Decrypt data using DES in CBC mode

plaintextDES = zeros(m*n/8,8);
for i = 1:m*n/8
    DESenctextss = DESenctext1(i,:);
    re_plaintext = decry(DESenctextss,r,keyi);
    if i == 1
        plaintext = bitxor(re_plaintext,CO);
    else
        plaintext = bitxor(re_plaintext,DESenctext1
        ((i-1),:));
    end;
    plaintextDES(i,:) = plaintext;
end;

function ciphertextDES = DESCFB(plaintext1,m,n,CO,r,
    keyi)
%Encrypt data using DES in CFB mode

ciphertextDES = zeros(m*n/8,8);
```

```
for i = 1:m*n/8
    plaintext = plaintext1(i,:);
    if i == 1
        cr1 = cipher(CO,r,keyi);

        ciphertext = bitxor(plaintext,cr1);
    else
        cr1 = cipher(ciphertext,r,keyi);

        ciphertext = bitxor(plaintext,cr1);
    end;
    ciphertextDES(i,:) = ciphertext;

end;

function plaintextDES = DESCFBDec(ciphertext1,m,n,CO,
    r,keyi)
%decrypt data using DES in CFB mode

plaintextDES = zeros(m*n/8,8);
for i = 1:m*n/8
    ciphertextss = ciphertext1(i,:);
    if i = =1
        cr1 = cipher(CO,r,keyi);

        re_plaintext = bitxor(ciphertextss,cr1);
    else
        cr1 = cipher(ciphertext1(i-1,:),r,keyi);

        re_plaintext = bitxor(ciphertextss,cr1);
    end;

        plaintextDES(i,:) = re_plaintext;

end;

function [xa1,Fs,nbits] = DesDecCBCAudio(x1,Fs,nbits,
    key,r,CO)
%Decryption Audio using DES in CBC mode

keyi = DESkeygen(key,r);
%CBC L = length(x1);
y1 = double(x1);
plaintextDES = zeros(L/8,8);
ciphertext1 = reshape(y1,L/8,8);
```

```matlab
for i = 1:L/8
    ciphertextss = ciphertext1(i,:);
    re_plaintext = decry(ciphertextss,r,keyi);
    if i == 1
        plaintext = bitxor(re_plaintext,CO);
    else
        plaintext = bitxor(re_plaintext,ciphertext1
        ((i-1),:));
    end;
    plaintextDES(i,:) = plaintext;
end;
xa1 = uint8(plaintextDES(:));

function xa1 = DesDecCBCImage(x1,key,r,CO)
%Decrypt Image using DES in CBC mode

keyi = DESkeygen(key,r);
%CBC
[m,n] = size(x1);
y1 = double(x1);
plaintextDES = zeros(m*n/8,8);
ciphertext1 = reshape(y1',8,m*n/8)';

for i = 1:m*n/8
    ciphertextss = ciphertext1(i,:);
    re_plaintext = decry(ciphertextss,r,keyi);
    if i == 1

        plaintext = bitxor(re_plaintext,CO);
    else
        plaintext = bitxor(re_plaintext,ciphertext1
        ((i-1),:));
    end;
    plaintextDES(i,:) = plaintext;

end;
ya1 = reshape(plaintextDES',n,m)';

xa1 = uint8(ya1);

function xa1 = DesDecCBCImageC(x,key,r,CO)
%Decrypt Colored Image using DES in CBC mode

keyi = DESkeygen(key,r);
%CBC
```

```
x1 = x(:,:,1);
x2 = x(:,:,2);
x3 = x(:,:,3);
y1 = double(x1);
y2 = double(x2);
y3 = double(x3);
[m,n] = size(x1);
plaintext(:,:,1) = reshape(y1',8,m*n/8)';

plaintext(:,:,2) = reshape(y2',8,m*n/8)';
plaintext(:,:,3) = reshape(y3',8,m*n/8)';
ciphertext = x;
for i = 1:3
plaintextDES = plaintext(:,:,i);
DESenctext = DESCBCDec(plaintextDES,m,n,CO,r,keyi);
DESenctext = reshape(DESenctext',n,m)';
ciphertext(:,:,i) = DESenctext;

end;
a1 = uint8(ciphertext);

function [xa1,Fs,nbits] = DesDecCFBAudio(x1,Fs,nbits,
    key,r,CO)
%Decrypt Audio using DES in CFB mode

keyi = DESkeygen(key,r);
%CFB L = length(x1);
y1 = double(x1);
plaintextDES = zeros(L/8,8);
ciphertext1 = reshape(y1,L/8,8);
for i = 1:L/8
    ciphertextss = ciphertext1(i,:);
    if i == 1
        cr1 = cipher(CO,r,keyi);

        re_plaintext = bitxor(ciphertextss,cr1);
    else
        cr1 = cipher(ciphertext1(i-1,:),r,keyi);

        re_plaintext = bitxor(ciphertextss,cr1);
    end;

    plaintextDES(i,:) = re_plaintext;

end;
```

```
xa1 = uint8(plaintextDES(:));

function xa1 = DesDecCFBImage(x1,key,r,CO)
%Decrypt Image using DES in CFB mode

keyi = DESkeygen(key,r);
%CFB
[m,n] = size(x1);
y1 = double(x1);
plaintextDES = zeros(m*n/8,8);
ciphertext1 = reshape(y1',8,m*n/8)';
for i = 1:m*n/8
    ciphertextss = ciphertext1(i,:);
    if i == 1
        cr1 = cipher(CO,r,keyi);

        re_plaintext = bitxor(ciphertextss,cr1);
    else
        cr1 = cipher(ciphertext1(i-1,:),r,keyi);

        re_plaintext = bitxor(ciphertextss,cr1);
    end;
    plaintextDES(i,:) = re_plaintext;
end;
ya1 = reshape(plaintextDES',n,m)';

xa1 = uint8(ya1);

function xa1 = DesDecCFBImageC(x,key,r,CO)
%Decrypt Colored Image using DES in CFB mode

keyi = DESkeygen(key,r);
%CFB
x1 = x(:,:,1);
x2 = x(:,:,2);
x3 = x(:,:,3);
y1 = double(x1);
y2 = double(x2);
y3 = double(x3);
[m,n] = size(x1);
plaintext(:,:,1) = reshape(y1',8,m*n/8)';
plaintext(:,:,2) = reshape(y2',8,m*n/8)';
plaintext(:,:,3) = reshape(y3',8,m*n/8)';
ciphertext = x;
for i = 1:3 plaintextDES = plaintext(:,:,i);
```

```
    DESenctext = DESCFBDec(plaintextDES,m,n,CO,r,keyi);
    DESenctext = reshape(DESenctext',n,m)';
    ciphertext(:,:,i) = DESenctext;
end

xa1 = uint8(ciphertext);

function [xa1,Fs,nbits] = DesDecECBAudio(x1,Fs,nbits,
    key,r)
%Decrypt Audio using DES in ECB mode

keyi = DESkeygen(key,r);
%ECB L = length(x1);
y1 = double(x1);
plaintextDES = zeros(L/8,8);
ciphertext1 = reshape(y1,L/8,8);
for i = 1:L/8 ciphertextss = ciphertext1(i,:);
    re_plaintext = decry(ciphertextss,r,keyi);
    plaintextDES(i,:) = re_plaintext;

end;
xa1 = uint8(plaintextDES(:));

function xa1 = DesDecECBImage(x1,key,r)
%Decrypt Image using DES in ECB mode

keyi = DESkeygen(key,r);
%ECB
[m,n] = size(x1);
y1 = double(x1);
plaintextDES = zeros(m*n/8,8);
ciphertext1 = reshape(y1',8,m*n/8)';
for i = 1:m*n/8
    ciphertextss = ciphertext1(i,:);
    re_plaintext = decry(ciphertextss,r,keyi);
    plaintextDES(i,:) = re_plaintext;
end;
ya1 = reshape(plaintextDES',n,m)';
xa1 = uint8(ya1);

function xa1 = DesDecECBImageC(x,key,r)
%Decrypt Colored Image using DES in ECB mode

keyi = DESkeygen(key,r);
%ECB x1 = x(:,:,1);
```

```
x2 = x(:,:,2);
x3 = x(:,:,3);
y1 = double(x1);
y2 = double(x2);
y3 = double(x3);
[m,n] = size(x1);
plaintext(:,:,1) = reshape(y1',8,m*n/8)';
plaintext(:,:,2) = reshape(y2',8,m*n/8)';
plaintext(:,:,3) = reshape(y3',8,m*n/8)';
ciphertext = x;
for i = 1:3 plaintextDES = plaintext(:,:,i);
    DESenctext = DESECBDec(plaintextDES,m,n,r,keyi);
    DESenctext = reshape(DESenctext',n,m)';
    ciphertext(:,:,i) = DESenctext;
end
xa1 = uint8(ciphertext);

function [xa1,Fs,nbits] = DesDecOFBAudio(x1,Fs,nbits,
    key,r,CO)
%Decrypt Audio using DES in OFB mode

keyi = DESkeygen(key,r);
%OFB L = length(x1);
y1 = double(x1);
plaintextDES = zeros(L/8,8);
ciphertext1 = reshape(y1,L/8,8);
for i = 1:L/8
    ciphertextss = ciphertext1(i,:);
    if i = =1
        cr = cipher(CO,r,keyi);
        re_plaintext = bitxor(ciphertextss,cr);
    else cr = cipher(cr,r,keyi);
        re_plaintext = bitxor(ciphertextss,cr);
    end;

        plaintextDES(i,:) = re_plaintext;

end;
xa1 = uint8(plaintextDES(:));

function xa1 = DesDecOFBImage(x1,key,r,CO)
%Decrypt Image using DES in OFB mode

keyi = DESkeygen(key,r);
%OFB [m,n] = size(x1);
```

```
y1 = double(x1);
plaintextDES = zeros(m*n/8,8);
ciphertext1 = reshape(y1',8,m*n/8)';
for i = 1:m*n/8
    ciphertextss = ciphertext1(i,:);
    if i == 1 cr = cipher(CO,r,keyi);
        re_plaintext = bitxor(ciphertextss,cr);
    else cr = cipher(cr,r,keyi);
        re_plaintext = bitxor(ciphertextss,cr);
    end;
    plaintextDES(i,:) = re_plaintext;
end;
ya1 = reshape(plaintextDES',n,m)';
xa1 = uint8(ya1);

function xa1 = DesDecOFBImageC(x,key,r,CO)
%Decrypt Colored Image using DES in OFB mode

keyi = DESkeygen(key,r);
%OFB x1 = x(:,:,1);
x2 = x(:,:,2);
x3 = x(:,:,3);
y1 = double(x1);
y2 = double(x2);
y3 = double(x3);
[m,n] = size(x1);
plaintext(:,:,1) = reshape(y1',8,m*n/8)';
plaintext(:,:,2) = reshape(y2',8,m*n/8)';
plaintext(:,:,3) = reshape(y3',8,m*n/8)';
ciphertext = x;
for i = 1:3 plaintextDES = plaintext(:,:,i);
    DESenctext = DESOFBDec(plaintextDES,m,n,CO,r,
      keyi);
    DESenctext = reshape(DESenctext',n,m)';
    ciphertext(:,:,i) = DESenctext;
end

xa1 = uint8(ciphertext);

function DESenctextDES = DESECB(plaintext1,m,n,r,keyi)
%Encrypt Data using DES in ECB mode

DESenctextDES = zeros(m*n/8,8);
for i = 1:m*n/8
    plaintext = plaintext1(i,:);
    DESenctext = cipher(plaintext,r,keyi);
```

```
    DESenctextDES(i,:) = DESenctext;

end;

function plaintextDES = DESECBDec(DESenctext1,m,n,r,
    keyi)
%Decrypt data using DES in ECB mode

plaintextDES = zeros(m*n/8,8);
for i = 1:m*n/8
    DESenctextss = DESenctext1(i,:);
    re_plaintext = decry(DESenctextss,r,keyi);
    plaintextDES(i,:) = re_plaintext;

end;

function [xa,Fs,nbits] = DesEncCBCAudio(x,Fs,nbits,
    key,r,CO)
%Encrypt Audio using DES in CBC mode

L = length(x);
y = double(x);
keyi = DESkeygen(key,r);
plaintext1 = reshape(y,L/8,8);
ciphertextDES = zeros(L/8,8);
    %CBC
for i = 1:L/8
    plaintext = plaintext1(i,:);
    if i == 1
            plaintexts = bitxor(plaintext,CO);
    else
            plaintexts = bitxor(plaintext,ciphertext);
    end;
    ciphertext = cipher(plaintexts,r,keyi);
    ciphertextDES(i,:) = ciphertext;
end;
xa = uint8(ciphertextDES(:));

function xa = DesEncCBCImage(x,key,r,CO)
%Encrypt Image using DES in CBC mode
[m,n] = size(x);
y = double(x);
keyi = DESkeygen(key,r);
plaintext1 = reshape(y',8,m*n/8)';
ciphertextDES = zeros(m*n/8,8);
```

```matlab
    %CBC
for i = 1:m*n/8
    plaintext = plaintext1(i,:);
    if i = =1
        plaintexts = bitxor(plaintext,CO);
    else
        plaintexts = bitxor(plaintext,ciphertext);
    end;
    ciphertext = cipher(plaintexts,r,keyi);
    ciphertextDES(i,:) = ciphertext;

end;
ya = reshape(ciphertextDES',n,m)';
xa = uint8(ya);

function xa = DesEncCBCImageC(x,key,r,CO)
%Encrypt Colored Image using DES in CBC mode
x1 = x(:,:,1);
x2 = x(:,:,2);
x3 = x(:,:,3);
[m,n] = size(x1);
y1 = double(x1);
y2 = double(x2);
y3 = double(x3);
keyi = DESkeygen(key,r);
plaintext(:,:,1) = reshape(y1',8,m*n/8)';
plaintext(:,:,2) = reshape(y2',8,m*n/8)';
plaintext(:,:,3) = reshape(y3',8,m*n/8)';
DESenctext = x;
  %CBC
for i = 1:3 plaintext1 = plaintext(:,:,i);
    DESenctextDES = DESCBC(plaintext1,m,n,CO,r,keyi);
    DESenctextDES = reshape(DESenctextDES',n,m)';
    DESenctext(:,:,i) = DESenctextDES;
end;
xa = uint8(DESenctext);

function [xa,Fs,nbits] = DesEncCFBAudio(x,Fs,nbits,key,
    r,CO)
%Encrypt Audio using DES in CFB mode
L = length(x);
y = double(x);
keyi = DESkeygen(key,r);
plaintext1 = reshape(y,L/8,8);
ciphertextDES = zeros(L/8,8);
```

```
    % CFB
    for i = 1:L/8
        plaintext = plaintext1(i,:);
        if i = =1
            cr1 = cipher(CO,r,keyi);
            ciphertext = bitxor(plaintext,cr1);
        else cr1 = cipher(ciphertext,r,keyi);
            ciphertext = bitxor(plaintext,cr1);
        end;
        ciphertextDES(i,:) = ciphertext;
    end;
    xa = uint8(ciphertextDES(:));


    function xa = DesEncCFBimage(x,key,r,CO)
    %Encrypt Image using DES in CFB mode
    [m,n] = size(x);
    y = double(x);
    keyi = DESkeygen(key,r);
    plaintext1 = reshape(y',8,m*n/8)';
    ciphertextDES = zeros(m*n/8,8);
    % CFB
    for i = 1:m*n/8
        plaintext = plaintext1(i,:);
        if i = =1
            cr1 = cipher(CO,r,keyi);
            ciphertext = bitxor(plaintext,cr1);
        else cr1 = cipher(ciphertext,r,keyi);
            ciphertext = bitxor(plaintext,cr1);
        end;
    ciphertextDES(i,:) = ciphertext;
    end;
    ya = reshape(ciphertextDES',n,m)';


    xa = uint8(ya);


    function xa = DesEncCFBImageC(x,key,r,CO)
    %Encrypt Colored Image using DES in CFB mode

    x1 = x(:,:,1); x2 = x(:,:,2); x3 = x(:,:,3);
    [m,n] = size(x1); y1 = double(x1);
    y2 = double(x2); y3 = double(x3);
    keyi = DESkeygen(key,r);
    plaintext(:,:,1) = reshape(y1',8,m*n/8)';
    plaintext(:,:,2) = reshape(y2',8,m*n/8)';
```

```
plaintext(:,:,3) = reshape(y3',8,m*n/8)';
DESenctext = x;
    % CFB
for i = 1:3 plaintext1 = plaintext(:,:,i);
    DESenctextDES = DESCFB(plaintext1,m,n,CO,r,keyi);
    DESenctextDES = reshape(DESenctextDES',n,m)';
    DESenctext(:,:,i) = DESenctextDES;
end;
xa = uint8(DESenctext);

function [xa,Fs,nbits] = DesEncECBAudio(x,Fs,nbits,
key,r)
%Encrypt Audio using DES in ECB mode
L = length(x);
y = double(x); keyi = DESkeygen(key,r);
plaintext1 = reshape(y,L/8,8);
ciphertextDES = zeros(L/8,8);
    % ECB
for i = 1:L/8 plaintext = plaintext1(i,:);
    ciphertext = cipher(plaintext,r,keyi);
    ciphertextDES(i,:) = ciphertext;
end;
xa = uint8(ciphertextDES(:));

function xa = DesEncECBImage(x,key,r)
%Encrypt Image using DES in ECB mode

[m,n] = size(x);
y = double(x); keyi = DESkeygen(key,r);
plaintext1 = reshape(y',8,m*n/8)';
ciphertextDES = zeros(m*n/8,8);
% ECB
for i = 1:m*n/8 plaintext = plaintext1(i,:);
    ciphertext = cipher(plaintext,r,keyi);
    ciphertextDES(i,:) = ciphertext;

end; ya = reshape(ciphertextDES',n,m)'; xa = uint8(ya);

function xa = DesEncECBImageC(x,key,r)
%Encrypt Colored Image using DES in ECB mode
x1 = x(:,:,1); x2 = x(:,:,2); x3 = x(:,:,3);
[m,n] = size(x1); y1 = double(x1);
y2 = double(x2);
y3 = double(x3); keyi = DESkeygen(key,r);
plaintext(:,:,1) = reshape(y1',8,m*n/8)';
```

```
plaintext(:,:,2) = reshape(y2',8,m*n/8)';
plaintext(:,:,3) = reshape(y3',8,m*n/8)';
DESenctext = x;

for i = 1:3 plaintext1 = plaintext(:,:,i);
    DESenctextDES = DESECB(plaintext1,m,n,r,keyi);
    DESenctextDES = reshape(DESenctextDES',n,m)';
    DESenctext(:,:,i) = DESenctextDES;
end;
xa = uint8(DESenctext);


function [xa,Fs,nbits] = DesEncOFBAudio(x,Fs,nbits,
    key,r,CO)
%Encrypt Audio using DES in OFB mode
L = length(x);
y = double(x); keyi = DESkeygen(key,r);
plaintext1 = reshape(y,L/8,8);
ciphertextDES = zeros(L/8,8);
    % OFB
for i = 1:L/8 plaintext = plaintext1(i,:); if i == 1
        cr = cipher(CO,r,keyi);
    else cr = cipher(cr,r,keyi);
    end; ciphertext = bitxor(plaintext,cr);
    ciphertextDES(i,:) = ciphertext;
end;
xa = uint8(ciphertextDES(:));


function xa = DesEncOFBImage(x,key,r,CO)
%Encrypt Image using DES in OFB mode

[m,n] = size(x);
y = double(x); keyi = DESkeygen(key,r);
plaintext1 = reshape(y',8,m*n/8)';
ciphertextDES = zeros(m*n/8,8);
% OFB
for i = 1:m*n/8 plaintext = plaintext1(i,:); if i == 1
        cr = cipher(CO,r,keyi);
    else cr = cipher(cr,r,keyi);
    end; ciphertext = bitxor(plaintext,cr);
    ciphertextDES(i,:) = ciphertext;
end; ya = reshape(ciphertextDES',n,m)'; xa = uint8(ya);


function xa = DesEncOFBImageC(x,key,r,CO)
%Encrypt Colored Image using DES in OFB mode
```

```
x1 = x(:,:,1);
x2 = x(:,:,2);
x3 = x(:,:,3);
[m,n] = size(x1);
y1 = double(x1);
y2 = double(x2);
y3 = double(x3);
keyi = DESkeygen(key,r);
plaintext(:,:,1) = reshape(y1',8,m*n/8)';
plaintext(:,:,2) = reshape(y2',8,m*n/8)';
plaintext(:,:,3) = reshape(y3',8,m*n/8)';
DESenctext = x;
% OFB
for i = 1:3 plaintext1 = plaintext(:,:,i);
    DESenctextDES = DESOFB(plaintext1,m,n,CO,r,keyi);
    DESenctextDES = reshape(DESenctextDES',n,m)';
    DESenctext(:,:,i) = DESenctextDES;
end;
xa = uint8(DESenctext);

function key1 = DESkeygen(Mainkey,r)
%DES Key Generation
% Inserting master key
% clc
% clear all

% Mainkey = ['A';'A';'B';'B';'0';'9';'1';'8';'2';'7';'
            3';'6';'C';'C';'D';'D'];
key2 = Convert256toHex(Mainkey);
key = zeros(r,4); for i = 1:r
key(i,:) = dec2binvecA(key2(i),4); end
key = rot90(key);
key = key(:);

    a = [57 49 41 33 25 17 9 1 58 50 42 34 26 18 10 2 59
         51 43 35 27 19 11 3 60 52 44 36 63 55 47 39 31
         23 15 7 62 54 46 38 30 22 14 6 61 53 45 37 29 21
         13 5 28 20 12 4];
    b = [14 17 11 24 1 5 3 28 15 6 21 10 23 19 12 4 26 8
         16 7 27 20 13 2 41 52 31 37 47 55 30 40 51 45 33
         48 44 49 39 56 34 53 46 42 50 36 29 32];
    %%%%%%%%%%%%%%%%%%%%%%%%%%%%%%%%%%%%%%%%%%
    output1 = key(a);
    key1 = zeros(r,48);
    left = output1(28:-1:1);
```

```
  right = output1(56:-1:29);
  for i = 1:r
  left = shiftleft(left,i);
  right = shiftleft(right,i);
  output(28:-1:1) = left;
  output(56:-1:29) = right;
  key1(i,:) = output(b);
end

function ciphertextDES = DESOFB(plaintext1,m,n,CO,r,
   keyi)
%Encrypt data using DES in OFB mode

ciphertextDES = zeros(m*n/8,8);
for i = 1:m*n/8
    plaintext = plaintext1(i,:);

    if i == 1
        cr = cipher(CO,r,keyi);

    else
       cr = cipher(cr,r,keyi);

    end;
    ciphertext = bitxor(plaintext,cr);
    ciphertextDES(i,:) = ciphertext;

end;

function plaintextDES = DESOFBDec(ciphertext1,m,n,CO,
   r,keyi)
%Decrypt data using DES in OFB mode

plaintextDES = zeros(m*n/8,8);
for i = 1:m*n/8
    ciphertextss = ciphertext1(i,:);
    if i = =1
        cr = cipher (CO,r,keyi);
        re_plaintext = bitxor(ciphertextss,cr);
    else cr = cipher(cr,r,keyi);
        re_plaintext = bitxor(ciphertextss,cr);
    end;

    plaintextDES(i,:) = re_plaintext;

end;
```

```
function y = sub(in)
%DES substitution
in1 = reshape(in',6,8)';
x = zeros(1,8);
for i = 1:8
    x(i) = binvec2decA(in1(i,end:-1:1));
end;
s = [14 0 4 15 13 7 1 4 2 14 15 2 11 13 8 1 3 10 10 6 6
     12 12 11 5 9 9 5 0 3 7 8 4 15 1 12 14 8 8 2 13 4 6
     9 2 1 11 7 15 5 12 11 9 3 7 14 3 10 10 0 5 6 0 13;
     15 3 1 13 8 4 14 7 6 15 11 2 3 8 4 14 9 12 7 0 2 1
     13 10 12 6 0 9 5 11 10 5 0 13 14 8 7 10 11 1 10 3 4
     15 13 4 1 2 5 11 8 6 12 7 6 12 9 0 3 5 2 14 15 9;
     10 13 0 7 9 0 14 9 6 3 3 4 15 6 5 10 1 2 13 8 12 5
     7 14 11 12 4 11 2 15 8 1 13 1 6 10 4 13 9 0 8 6 15
     9 3 8 0 7 11 4 1 5 2 14 12 3 5 11 10 5 14 2 7 12;
     7 13 13 8 14 11 3 5 0 6 6 15 9 0 10 3 1 4 2 7 8 2 5
     12 11 1 12 10 4 14 15 9 10 3 6 15 9 0 0 6 12 10 11
     1 7 13 13 8 15 9 1 4 3 5 14 11 5 12 2 7 8 2 4 14;
     2 14 12 11 4 2 1 12 7 4 10 7 11 13 6 1 8 5 5 0 3
     15 15 10 13 3 0 9 14 8 9 6 4 11 2 8 1 12 11 7
     10 1 13 14 7 2 8 13 15 6 9 15 12 0 5 9 6 10 3 4
     0 5 14 3;
     12 10 1 15 10 4 15 2 9 7 2 12 6 9 8 5 0 6 13 1 3 13
     4 14 14 0 7 11 5 3 11 8 9 4 14 3 15 2 5 12 2 9 8 5
     12 15 3 10 7 11 0 14 4 1 10 7 1 6 13 0 11 8 6 13;
     4 13 11 0 2 11 14 7 15 4 0 9 8 1 13 10 3 14 12 3 9
     5 7 12 5 2 10 15 6 8 1 6 1 6 4 11 11 13 13 8 12 1 3
     4 7 10 14 7 10 9 15 5 6 0 8 15 0 14 5 2 9 3 2 12;
     13 1 2 15 8 13 4 8 6 10 15 3 11 7 1 4 10 12 9 5 3 6
     14 11 5 0 0 14 12 9 7 2 7 2 11 1 4 14 1 7 9 4 12 10
     14 8 2 13 0 15 6 12 10 9 13 0 15 3 3 5 5 6 8 11];
out1 = zeros(1,8);
for i = 1:8
out1(i) = s(i,x(i)+1);
end
out = zeros(8,4);
for i = 1:8
    out(i,:) = dec2binvecA(out1(i),4);
end
out = rot90(out);
y = out(:);

function expansion = exp1(right)
%DES expansion
```

```
a = [32,1,2,3,4,5,4,5,6,7,8,9,8,9,10,11,12,13,12,13,14,
     15,16,17,16,17,18,19,20,21,20,21,22,23,24,25,24,25,
     26,27,28,29,28,29,30,31,32,1];
expansion = right(a);

function y = finalpermutation(x)
%DES final permutation

a = [40 8 48 16 56 24 64 32 39 7 47 15 55 23 63 31 38
     6 46 14 54 22 62 30 37 5 45 13 53 21 61 29 36 4 44
     12 52 20 60 28 35 3 43 11 51 19 59 27 34 2 42 10
     50 18 58 26 33 1 41 9 49 17 57 25];
y = x(a);

function y = frombase256(x)
%Convert 256 base number to decimal
y = 0;
for i = 1:4
    y = y+(x(i)*256^(i-1));
end

function ip = InitialPermutation(plaindes)
%DES Initial Permutation
a = [58 50 42 34 26 18 10 2 60 52 44 36 28 20 12 4 62
     54 46 38 30 22 14 6 64 56 48 40 32 24 16 8 57 49
     41 33 25 17 9 1 59 51 43 35 27 19 11 3 61 53 45 37
     29 21 13 5 63 55 47 39 31 23 15 7];
ip = plaindes(a);

function y = LSB5(x)
%return the 5 least significant bits of x
y = dec2bin(x(1),8);
u = y(8:-1:3); u = u(end:-1:1);
y = bin2dec(u);

function y = CHaoticCipher(im)
% This function encrypts square image using baker map
im = double(im);
n = [10,5,12,5,10,8,14,10,5,12,5,10,8,14,10,5,12,5,10,8,
     14,10,5,12,5,10,8,14,10,5,12,5,10,8,14,10,5,12,5,10,
     8,14,10,5,12,5,10,8,14,10,5,12,5,10,8,14,];
[pr,pc] = chaomat(n);
pim = chaoperm(im,pr,pc,3,'forward');
y = uint8(pim);
imshow(y);
```

```
    function [pr,pc] = chaomat(n)
%
I = sum(n);
k = size(n,2);
for i = 1:k
    N(i+1) = 1;
    for j = 1:i
        N(i+1) = N(i+1)+n(j);
    end
end

N(1) = 1;

for cb = 1:k
  for rb = 1:n(cb)
    rbstartcol(rb) = mod((rb-1)*I,n(cb));
    rbendcol(rb) = mod((rb*I-1),n(cb));
    rbstartrow(rb) = fix(((rb-1)*I)/n(cb));
    rbendrow(rb) = fix((rb*I-1)/n(cb));
    mincol(rb) = min([rbendcol(rb)+1,rbstartcol(rb)]);
    maxcol(rb) = max([rbendcol(rb),rbstartcol(rb)-1]);
  end
for i = 1:I
  for j = N(cb):N(cb+1)-1
    newindex(i,j-N(cb)+1) = (i-1)*n(cb)+(n(cb)-
    j+N(cb)-1);
    newindexmod(i,j-N(cb)+1) = mod(newindex(i,j-
    N(cb)+1),n(cb));
    newindexquotient(i,j-N(cb)+1) = fix(newindex(i,j-
    N(cb)+1)/n(cb));
    rowblockindex(i,j-N(cb)+1) = fix(newindex(i,j-
    N(cb)+1)/I)+1;
  end
end
for i = 1:I
  for j = 1:n(cb)
    for rb = 1:n(cb)
      if rowblockindex(i,j) = =rb;
        if newindexmod(i,j)>maxcol(rb)
          col = rbendrow(rb)-
          newindexquotient(i,j)+(n(cb)-1-
          newindexmod(i,j))*(rbendrow(rb)-
          rbstartrow(rb));
        elseif newindexmod(i,j)> = mincol(rb) &
        newindexmod(i,j)< = maxcol(rb)
```

```
                    if rbstartcol(rb)>rbendcol(rb)
                        c = 0;
                        d = -1;
                    else
                        c = 1;
                        d = 1;
                    end
                    col = (rbendrow(rb)-
                     rbstartrow(rb))*(n(cb)-1-
                     maxcol(rb))+(rbendrow(rb)-newindexquot
                     ient(i,j)+c)+(maxcol(rb)-
                     newindexmod(i,j))*(rbendrow(rb)-
                     rbstartrow(rb)+d);
                else%if newindexmod(i,j)< = mincol(rb)
                    col = I-mincol(rb)*(rbendrow(rb)-
                     rbstartrow(rb))+(rbendrow(rb)-newindex
                     quotient(i,j)+1)+(mincol(rb)-1-
                     newindexmod(i,j))*(rbendrow(rb)-
                     rbstartrow(rb));
                end
                    row = 1+I-N(cb+1)+rowblockindex(i,j);
            end
        end
                    pr(i,j+N(cb)-1) = row;
                    pc(i,j+N(cb)-1) = col;
        end
    end
end

function out = chaoperm(im,pr,pc,num,forward)

%
[rows,cols] = size(im);
mat = zeros([rows,cols,num+1]);
mat(:,:,1) = im(:,:);
for loc = 2:num+1
if(strcmp(forward,'forward'))
        for i = 1:rows
            for j = 1:cols
                mat(pr(i,j),pc(i,j),loc) =
                 mat(i,j,loc-1);
            end
        end
elseif(strcmp(forward,'backward'))
        for i = 1:rows
```

```
            for j = 1:cols
                mat(i,j,loc) = mat(pr(i,j),pc(i,j),loc-1);
            end
        end
    end
end
out = mat(:,:,num+1);

function ID = IDMF(x,y)
%this fuunction(irregular Deviation Measuring Factor) is
based on how much the deviation cased by encryption is
%irregular.
%x:-Original Image
%y:-Encrypted Image
%D:-Maximum Deviation Measuring Factor
x = double(x);
y = double(y);
%first,calculate the difference between each pixel
value before and after
%encryption
D = uint8(abs(x-y));
%calculate the H histogram of the difference
H = imhist(D)
%calculate the average of H
DC = 0;
for i = 1:256
    DC = DC+H(i);
end
DC = DC/256;
%subtract the DC from H at every point
for i = 1:256
    AC(i) = abs(H(i)-DC);
end
%calculate the sum of AC
ID = 0;
for i = 1:256
    ID = ID+AC(i);
End;
%the lower the value, the better the encryption

function D = MDMF(x,y)
%this fuunction calculates the Maximum Deviation
  Measuring Factor which
%calculates the deviation between the original and
  encrypted image.
```

```
%x:-Original Image
%y:-Encrypted Image
%D:-Maximum Deviation Measuring Factor
%first,calculate the histogram between the original
  and encrypted image
x1 = imhist(x);
y1 = imhist(y);
%then calculate the difference between the two
diff = abs(y1-x1);
%then calculate D as follows
D1 = 0;
for i = 2:255
D1 = D1+diff(i);
end;
D2 = (diff(1)+diff(255))/2;
D = D1+D2;
function y = psnr(im1,im2)
%Peak Signal to Noise Ratio
MSEa = 0;
[m,n] = size(im1);
im1 = double(im1);
im2 = double(im2);
for i = 1:m
for j = 1:n
x = double(im1(i,j)-im2(i,j));
x1 = x^2/(m*n);
MSEa = MSEa+x1;
end;
end;
MSEa = double(MSEa);
y = 10*log10((255^2)/MSEa);
```

Appendix B

```
%Image with no Encryption over OFDM system + AWGN ....
%Transmitter..............................................
%Data Generation ........................................

f = imread ('Cameraman.tif');
[M,N] = size(f) ;
g1 = im2col(f, [M,N], [M,N], 'distinct');
h1 = dec2bin(double(g1));
[M1,N1] = size(h1) ;
z1 = zeros (M1,N1) ;
for i = 1
M1
for j = 1: N1
z1(i,j) = str2num(h1(i,j));
end;
end;
[R1,T1] = size(z1) ;
zz1 = reshape(z1,R1*T1, 1);
% The transmitted data ...............................
zz = [zz1];
        trel = poly2trellis(7,[171 133]);% Trellis
        data1 = zz;
        zz = convenc(data1,trel);
s1 = length(zz);
  para = 128;
```

```
nd = 6; %number of information OFDM symbol for one
        loop
ml = 2; %Modulation level: QPSK
sr = 250000; %symbol rate
  br = sr.*ml;      %Bit rat per carrier
Ipoint = 8;%Number of over samples
gilen = 32;
flat = 1;
fd = 600;

        fftlen = para;
        noc = para;
        ofdm_length = para*nd*ml; %Total no for one loop

%Dividing the image into blocks ......................
nloops = ceil((length(zz))/ofdm_length);
new_data = nloops*ofdm_length ;
nzeros = new_data - length(zz);
input_data = [zz;zeros(nzeros,1)];
input_data2 = reshape(input_data,ofdm_length,nloops);

%Transmission ON FFT_OFDM ...................... ......

    for ebno = [0,2,4,6,8,10];

demodata1 = zeros(ofdm_length,nloops);
for jj = 1: nloops % loop for columns
  serdata1 = input_data2(:,jj)';
  demodata = fft_channel(serdata1,para,nd,ml,gilen,fftl
  en,sr,ebno, br);
            demodata1(:,jj) = demodata(: ); % the
            output of ofdm columns
end

%Received image .......................................
  [Mr,Nr] = size(demodata1);
% demodata2 = demodata1(: );
  yy = reshape (demodata1,Mr*Nr,1);%
part1 = yy(1: s1);
  yy11 = part1;
%%%%%%%%%%%%%%%%%%%%%%%%%%%%%%%%%%%%DECODING ..........
  yy11 = vitdec(yy11',trel,1,'term','hard');% Decode.

yy1 = reshape(yy11,[R1,T1]);
for i = 1: M1
```

```
for j = 1: N1
zn1(i,j) = num2str(yy1(i,j));
end;
end;
hn1 = bin2dec(zn1);
gn1 = col2im(hn1, [M,N], [M,N], 'distinct');
 %The Error between Trans ...........................
    output_image = gn1/255;

output_image1 = medfilt2(output_image);

%The Error between Trans ...........................
 MSE1 = sum(sum((double(f)/255-output_image).^2))/
        prod(size(f));
PSNR = 10*log(1/MSE1)/log(10)

 MSE11 = sum(sum((double(f)/255-output_image1).^2))/
        prod(size(f));
PSNR1 = 10*log(1/MSE11)/log(10);

    end

%Transmission ON DCT_OFDM ...........................

    for ebno = [0,2,4,6,8,10];

demodata1 = zeros(ofdm_length,nloops);
for jj = 1: nloops % loop for columns
    serdata1 = input_data2(:,jj)';
    demodata = dct_channel(serdata1,para,nd,ml,gilen,f
            ftlen,sr,ebno, br);
    demodata1(:,jj) = demodata(: ); % the output of
                    ofdm columns
end

%Received image ....................................
 [Mr,Nr] = size(demodata1);
% demodata2 = demodata1(: );
  yy = reshape (demodata1,Mr*Nr,1);%
 part1 = yy(1: s1);
yy11 = part1;
%%%%%%%%%%%%%%%%%%%%%%%%%%%%%%%%%%%%%DECODING ...........
 yy11 = vitdec(yy11',trel,1,'term','hard');% Decode.

yy1 = reshape(yy11,[R1,T1]);
```

```
for i = 1: M1
for j = 1: N1
zn1(i,j) = num2str(yy1(i,j));
end;
end;
hn1 = bin2dec(zn1);
gn1 = col2im(hn1, [M,N], [M,N], 'distinct');
  %The Error between Trans .......................
    output_image = gn1/255;

output_image1 = medfilt2(output_image);

%The Error between Trans .............................
  MSE1 = sum(sum((double(f)/255-output_image).^2))/
        prod(size(f));
PSNR = 10*log(1/MSE1)/log(10)
  MSE11 = sum(sum((double(f)/255-output_image1).^2))/
          prod(size(f));
PSNR1 = 10*log(1/MSE11)/log(10);

end
%Transmission ON DWT_OFDM ...........................

  for ebno = [0,2,4,6,8,10];

demodata1 = zeros(ofdm_length,nloops);
for jj = 1: nloops % loop for columns
        serdata1 = input_data2(:,jj)';
        demodata = dwt_channel(serdata1,para,nd,ml,gilen
                ,fftlen,sr,ebno, br);
                demodata1(:,jj) = demodata(: ); % the
                                    output of ofdm
                                    columns
end

%Received image .......................................
  [Mr,Nr] = size(demodata1);
% demodata2 = demodata1(: );
  yy = reshape (demodata1,Mr*Nr,1);%
 part1 = yy(1: s1);
yy11 = part1;
%%%%%%%%%%%%%%%%%%%%%%%%%%%%%%%%%%%%%%%DECODING ...........
  yy11 = vitdec(yy11',trel,1,'term','hard');% Decode.

yy1 = reshape(yy11,[R1,T1]);
```

```
for i = 1: M1
for j = 1: N1
zn1(i,j) = num2str(yy1(i,j));
end;
end;
hn1 = bin2dec(zn1);
gn1 = col2im(hn1, [M,N], [M,N], 'distinct');
  %The Error between Trans ...........................
    output_image = gn1/255;

output_image1 = medfilt2(output_image);

%The Error between Trans ...........................
  MSE1 = sum(sum((double(f)/255-output_image).^2))/
        prod(size(f));
PSNR = 10*log(1/MSE1)/log(10)
  MSE11 = sum(sum((double(f)/255-output_image1).^2))/
          prod(size(f));
PSNR1 = 10*log(1/MSE11)/log(10);

   end

%Original clipping + companding + AWGN ..............
%Transmitter ........................................
%Data Generation ....................................

f = imread ('Cameraman.tif');
[M,N] = size(f) ;
g1 = im2col(f, [M,N], [M,N], 'distinct');
h1 = dec2bin(double(g1));
[M1,N1] = size(h1) ;
z1 = zeros (M1,N1) ;
for i = 1: M1
for j = 1: N1
z1(i,j) = str2num(h1(i,j));
end;
end;
[R1,T1] = size(z1) ;
zz1 = reshape(z1,R1*T1, 1);
% The transmitted data ...............................
zz = [zz1];

%ENCODING ...........................................
      trel = poly2trellis(7,[171 133]);% Trellis
      data1 = zz;
```

```
        zz = convenc(data1,trel);
s1 = length(zz);
 para = 128;
nd = 6; %number of information OFDM symbol for one loop
ml = 2; %Modulation level: QPSK
sr = 250000; %symbol rate
 br = sr.*ml; %Bit rat per carrier
Ipoint = 8;%Number of over samples
gilen = 32;
flat = 1;
fd = 600;


        dctlen = para;
        noc = para;
        ofdm_length = para*nd*ml; %Total no for one loop

%Dividing the image into blocks .....................
nloops = ceil((length(zz))/ofdm_length);
new_data = nloops*ofdm_length ;
nzeros = new_data - length(zz);
input_data = [zz;zeros(nzeros,1)];
input_data2 = reshape(input_data,ofdm_length,nloops);

%Transmission ON FFT_OFDM ............................

        for ebno = [0,2,4,6,8,10];

demodata1 = zeros(ofdm_length,nloops);
for jj = 1: nloops % loop for columns
        serdata1 = input_data2(:,jj)';
        demodata = fft_channel_clipping_companding(serda
                ta1,para,nd,ml,gilen,dctlen,sr,ebno,
                br);
                demodata1(:,jj) = demodata(: ); % the
                                 output of ofdm
                                 columns

end

%Received image ...................... .......
 [Mr,Nr] = size(demodata1);
% demodata2 = demodata1(: );
 yy = reshape (demodata1,Mr*Nr,1);%
 part1 = yy(1: s1);
yy11 = part1;
```

```
%DECODING ........................................
  yy11 = vitdec(yy11',trel,1,'term','hard');% Decode.

yy1 = reshape(yy11,[R1,T1]);
for i = 1: M1
for j = 1: N1
zn1(i,j) = num2str(yy1(i,j));
end;
end;
hn1 = bin2dec(zn1);
gn1 = col2im(hn1, [M,N], [M,N], 'distinct');
%The Error between Trans ..............................
        output_image = gn1/255;

output_image1 = medfilt2(output_image);

%The Error between Trans ..............................
  MSE1 = sum(sum((double(f)/255-output_image).^2))/
prod(size(f));
PSNR = 10*log(1/MSE1)/log(10)

  MSE11 = sum(sum((double(f)/255-output_image1).^2))/
prod(size(f));
PSNR1 = 10*log(1/MSE11)/log(10);

        end

%Transmission ON DCT_OFDM ...........................

        for ebno = [0,2,4,6,8,10];

demodata1 = zeros(ofdm_length,nloops);
for jj = 1: nloops           % loop for columns
        serdata1 = input_data2(:,jj)';
        demodata = dct_channel_clipping_companding(serda
                ta1,para,nd,ml,gilen,dctlen,sr,ebno,
                br);
                demodata1(:,jj) = demodata(: ); % the
                output of ofdm columns

end

%Received image ..................................
  [Mr,Nr] = size(demodata1);
% demodata2 = demodata1(: );
```

```
 yy = reshape (demodata1,Mr*Nr,1);%
 part1 = yy(1: s1);
yy11 = part1;
%DECODING ..........................................
 yy11 = vitdec(yy11',trel,1,'term','hard');% Decode.

yy1 = reshape(yy11,[R1,T1]);
for i = 1: M1
for j = 1: N1
zn1(i,j) = num2str(yy1(i,j));
end;
end;
hn1 = bin2dec(zn1);
gn1 = col2im(hn1, [M,N], [M,N], 'distinct');
        %The Error between Trans .......................
          output_image = gn1/255;
output_image1 = medfilt2(output_image);

%The Error between Trans ..............................
 MSE1 = sum(sum((double(f)/255-output_image).^2))/
        prod(size(f));
PSNR = 10*log(1/MSE1)/log(10)

 MSE11 = sum(sum((double(f)/255-output_image1).^2))/
         prod(size(f));
PSNR1 = 10*log(1/MSE11)/log(10);

end

%Transmission ON DWT_OFDM ............................

for ebno = [0,2,4,6,8,10];

demodata1 = zeros(ofdm_length,nloops);
for jj = 1: nloops % loop for columns
       serdata1 = input_data2(:,jj)';
       demodata = dwt_channel_clipping_companding(serda
                  ta1,para,nd,ml,gilen,dctlen,sr,ebno,
                  br);
                  demodata1(:,jj) = demodata(: ); % the
                                    output of ofdm
                                    columns

end
```

```
%Received image .......................................
 [Mr,Nr] = size(demodata1);
% demodata2 = demodata1(: );
 yy = reshape (demodata1,Mr*Nr,1);%
 part1 = yy(1: s1);
yy11 = part1;
%DECODING ...........................................
 yy11 = vitdec(yy11',trel,1,'term','hard');% Decode.

yy1 = reshape(yy11,[R1,T1]);
for i = 1: M1
for j = 1: N1
zn1(i,j) = num2str(yy1(i,j));
end;
end;
hn1 = bin2dec(zn1);
gn1 = col2im(hn1, [M,N], [M,N], 'distinct');
%The Error between Trans ............................
        output_image = gn1/255;

output_image1 = medfilt2(output_image);

%The Error between Trans ............................
 MSE1 = sum(sum((double(f)/255-output_image).^2))/
        prod(size(f));
PSNR = 10*log(1/MSE1)/log(10)

 MSE11 = sum(sum((double(f)/255-output_image1).^2))/
        prod(size(f));
PSNR1 = 10*log(1/MSE11)/log(10);

end

%Image without Encryption + clipping over
estimated channel ................ ...........
%Transmitter ...........................................
%Data Generation ......................................
f = imread ('Cameraman.tif');
[M,N] = size(f) ;
g1 = im2col(f, [M,N], [M,N], 'distinct');
h1 = dec2bin(double(g1));
[M1,N1] = size(h1) ;
z1 = zeros (M1,N1) ;
for i = 1: M1
for j = 1: N1
```

```
z1(i,j) = str2num(h1(i,j));
end;
end;
[R1,T1] = size(z1) ;
zz1 = reshape(z1,R1*T1, 1);
% The transmitted data ................ ................
zz = [zz1];

%ENCODING ...........................................
        trel = poly2trellis(7,[171 133]);% Trellis
        data1 = zz;
        zz = convenc(data1,trel);
s1 = length(zz);
 para = 128;
nd = 6; %number of information OFDM symbol for one
        loop
m1 = 2; %Modulation level: QPSK
sr = 250000; %symbol rate
 br = sr.*m1; %Bit rat per carrier
Ipoint = 8;%Number of over samples
gilen = 32;
flat = 1;
fd = 600;
        fftlen = para;
        noc = para;
        ofdm_length = para*nd*m1; %Total no for one
                    loop

%Dividing the image into blocks .....................
nloops = ceil((length(zz))/ofdm_length);
new_data = nloops*ofdm_length ;
nzeros = new_data - length(zz);
input_data = [zz;zeros(nzeros,1)];
input_data2 = reshape(input_data,ofdm_length,nloops);

%Transmission ON FFT_OFDM ...........................

        for ebno = [0,2,4,6,8,10];

demodata1 = zeros(ofdm_length,nloops);
for jj = 1: nloops % loop for columns
        serdata = input_data2(:,jj)';
        demodata = fft_channel_estimation_no_mapping_cli
                    pping(serdata,para,nd,m1,gilen,fftlen
                    ,sr,ebno, br,fd,flat);
```

```
          demodata1(:,jj) = demodata(: ); % the
                                output of ofdm
                                columns

end

%Received image .......................................
  [Mr,Nr] = size(demodata1);
% demodata2 = demodata1(: );
  yy = reshape (demodata1,Mr*Nr,1);%
  part1 = yy(1: s1);
yy11 = part1;
%DECODING .............................................
yy11 = vitdec(yy11',trel,1,'term','hard');% Decode.

yy1 = reshape(yy11,[R1,T1]);
for i = 1: M1
for j = 1: N1
zn1(i,j) = num2str(yy1(i,j));
end;
end;
hn1 = bin2dec(zn1);
gn1 = col2im(hn1, [M,N], [M,N], 'distinct');
        %The Error between Trans ......................
          output_image = gn1/255;

output_image1 = medfilt2(output_image);

%The Error between Trans ............................
  MSE1 = sum(sum((double(f)/255-output_image).^2))/
          prod(size(f));
PSNR = 10*log(1/MSE1)/log(10)

  MSE11 = sum(sum((double(f)/255-output_image1).^2))/
          prod(size(f));
PSNR1 = 10*log(1/MSE11)/log(10);

end

        %Transmission ON DCT_OFDM .....................

for ebno = [0,2,4,6,8,10];

demodata1 = zeros(ofdm_length,nloops);
for jj = 1: nloops          % loop for columns
```

```
        serdata = input_data2(:,jj)';
        demodata = dct_channel_estimation_no_mapping_cli
                pping(serdata,para,nd,m1,gilen,fftlen
                ,sr,ebno, br,fd,flat);
                    demodata1(:,jj) = demodata(: ); % the
                                        output of ofdm
                                        columns


end

%Received image ......................................
[Mr,Nr] = size(demodata1);
% demodata2 = demodata1(: );
yy = reshape (demodata1,Mr*Nr,1);%
part1 = yy(1: s1);
yy11 = part1;
%DECODING ...........................................
 yy11 = vitdec(yy11',trel,1,'term','hard');% Decode.

yy1 = reshape(yy11,[R1,T1]);
for i = 1: M1
for j = 1: N1
zn1(i,j) = num2str(yy1(i,j));
end;
end;
hn1 = bin2dec(zn1);
gn1 = col2im(hn1, [M,N], [M,N], 'distinct');
 %The Error between Trans ...........................
   output_image = gn1/255;

output_image1 = medfilt2(output_image);

%The Error between Trans ...........................
 MSE1 = sum(sum((double(f)/255-output_image).^2))/
        prod(size(f));
PSNR = 10*log(1/MSE1)/log(10)

 MSE11 = sum(sum((double(f)/255-output_image1).^2))/
        prod(size(f));
PSNR1 = 10*log(1/MSE11)/log(10);

    end

    %Transmission ON DWT_OFDM ......................
```

```
      for ebno = [0,2,4,6,8,10];

demodata1 = zeros(ofdm_length,nloops);
for jj = 1: nloops % loop for columns
      serdata = input_data2(:,jj)';
      demodata = dwt_channel_estimation_no_mapping_cli
            pping(serdata,para,nd,m1,gilen,fftlen
            ,sr,ebno, br,fd,flat);
            demodata1(:,jj) = demodata(: ); % the
                              output of ofdm
                              columns

end

%Received image ...................................
 [Mr,Nr] = size(demodata1);
% demodata2 = demodata1(: );
 yy = reshape (demodata1,Mr*Nr,1);%
 part1 = yy(1: s1);
yy11 = part1;
%DECODING ...........................................
 yy11 = vitdec(yy11',trel,1,'term','hard');% Decode.

yy1 = reshape(yy11,[R1,T1]);
for i = 1: M1
for j = 1: N1
zn1(i,j) = num2str(yy1(i,j));
end;
end;
hn1 = bin2dec(zn1);
gn1 = col2im(hn1, [M,N], [M,N], 'distinct');
 %The Error between Trans ............................
   output_image = gn1/255;

output_image1 = medfilt2(output_image);

%The Error between Trans .............................
 MSE1 = sum(sum((double(f)/255-output_image).^2))/
        prod(size(f));
PSNR = 10*log(1/MSE1)/log(10)

 MSE11 = sum(sum((double(f)/255-output_image1).^2))/
        prod(size(f));
PSNR1 = 10*log(1/MSE11)/log(10);

end
```

```
%Image with no Encryption over OFDM system +
offset + cyclic prefix + AWGN ...............
%Transmitter .........................................
%Data Generation .....................................
%Transmitter .............................................

%Data Generation ....................................
f = imread ('Cameraman.tif');
[M,N] = size(f) ;
g1 = im2col(f, [M,N], [M,N], 'distinct');
h1 = dec2bin(double(g1));
[M1,N1] = size(h1) ;
z1 = zeros (M1,N1) ;
for i = 1: M1
for j = 1: N1
z1(i,j) = str2num(h1(i,j));
end;
end;
[R1,T1] = size(z1) ;
zz1 = reshape(z1,R1*T1, 1);
% The transmitted data ..............................
zz = [zz1];
        trel = poly2trellis(7,[171 133]);% Trellis
        data1 = zz;
        zz = convenc(data1,trel);
s1 = length(zz);
 para = 128;
nd = 6; %number of information OFDM symbol for one loop
m1 = 2; %Modulation level: QPSK
sr = 250000; %symbol rate
 br = sr.*ml; %Bit rate per carrier
Ipoint = 8;%Number of over samples
gilen = 32;
flat = 1;
fd = 600;
epsilon = 0.1;

        fftlen = para;
        noc = para;
        ofdm_length = para*nd*ml; %Total no for one loop

%Dividing the image into blocks .....................

nloops = ceil((length(zz))/ofdm_length);
new_data = nloops*ofdm_length ;
```

```
nzeros = new_data - length(zz);
input_data = [zz;zeros(nzeros,1)];
input_data2 = reshape(input_data,ofdm_length,nloops);

%Transmission ON FFT_OFDM ............................

        for ebno = [0,2,4,6,8,10];

demodata1 = zeros(ofdm_length,nloops);
for jj = 1: nloops % loop for columns
        serdata = input_data2(:,jj)';
        demodata = fft_channel_offset_g(serdata,para,nd,
                m1,gilen,fftlen,sr,ebno, br,epsilon);
                demodata1(:,jj) = demodata(: ); % the
                                        output of ofdm
                                        columns

end

%Received image ......................................
  [Mr,Nr] = size(demodata1);
% demodata2 = demodata1(: );
  yy = reshape (demodata1,Mr*Nr,1);%
  part1 = yy(1: s1);
yy11 = part1;
%%%%%%%%%%%%%%%%%%%%%%%%%%%%%%%%%%%%%DECODING ...........
  yy11 = vitdec(yy11',trel,1,'term','hard');% Decode.

yy1 = reshape(yy11,[R1,T1]);
for i = 1: M1
for j = 1: N1
zn1(i,j) = num2str(yy1(i,j));
end;
end;
hn1 = bin2dec(zn1);
gn1 = col2im(hn1, [M,N], [M,N], 'distinct');
  %The Error between Trans ...........................
    output_image = gn1/255;

output_image1 = medfilt2(output_image);

%The Error between Trans ............................
  MSE1 = sum(sum((double(f)/255-output_image).^2))/
        prod(size(f));
PSNR = 10*log(1/MSE1)/log(10)
```

```
MSE11 = sum(sum((double(f)/255-output_image1).^2))/
        prod(size(f));
PSNR1 = 10*log(1/MSE11)/log(10);


        end

        %Transmission ON DCT_OFDM ......................

        for ebno = [0,2,4,6,8,10];

demodata1 = zeros(ofdm_length,nloops);
for jj = 1: nloops % loop for columns
        serdata1 = input_data2(:,jj)';
        demodata = dct_channel_offset_g(serdata1,para,
                    nd,m1,gilen,fftlen,sr,ebno,
                    br,epsilon);
                    demodata1(:,jj) = demodata(: ); % the
                                        output of ofdm
                                        columns

end

%Received image .......................................
 [Mr,Nr] = size(demodata1);
% demodata2 = demodata1(: );
 yy = reshape (demodata1,Mr*Nr,1);%
 part1 = yy(1: s1);
yy11 = part1;
%%%%%%%%%%%%%%%%%%%%%%%%%%%%%%%%%%DECODING ...........
 yy11 = vitdec(yy11',trel,1,'term','hard');% Decode.

yy1 = reshape(yy11,[R1,T1]);
for i = 1: M1
for j = 1: N1
zn1(i,j) = num2str(yy1(i,j));
end;
end;
hn1 = bin2dec(zn1);
gn1 = col2im(hn1, [M,N], [M,N], 'distinct');
 %The Error between Trans ...........................
    output_image = gn1/255;

output_image1 = medfilt2(output_image);

%The Error between Trans ...............................
```

```
  MSE1 = sum(sum((double(f)/255-output_image).^2))/
         prod(size(f));
PSNR = 10*log(1/MSE1)/log(10)

  MSE11 = sum(sum((double(f)/255-output_image1).^2))/
          prod(size(f));
PSNR1 = 10*log(1/MSE11)/log(10);

    end

    %Transmission ON DWT_OFDM ......................

    for ebno = [0,2,4,6,8,10];

demodata1 = zeros(ofdm_length,nloops);
for jj = 1: nloops % loop for columns
       serdata1 = input_data2(:,jj)';
       demodata = dwt_channel_offset_g(serdata1,para,
                  nd,m1,gilen,fftlen,sr,ebno,
                  br,epsilon);
                  demodata1(:,jj) = demodata(: ); % the
                                    output of ofdm
                                    columns

end

%Received image ......................................
  [Mr,Nr] = size(demodata1);
% demodata2 = demodata1(: );
 yy = reshape (demodata1,Mr*Nr,1);%
 part1 = yy(1: s1);
yy11 = part1;
%%%%%%%%%%%%%%%%%%%%%%%%%%%%%%%%%%%%%%DECODING ...........
 yy11 = vitdec(yy11',trel,1,'term','hard');% Decode.

yy1 = reshape(yy11,[R1,T1]);
for i = 1: M1
for j = 1: N1
zn1(i,j) = num2str(yy1(i,j));
end;
end;
hn1 = bin2dec(zn1);
gn1 = col2im(hn1, [M,N], [M,N], 'distinct');
 %The Error between Trans ...........................
   output_image = gn1/255;
```

```
output_image1 = medfilt2(output_image);

%The Error between Trans ............................
 MSE1 = sum(sum((double(f)/255-output_image).^2))/
        prod(size(f));
PSNR = 10*log(1/MSE1)/log(10)

 MSE11 = sum(sum((double(f)/255-output_image1).^2))/
         prod(size(f));
PSNR1 = 10*log(1/MSE11)/log(10);

end

%Image with no Encryption over OFDM system + offset +
zero padding + AWGN ...............
%Transmitter ...........................................
%Data Generation .......................................

f = imread ('Cameraman.tif');
[M,N] = size(f) ;
g1 = im2col(f, [M,N], [M,N], 'distinct');
h1 = dec2bin(double(g1));
[M1,N1] = size(h1) ;
z1 = zeros (M1,N1) ;
for i = 1: M1
for j = 1: N1
z1(i,j) = str2num(h1(i,j));
end;
end;
[R1,T1] = size(z1) ;
zz1 = reshape(z1,R1*T1, 1);
% The transmitted data ............... ...............
............... .......
zz = [zz1];
      trel = poly2trellis(7,[171 133]);% Trellis
      data1 = zz;
      zz = convenc(data1,trel);
s1 = length(zz);
 para = 128;
nd = 6; %number of information OFDM symbol for one loop
m1 = 2; %Modulation level: QPSK
sr = 250000; %symbol rate
 br = sr.*ml; %Bit rate per carrier
Ipoint = 8;%Number of over samples
gilen = 32;
```

```
flat = 1;
fd = 600;
epsilon = 0.1;
      fftlen = para;
      noc = para;
      ofdm_length = para*nd*ml; %Total no for one loop

%Dividing the image into blocks ......................
nloops = ceil((length(zz))/ofdm_length);
new_data = nloops*ofdm_length ;
nzeros = new_data - length(zz);
input_data = [zz;zeros(nzeros,1)];
input_data2 = reshape(input_data,ofdm_length,nloops);


%Transmission ON FFT_OFDM ..............................


      for ebno = [0,2,4,6,8,10];

demodata1 = zeros(ofdm_length,nloops);
for jj = 1: nloops % loop for columns
      serdata = input_data2(:,jj)';
      demodata = fft_channel_offset_zg(serdata,para,
               nd,ml,gilen,fftlen,sr,ebno,
               br,epsilon);
               demodata1(:,jj) = demodata(: ); % the
                                output of ofdm
                                columns


end

%Received image .......................................
[Mr,Nr] = size(demodata1);
% demodata2 = demodata1(: );
 yy = reshape (demodata1,Mr*Nr,1);%
 part1 = yy(1: s1);
yy11 = part1;
%%%%%%%%%%%%%%%%%%%%%%%%%%%%%%%%%%%%%DECODING ..........
 yy11 = vitdec(yy11',trel,1,'term','hard');% Decode.

yy1 = reshape(yy11,[R1,T1]);
for i = 1: M1
for j = 1: N1
zn1(i,j) = num2str(yy1(i,j));
end;
end;
```

```
hn1 = bin2dec(zn1);
gn1 = col2im(hn1, [M,N], [M,N], 'distinct');
 %The Error between Trans ...........................
   output_image = gn1/255;


output_image1 = medfilt2(output_image);

%The Error between Trans ...........................
 MSE1 = sum(sum((double(f)/255-output_image).^2))/
        prod(size(f));
PSNR = 10*log(1/MSE1)/log(10)

 MSE11 = sum(sum((double(f)/255-output_image1).^2))/
        prod(size(f));
PSNR1 = 10*log(1/MSE11)/log(10);

end

%Transmission ON DCT_OFDM ...........................

      for ebno = [0,2,4,6,8,10];

demodata1 = zeros(ofdm_length,nloops);
for jj = 1: nloops % loop for columns
      serdata1 = input_data2(:,jj)';
      demodata = dct_channel_offset_zg(serdata1,para,n
                 d,m1,gilen,fftlen,sr,ebno,
                 br,epsilon);
                 demodata1(:,jj) = demodata(: ); % the
                                    output of ofdm
                                    columns

end

%Received image .....................................
 [Mr,Nr] = size(demodata1);
% demodata2 = demodata1(: );
 yy = reshape (demodata1,Mr*Nr,1);%
 part1 = yy(1: s1);
yy11 = part1;
%%%%%%%%%%%%%%%%%%%%%%%%%%%%%%%%%%%%%%%DECODING ...........
yy11 = vitdec(yy11',trel,1,'term','hard');% Decode.

yy1 = reshape(yy11,[R1,T1]);
```

```
for i = 1: M1
for j = 1: N1
zn1(i,j) = num2str(yy1(i,j));
end;
end;
hn1 = bin2dec(zn1);
gn1 = col2im(hn1, [M,N], [M,N], 'distinct');
 %The Error between Trans ..........................
   output_image = gn1/255;

output_image1 = medfilt2(output_image);

%The Error between Trans ..............................
 MSE1 = sum(sum((double(f)/255-output_image).^2))/
        prod(size(f));
PSNR = 10*log(1/MSE1)/log(10)

 MSE11 = sum(sum((double(f)/255-output_image1).^2))/
         prod(size(f));
PSNR1 = 10*log(1/MSE11)/log(10);

        end

%Transmission ON DWT_OFDM ...........................

        for ebno = [0,2,4,6,8,10];

demodata1 = zeros(ofdm_length,nloops);
for jj = 1: nloops % loop for columns
       serdata1 = input_data2(:,jj)';
       demodata = dwt_channel_offset_zg(serdata1,para,n
                  d,m1,gilen,fftlen,sr,ebno,
                  br,epsilon);
                  demodata1(:,jj) = demodata(: ); % the
                                    output of ofdm
                                    columns

end

%Received image ......................................
 [Mr,Nr] = size(demodata1);
% demodata2 = demodata1(: );
 yy = reshape (demodata1,Mr*Nr,1);%
 part1 = yy(1: s1);
yy11 = part1;
```

```
%%%%%%%%%%%%%%%%%%%%%%%%%%%%%%%%%DECODING ...........
  yy11 = vitdec(yy11',trel,1,'term','hard');% Decode.

yy1 = reshape(yy11,[R1,T1]);
for i = 1: M1
for j = 1: N1
zn1(i,j) = num2str(yy1(i,j));
end;
end;
hn1 = bin2dec(zn1);
gn1 = col2im(hn1, [M,N], [M,N], 'distinct');
 %The Error between Trans ...........................
   output_image = gn1/255;

output_image1 = medfilt2(output_image);

%The Error between Trans ...........................
 MSE1 = sum(sum((double(f)/255-output_image).^2))/
       prod(size(f));
PSNR = 10*log(1/MSE1)/log(10)

 MSE11 = sum(sum((double(f)/255-output_image1).^2))/
        prod(size(f));
PSNR1 = 10*log(1/MSE11)/log(10);

end

%Image with no Encryption + OFDM + offset + cyclic
prefix + over estimated channel...
%Transmitter .........................................
%Data Generation .....................................

%Transmitter .........................................

%Data Generation .....................................
f = imread ('Cameraman.tif');
[M,N] = size(f) ;
g1 = im2col(f, [M,N], [M,N], 'distinct');
h1 = dec2bin(double(g1));
[M1,N1] = size(h1) ;
z1 = zeros (M1,N1) ;
for i = 1: M1
for j = 1: N1
z1(i,j) = str2num(h1(i,j));
end;
```

```
end;
[R1,T1] = size(z1) ;
zz1 = reshape(z1,R1*T1, 1);
% The transmitted data ...............................
zz = [zz1];


%ENCODING ..............................................
       trel = poly2trellis(7,[171 133]);% Trellis
       data1 = zz;
       zz = convenc(data1,trel);
s1 = length(zz);
 para = 128;
nd = 6; %number of information OFDM symbol for one
        loop
m1 = 2; %Modulation level: QPSK
sr = 250000; %symbol rate
 br = sr.*m1; %Bit rate per carrier
Ipoint = 8;%Number of over samples
gilen = 32;
flat = 1;
fd = 600;


       fftlen = para;
       noc = para;
       ofdm_length = para*nd*m1; %Total no for one loop

%Dividing the image into blocks ....................
nloops = ceil((length(zz))/ofdm_length);
new_data = nloops*ofdm_length ;
nzeros = new_data - length(zz);
input_data = [zz;zeros(nzeros,1)];
input_data2 = reshape(input_data,ofdm_length,nloops);


%Transmission ON FFT_OFDM ...........................

       for ebno = [0,2,4,6,8,10];


demodata1 = zeros(ofdm_length,nloops);
for jj = 1: nloops % loop for columns
       serdata = input_data2(:,jj)';
       demodata = fft_channel_estimation_no_mapping_g(s
                  erdata,para,nd,m1,gilen,fftlen,sr,e
                  bno, br,fd,flat);
```

```
              demodata1(:,jj) = demodata(: ); % the
                               output of ofdm
                               columns

end

%Received image .......................................
  [Mr,Nr] = size(demodata1);
% demodata2 = demodata1(: );
  yy = reshape (demodata1,Mr*Nr,1);%
  part1 = yy(1: s1);
yy11 = part1;
%DECODING ............................................
  yy11 = vitdec(yy11',trel,1,'term','hard');% Decode.

yy1 = reshape(yy11,[R1,T1]);
for i = 1: M1
for j = 1: N1
zn1(i,j) = num2str(yy1(i,j));
end;
end;
hn1 = bin2dec(zn1);
gn1 = col2im(hn1, [M,N], [M,N], 'distinct');
  %The Error between Trans ...........................
    output_image = gn1/255;

output_image1 = medfilt2(output_image);

%The Error between Trans ...........................
  MSE1 = sum(sum((double(f)/255-output_image).^2))/
        prod(size(f));
PSNR = 10*log(1/MSE1)/log(10)

  MSE11 = sum(sum((double(f)/255-output_image1).^2))/
         prod(size(f));
PSNR1 = 10*log(1/MSE11)/log(10);

end

%Transmission ON DCT_OFDM .............................

for ebno = [0,2,4,6,8,10];

demodata1 = zeros(ofdm_length,nloops);
for jj = 1: nloops % loop for columns
```

```
        serdata = input_data2(:,jj)';
        demodata = dct_channel_estimation_no_mapping_g(s
                erdata,para,nd,m1,gilen,fftlen,sr,e
                bno, br,fd,flat);
                demodata1(:,jj) = demodata(: ); % the
                                        output of ofdm
                                        columns


end

%Received image ......................................
  [Mr,Nr] = size(demodata1);
% demodata2 = demodata1(: );
  yy = reshape (demodata1,Mr*Nr,1);%
  part1 = yy(1: s1);
yy11 = part1;
%DECODING ...........................................
  yy11 = vitdec(yy11',trel,1,'term','hard');% Decode.

yy1 = reshape(yy11,[R1,T1]);
for i = 1: M1
for j = 1: N1
zn1(i,j) = num2str(yy1(i,j));
end;
end;
hn1 = bin2dec(zn1);
gn1 = col2im(hn1, [M,N], [M,N], 'distinct');
  %The Error between Trans ...........................
    output_image = gn1/255;

output_image1 = medfilt2(output_image);

%The Error between Trans ...........................
  MSE1 = sum(sum((double(f)/255-output_image).^2))/
        prod(size(f));
PSNR = 10*log(1/MSE1)/log(10)

  MSE11 = sum(sum((double(f)/255-output_image1).^2))/
        prod(size(f));
PSNR1 = 10*log(1/MSE11)/log(10);


        end

        %Transmission ON DWT_OFDM ....................
```

```matlab
    for ebno = [0,2,4,6,8,10];

demodata1 = zeros(ofdm_length,nloops);
for jj = 1: nloops % loop for columns
      serdata = input_data2(:,jj)';
      demodata = dwt_channel_estimation_no_mapping_g(s
               erdata,para,nd,m1,gilen,fftlen,sr,e
               bno, br,fd,flat);
               demodata1(:,jj) = demodata(: ); % the
                                output of ofdm
                                columns

end

%Received image ......................................
 [Mr,Nr] = size(demodata1);
% demodata2 = demodata1(: );
 yy = reshape (demodata1,Mr*Nr,1);%
 part1 = yy(1: s1);
yy11 = part1;
%DECODING ............. ...........................
 yy11 = vitdec(yy11',trel,1,'term','hard');% Decode.

yy1 = reshape(yy11,[R1,T1]);
for i = 1: M1
for j = 1: N1
zn1(i,j) = num2str(yy1(i,j));
end;
end;
hn1 = bin2dec(zn1);
gn1 = col2im(hn1, [M,N], [M,N], 'distinct');
 %The Error between Trans ...........................
   output_image = gn1/255;

output_image1 = medfilt2(output_image);

%The Error between Trans ...........................
 MSE1 = sum(sum((double(f)/255-output_image).^2))/
       prod(size(f));
PSNR = 10*log(1/MSE1)/log(10)

 MSE11 = sum(sum((double(f)/255-output_image1).^2))/
        prod(size(f));
PSNR1 = 10*log(1/MSE11)/log(10);

end
```

```
%Image with no Encryption + OFDM + offset + zero
padding + over estimated channel........
%Transmitter ..........................................
%Data Generation ......................................
%Transmitter ..........................................
%Data Generation ......................................
f = imread ('Cameraman.tif');
[M,N] = size(f) ;
g1 = im2col(f, [M,N], [M,N], 'distinct');
h1 = dec2bin(double(g1));
[M1,N1] = size(h1) ;
z1 = zeros (M1,N1) ;
for i = 1: M1
for j = 1: N1
z1(i,j) = str2num(h1(i,j));
end;
end;
[R1,T1] = size(z1) ;
zz1 = reshape(z1,R1*T1, 1);
% The transmitted data .............................
zz = [zz1];

%ENCODING ............................................
        trel = poly2trellis(7,[171 133]);% Trellis
        data1 = zz;
        zz = convenc(data1,trel);
s1 = length(zz);
 para = 128;
nd = 6; %number of information OFDM symbol for one
loop
m1 = 2; %Modulation level: QPSK
sr = 250000; %symbol rate
 br = sr.*m1; %Bit rate per carrier
Ipoint = 8;%Number of over samples
gilen = 32;
flat = 1;
fd = 600;

        fftlen = para;
        noc = para;
        ofdm_length = para*nd*m1; %Total no for one loop

%Dividing the image into blocks .....................

nloops = ceil((length(zz))/ofdm_length);
```

```
new_data = nloops*ofdm_length ;
nzeros = new_data - length(zz);
input_data = [zz;zeros(nzeros,1)];
input_data2 = reshape(input_data,ofdm_length,nloops);

%Transmission ON FFT_OFDM ...............................

        for ebno = [0,2,4,6,8,10];

demodata1 = zeros(ofdm_length,nloops);
for jj = 1: nloops % loop for columns
        serdata = input_data2(:,jj)';
        demodata = fft_channel_estimation_no_mapping_zg
                   (serdata,para,nd,m1,gilen,fftlen,sr,e
                   bno, br,fd,flat);
                   demodata1(:,jj) = demodata(: ); % the
                                     output of ofdm
                                     columns

end

%Received image .......................................
  [Mr,Nr] = size(demodata1);
% demodata2 = demodata1(: );
  yy = reshape (demodata1,Mr*Nr,1);%
  part1 = yy(1: s1);
yy11 = part1;
%DECODING ...........................................
  yy11 = vitdec(yy11',trel,1,'term','hard');% Decode.

yy1 = reshape(yy11,[R1,T1]);
for i = 1: M1
for j = 1: N1
zn1(i,j) = num2str(yy1(i,j));
end;
end;
hn1 = bin2dec(zn1);
gn1 = col2im(hn1, [M,N], [M,N], 'distinct');
  %The Error between Trans ..........................
    output_image = gn1/255;

output_image1 = medfilt2(output_image);

%The Error between Trans ...............................
```

```
MSE1 = sum(sum((double(f)/255-output_image).^2))/
        prod(size(f));
PSNR = 10*log(1/MSE1)/log(10)

MSE11 = sum(sum((double(f)/255-output_image1).^2))/
        prod(size(f));
PSNR1 = 10*log(1/MSE11)/log(10);

    end

    %Transmission ON DCT_OFDM .........................

    for ebno = [0,2,4,6,8,10];

demodata1 = zeros(ofdm_length,nloops);
for jj = 1: nloops % loop for columns
      serdata = input_data2(:,jj)';
      demodata = dct_channel_estimation_no_mapping_zg
                  (serdata,para,nd,m1,gilen,fftlen,sr,e
                  bno, br,fd,flat);
                  demodata1(:,jj) = demodata(: ); % the
                                    output of ofdm
                                    columns

end

%Received image .......................................
  [Mr,Nr] = size(demodata1);
% demodata2 = demodata1(: );
 yy = reshape (demodata1,Mr*Nr,1);%
 part1 = yy(1: s1);
yy11 = part1;
%DECODING .............................................
 yy11 = vitdec(yy11',trel,1,'term','hard');% Decode.

yy1 = reshape(yy11,[R1,T1]);
for i = 1: M1
for j = 1: N1
zn1(i,j) = num2str(yy1(i,j));
end;
end;
hn1 = bin2dec(zn1);
gn1 = col2im(hn1, [M,N], [M,N], 'distinct');
 %The Error between Trans .............................
```

```
    output_image = gn1/255;

output_image1 = medfilt2(output_image);

%The Error between Trans ..............................
 MSE1 = sum(sum((double(f)/255-output_image).^2))/
        prod(size(f));
PSNR = 10*log(1/MSE1)/log(10)

 MSE11 = sum(sum((double(f)/255-output_image1).^2))/
         prod(size(f));
PSNR1 = 10*log(1/MSE11)/log(10);

    end

    %Transmission ON DWT_OFDM .........................

    for ebno = [0,2,4,6,8,10];

demodata1 = zeros(ofdm_length,nloops);
for jj = 1: nloops % loop for columns
        serdata = input_data2(:,jj)';
        demodata = dwt_channel_estimation_no_mapping_zg(
                   serdata,para,nd,m1,gilen,fftlen,sr,e
                   bno, br,fd,flat);
                   demodata1(:,jj) = demodata(: ); % the
                                     output of ofdm
                                     columns

end

%Received image ......................................
 [Mr,Nr] = size(demodata1);
% demodata2 = demodata1(: );
 yy = reshape (demodata1,Mr*Nr,1);%
 part1 = yy(1: s1);
yy11 = part1;
%DECODING ............................................
 yy11 = vitdec(yy11',trel,1,'term','hard');% Decode.

yy1 = reshape(yy11,[R1,T1]);
for i = 1: M1
for j = 1: N1
zn1(i,j) = num2str(yy1(i,j));
end;
```

```
end;
hn1 = bin2dec(zn1);
gn1 = col2im(hn1, [M,N], [M,N], 'distinct');
 %The Error between Trans ...........................
   output_image = gn1/255;

output_image1 = medfilt2(output_image);

%The Error between Trans ...........................
 MSE1 = sum(sum((double(f)/255-output_image).^2))/
       prod(size(f));
PSNR = 10*log(1/MSE1)/log(10)

 MSE11 = sum(sum((double(f)/255-output_image1).^2))/
       prod(size(f));
PSNR1 = 10*log(1/MSE11)/log(10);

end

%%%%%%%%%%%%%%%%%%%%%%%%%%%%%%%%%%%%%%%%%%%%%%%%%%%%%%%%%%%%
function outdemodata = fft_channel(serdata,para,nd,m1,
                     gilen,fftlen,sr,ebno, br)

%serial to parallel conversion ......................

paradata = reshape(serdata,para,nd*m1);
%QPSK modulation .....................................
[ich,qch] = qpskmod(paradata,para,nd,m1);
% [ich0,qch0] = compoversamp(ich01,qch01,length(ich01)
              ,Ipoint);
kmod = 1/sqrt(2);
ich1 = ich.*kmod;
qch1 = qch.*kmod;
%IFFT ............. ..................................
x = ich1 + qch1.*j;
y = ifft(x);
ich2 = real (y);
qch2 = imag (y);

%Guard interval insertion .............................

[ich3,qch3] = giins(ich2,qch2,fftlen,gilen,nd);
fftlen2 = fftlen + gilen;

%Attenuation Calculation .............................
```

```
spow = sum(ich3.^2+qch3.^2)/nd./para;
attn = 0.5*spow*sr/br*10.^(-ebno/10);
attn = sqrt (attn);


%Receiver ............................................
%AWGN addition ......................................

  [ich4,qch4] = comb(ich3,qch3,attn);

%Guard interval removal ...............................
  [ich5,qch5] = girem (ich4,qch4,fftlen2,gilen,nd);
%FFT ............ ...............................
rx = ich5 + qch5.*j;
ry = fft(rx);
ich6 = real (ry);
qch6 = imag (ry);


%Demodulation ......................................
  ich7 = ich6./kmod;
  qch7 = qch6./kmod;
  outdemodata = qpskdemod (ich7,qch7,para,nd,m1);
function outdemodata = dct_channel(serdata,para,nd,m1,
                        gilen,dctlen,sr,ebno, br)


%serial to parallel conversion ......................

paradata = reshape(serdata,para,nd*m1);
%QPSK modulation ....................................
[ich,qch] = qpskmod(paradata,para,nd,m1);
% [ich0,qch0] =
compoversamp(ich01,qch01,length(ich01),Ipoint);
kmod = 1/sqrt(2);
ich1 = ich.*kmod;
qch1 = qch.*kmod;
%IDCT ..........................................
x = ich1 + qch1.*j;
y = idct(x);
ich2 = real (y);
qch2 = imag (y);


%Guard interval insertion ...........................

[ich3,qch3] = giins(ich2,qch2,dctlen,gilen,nd);
dctlen2 = dctlen + gilen;
```

```
%Attenuation Calculation ............................

spow = sum(ich3.^2+qch3.^2)/nd./para;
attn = 0.5*spow*sr/br*10.^(-ebno/10);
attn = sqrt (attn);


%Receiver ............................................

%AWGN addition ......................................

  [ich4,qch4] = comb(ich3,qch3,attn);

%Guard interval removal ..............................
[ich5,qch5] = girem (ich4,qch4,dctlen2,gilen,nd);
%DCT ................................................
rx = ich5 + qch5.*j;
ry = dct(rx);
ich6 = real (ry);
qch6 = imag (ry);


%Demodulation .......................................
  ich7 = ich6./kmod;
  qch7 = qch6./kmod;
  outdemodata = qpskdemod (ich7,qch7,para,nd,m1);


function outdemodata = dwt_channel(serdata,para,nd,m1,
                       gilen,dctlen,sr,ebno, br)


%serial to parallel conversion ......................

paradata = reshape(serdata,para,nd*m1);
%QPSK modulation ....................................
[ich,qch] = qpskmod(paradata,para,nd,m1);
% [ich0,qch0] = compoversamp(ich01,qch01,length
               (ich01),Ipoint);
kmod = 1/sqrt(2);
ich1 = ich.*kmod;
qch1 = qch.*kmod;
%IDWT ...............................................
x = ich1 + qch1.*j;
y = wavelet('D6',-1,x,'zpd'); % Invert 5 stages
%; % 2D wavelet transform
% R = wavelet('2D CDF 9/7',-2,Y); % Recover X from Y%
Forward transform with 5 stages
```

```
ich2 = real (y);
qch2 = imag (y);

%Guard interval insertion .............................

[ich3,qch3] = giins(ich2,qch2,dctlen,gilen,nd);
dctlen2 = dctlen + gilen;

%Attenuation Calculation .............................

spow = sum(ich3.^2+qch3.^2)/nd./para;
attn = 0.5*spow*sr/br*10.^(-ebno/10);
attn = sqrt (attn);

%Receiver .........................................

%AWGN addition ........................................

 [ich4,qch4] = comb(ich3,qch3,attn);

%Guard interval removal ..............................
 [ich5,qch5] = girem (ich4,qch4,dctlen2,gilen,nd);
%DWT .................................................
rx = ich5 + qch5.*j;
ry = wavelet('D6',1,rx,'zpd');; % 2D wavelet transform
% R = wavelet('2D CDF 9/7',-2,Y); % Recover X from Y;
ich6 = real (ry);
qch6 = imag (ry);

%Demodulation ........................................
 ich7 = ich6./kmod;
 qch7 = qch6./kmod;
 outdemodata = qpskdemod (ich7,qch7,para,nd,m1);

function outdemodata = fft_channel_clipping_companding
                        (serdata,para,nd,m1,gilen,fftle
                        n,sr,ebno, br)

paradata = reshape(serdata,para,nd*m1);
%QPSK modulation ......................................
[ich,qch] = qpskmod(paradata,para,nd,m1);
% [ich0,qch0] = compoversamp(ich01,qch01,length
                (ich01),Ipoint);
kmod = 1/sqrt(2);
ich1 = ich.*kmod;
```

```
qch1 = qch.*kmod;

%IFFT ............... ...................................
x = ich1 + qch1.*i;
y = ifft(x);
CR = 4;
        clipping_threshold = (10^(CR/10))*sqrt(mean(ab
                            s(y).^2));
        tx_signal_Ang = angle(y);

        for ii = 1: length(y)
         if y(ii) = = y(ii);
             y(ii) = y(ii);

         elseif abs(y(ii)).^2> clipping_threshold
             y(ii) = clipping_threshold.*exp(sqrt
                      (-1)*tx_signal_Ang(ii));

        end
        end
%companding .........................
u = 4;

        tx_lfdma_max = max(abs(y(1: nd*para)));
        tx_lfdma_Abs = abs(y(1: nd*para)); % tx data
                     amplitude

        TxSamples_lfdma1 = tx_lfdma_max*((log10(1+u*(tx_
                          lfdma_Abs./tx_lfdma_max)))/
                          log10(u+1)).*sign(y(1:
                          nd*para));

        tx_ifdma_max = max(abs(y(1: nd*para)));
        tx_ifdma_Abs = abs(y(1: nd*para)); % tx data
                     amplitude
        yy = tx_ifdma_max*((log10(1+u*(tx_ifdma_Abs./
            tx_ifdma_max)))/log10(u+1)).*sign(y
            (1: nd*para));

ich3 = real (yy);
qch3 = imag (yy);

%Guard interval insertion ..........................

[ich4,qch4] = giins(ich3,qch3,fftlen,gilen,nd);
```

```
fftlen2 = fftlen + gilen;

%Attenuation Calculation ...............  .............
spow = sum(ich4.^2+qch4.^2)/nd./para;
attn = 0.5*spow*sr/br*10.^(-ebno/10);
attn = sqrt (attn);
%fading ...............................................
%******* Create Rayleigh fading channel object.*******
%****************** AWGN addition ******************
%Receiver ............................................

%AWGN addition ..................... ...................

  [ich5,qch5] = comb(ich4,qch4,attn);
%perfect fading compensation ..........................

%Guard interval removal ..............................
  [ich6,qch6] = girem (ich5,qch5,fftlen2,gilen,nd);
%FFT ................................................
rx = ich6 + qch6.*j;

         %%%%%%%%%%% Expanding
         rx_lfdma_Abs = abs(rx);
       r_lfdma_max = tx_lfdma_max;

RxSamples_lfdma = (r_lfdma_max/u)*(exp(log10(1+u)*2.30
2585093*rx_lfdma_Abs./r_lfdma_max)-1).*sign(rx);

       rx_ifdma_Abs = abs(rx);
       r_ifdma_max = tx_ifdma_max;

rxx = (r_ifdma_max/u)*(exp(log10(1+u)*2.302585093
*rx_ifdma_Abs./r_ifdma_max)-1).*sign(rx);

ry = fft(rxx);
ich7 = real (ry);
qch7 = imag (ry);

%Demodulation .........................................
  ich10 = ich7./kmod;
  qch10 = qch7./kmod;
  outdemodata = qpskdemod (ich10,qch10,para,nd,m1);

function outdemodata = dct_channel_clipping_companding
(serdata,para,nd,m1,gilen,dctlen,sr,ebno, br)
```

```
%serial to parallel conversion .......................

paradata = reshape(serdata,para,nd*m1);
%QPSK modulation ......................................
[ich,qch] = qpskmod(paradata,para,nd,m1);
% [ich0,qch0] = compoversamp(ich01,qch01,length
                (ich01),Ipoint);
kmod = 1/sqrt(2);
ich1 = ich.*kmod;
qch1 = qch.*kmod;
%IDCT .................................................
x = ich1 + qch1.*j;
y = idct(x);

CR = 4;
  clipping_threshold = (10^(CR/10))*sqrt(mean(ab
                     s(y).^2));
   tx_signal_Ang = angle(y);

   for ii = 1: length(y)
    if y(ii) = = y(ii);
        y(ii) = y(ii);

      elseif abs(y(ii)).^2> clipping_threshold
          y(ii) = clipping_threshold.*exp(sqrt(-1)*tx_
                 signal_Ang(ii));

      end
end
%companding ...............................
u = 4;

        tx_lfdma_max = max(abs(y(1: nd*para)));
        tx_lfdma_Abs = abs(y(1: nd*para)); % tx data
                    amplitude

        TxSamples_lfdma1 = tx_lfdma_max*((log10(1+u*
                         (tx_lfdma_Abs./tx_lfdma_
                         max)))/log10(u+1)).*sign(y
                         (1: nd*para));

        tx_ifdma_max = max(abs(y(1: nd*para)));
        tx_ifdma_Abs = abs(y(1: nd*para)); % tx data
                    amplitude
```

```
        yy = tx_ifdma_max*((log10(1+u*(tx_ifdma_Abs./
             tx_ifdma_max)))/log10(u+1)).*sign(y(1:
             nd*para));

ich2 = real (yy);
qch2 = imag (yy);

%Guard interval insertion ...........................

[ich3,qch3] = giins(ich2,qch2,dctlen,gilen,nd);
dctlen2 = dctlen + gilen;

%Attenuation Calculation ...........................

spow = sum(ich3.^2+qch3.^2)/nd./para;
attn = 0.5*spow*sr/br*10.^(-ebno/10);
attn = sqrt (attn);

%Receiver .........................................

%AWGN addition ....................................

  [ich4,qch4] = comb(ich3,qch3,attn);

%Guard interval removal ...........................
  [ich5,qch5] = girem (ich4,qch4,dctlen2,gilen,nd);
%DCT ..............................................
rx = ich5 + qch5.*j;

        %%%%%%%%%% Expanding
        rx_lfdma_Abs = abs(rx);
        r_lfdma_max = tx_lfdma_max;

RxSamples_lfdma = (r_lfdma_max/u)*(exp(log10(1+u)*2.30
                  2585093*rx_lfdma_Abs./r_lfdma_max)-
                  1).*sign(rx);

        rx_ifdma_Abs = abs(rx);
        r_ifdma_max = tx_ifdma_max;

rxx = (r_ifdma_max/u)*(exp(log10(1+u)*2.302585093
*rx_ifdma_Abs./r_ifdma_max)-1).*sign(rx);

ry = dct(rxx);
ich6 = real (ry);
```

```
qch6 = imag (ry);

%Demodulation .........................................

  ich7 = ich6./kmod;
  qch7 = qch6./kmod;
  outdemodata = qpskdemod (ich7,qch7,para,nd,m1);


function outdemodata = dwt_channel_clipping_companding
(serdata,para,nd,m1,gilen,dctlen,sr,ebno, br)


%serial to parallel conversion ......................

paradata = reshape(serdata,para,nd*m1);
%QPSK modulation ...................................
[ich,qch] = qpskmod(paradata,para,nd,m1);
% [ich0,qch0] = compoversamp(ich01,qch01,length
               (ich01),Ipoint);
kmod = 1/sqrt(2);
ich1 = ich.*kmod;
qch1 = qch.*kmod;
%IDWT ..............................................
x = ich1 + qch1.*j;
y = wavelet('D6',-1,x,'zpd'); % Invert 5 stages
%; % 2D wavelet transform
% R = wavelet('2D CDF 9/7',-2,Y); % Recover X from Y%
Forward transform with 5 stages
CR = 4;
        clipping_threshold = (10^(CR/10))*sqrt(mean(ab
                        s(y).^2));
      tx_signal_Ang = angle(y);
      for ii = 1: length(y)
       if y(ii) = = y(ii);
            y(ii) = y(ii);

        elseif abs(y(ii)).^2> clipping_threshold
             y(ii) = clipping_threshold.*exp(sqrt
                     (-1)*tx_signal_Ang(ii));


        end
end
%companding ...............................
u = 4;
```

```
          tx_lfdma_max = max(abs(y(1: nd*para)));
          tx_lfdma_Abs = abs(y(1: nd*para)); % tx data
                         amplitude
          TxSamples_lfdma1 = tx_lfdma_max*((log10(1+u*
                             (tx_lfdma_Abs./tx_lfdma_
                             max)))/log10(u+1)).*sign
                             (y(1: nd*para));
          tx_ifdma_max = max(abs(y(1: nd*para)));
          tx_ifdma_Abs = abs(y(1: nd*para)); % tx data
                         amplitude
          yy = tx_ifdma_max*((log10(1+u*(tx_ifdma_Abs./
               tx_ifdma_max)))/log10(u+1)).*sign(y(1:
               nd*para));

ich2 = real (yy);
qch2 = imag (yy);

%Guard interval insertion ............................

[ich3,qch3] = giins(ich2,qch2,dctlen,gilen,nd);
dctlen2 = dctlen + gilen;

%Attenuation Calculation ............................

spow = sum(ich3.^2+qch3.^2)/nd./para;
attn = 0.5*spow*sr/br*10.^(-ebno/10);
attn = sqrt (attn);

%Receiver ..........................................
%AWGN addition .....................................
  [ich4,qch4] = comb(ich3,qch3,attn);

%Guard interval removal .............................

  [ich5,qch5] = girem (ich4,qch4,dctlen2,gilen,nd);
%DWT ...............................................
rx = ich5 + qch5.*j;

          %%%%%%%%%% Expanding
          rx_lfdma_Abs = abs(rx);
          r_lfdma_max = tx_lfdma_max;

RxSamples_lfdma = (r_lfdma_max/u)*(exp(log10(1+u)*2.30
                  2585093*rx_lfdma_Abs./r_lfdma_max)
                  -1).*sign(rx);
```

```
      rx_ifdma_Abs = abs(rx);
      r_ifdma_max = tx_ifdma_max;

rxx = (r_ifdma_max/u)*(exp(log10(1+u)*2.302585093
      *rx_ifdma_Abs./r_ifdma_max)-1).*sign(rx);

ry = wavelet('D6',1,rxx,'zpd');; % 2D wavelet transform
% R = wavelet('2D CDF 9/7',-2,Y); % Recover X from Y;
ich6 = real (ry);
qch6 = imag (ry);

%Demodulation ......................................
ich7 = ich6./kmod;
qch7 = qch6./kmod;
outdemodata = qpskdemod (ich7,qch7,para,nd,m1);

function outdemodata = fft_channel_estimation_no_
mapping_clipping(serdata,para,nd,m1,gilen,fftlen,sr,e
bno, br,fd,flat)

paradata = reshape(serdata,para,nd*m1);
%QPSK modulation .....................................
[ich,qch] = qpskmod(paradata,para,nd,m1);
% [ich0,qch0] = compoversamp(ich01,qch01,length
(ich01),Ipoint);
kmod = 1/sqrt(2);
ich1 = ich.*kmod;
qch1 = qch.*kmod;
%channel estimation data generation .................
kndata = zeros(1,fftlen);
kndata0 = 2.*(rand(1,para)<0.5)-1;
kndata(1: para/2) = kndata0(1: para/2);
kndata((para/2)+1: para) = kndata0((para/2)+1: para);
ceich = kndata;
ceqch = zeros(1,para);
%data mapping ........................................
ich2 = [ceich.' ich1];
qch2 = [ceqch.' qch1];
%IFFT ................................................
x = ich2 + qch2.*i;

y = ifft(x);
CR = 4;
      clipping_threshold = (10^(CR/10))*sqrt(mean(ab
                          s(y).^2));
```

```
        tx_signal_Ang = angle(y);
        for ii = 1: length(y)
          if y(ii) = = y(ii);
              y(ii) = y(ii);

          elseif abs(y(ii)).^2> clipping_threshold
              y(ii) = clipping_threshold.*exp(sqrt(-1)*tx_
                      signal_Ang(ii));

          end
      end

ich3 = real (y);
qch3 = imag (y);

%Guard interval insertion ...........................
[ich4,qch4] = giins(ich3,qch3,fftlen,gilen,nd+1);
fftlen2 = fftlen + gilen;

%Attenuation Calculation ..........................
spow = sum(ich4.^2+qch4.^2)/nd./para;
attn = 0.5*spow*sr/br*10.^(-ebno/10);
attn = sqrt (attn);
%fading ...........................................
%******* Create Rayleigh fading channel object.*******
tstp = 1/sr/(fftlen+gilen);
itau = [0,2,3,4];
dlvll = [0,10,20,25];
n0 = [6,7,6,7];
th1 = [0,0,0,0];
itnd1 = [1000,2000,3000,4000];
now1 = 4;
itnd0 = nd*(fftlen+gilen)*20;
[ifade,qfade,ramp,rcos,rsin] = sefade(ich4,qch4,itau,d
lvll,th1,n0,itnd1,now1,length(ich4),tstp,fd,flat);
itnd1 = itnd1+itnd0;
ich4 = ifade;
qch4 = qfade;

%****************** AWGN addition ******************
%Receiver .........................................

%AWGN addition ....................................
[ich5,qch5] = comb(ich4,qch4,attn);
```

```
%perfect fading compensation .........................
  ifade2 = 1./ramp.*(rcos(1,: ).*ich5+rsin
          (1,: ).*qch5);
  qfade2 = 1./ramp.*(-rsin(1,: ).*ich5+rcos
          (1,:).*qch5);
  ich5 = ifade2;
  qch5 = qfade2;

%Guard interval removal ...........................
  [ich6,qch6] = girem (ich5,qch5,fftlen2,gilen,nd+1);
%FFT ...............................................
rx = ich6 + qch6.*j;
ry = fft(rx);
ich7 = real (ry);
qch7 = imag (ry);
%fading compensation by channel estimation symbol ....
ce = 1;
ice0 = ich2(:,ce);
qce0 = qch2(:,ce);
ice1 = ich7(:,ce);
qce1 = qch7(:,ce);
%calculate reverse rotation .........................
iv = real((1./(ice1.^2+qce1.^2)).*(ice0+i.*qce0).
     *(ice1-i.*qce1));
qv = imag((1./(ice1.^2+qce1.^2)).*(ice0+i.*qce0).
     *(ice1-i.*qce1));
%matrix for reverse rotation ........................
ieqv1 = [iv iv iv iv iv iv iv];
qeqv1 = [qv qv qv qv qv qv qv];
%reverse rotation ................ ........
icompen = real((ich7+i.*qch7).*(ieqv1+i.*qeqv1));
qcompen = imag((ich7+i.*qch7).*(ieqv1+i.*qeqv1));
ich7 = icompen;
qch7 = qcompen;
%channel estimation symbol removal ..................
knd = 1;      %number of known channel estimation ofdm
              symbol
ich9 = ich7(:,knd+1: nd+1);
qch9 = qch7(:,knd+1: nd+1);

%Demodulation ......................................
ich10 = ich9./kmod;
qch10 = qch9./kmod;
outdemodata = qpskdemod (ich10,qch10,para,nd,m1);
```

```
function outdemodata = dct_channel_estimation_no_
                       mapping_clipping(serdata,para,n
                       d,m1,gilen,fftlen,sr,ebno,
                       br,fd,flat)

paradata = reshape(serdata,para,nd*m1);

%QPSK modulation ...................................
[ich,qch] = qpskmod(paradata,para,nd,m1);
% [ich0,qch0] = compoversamp(ich01,qch01,length(ich01)
            ,Ipoint);
kmod = 1/sqrt(2);
ich1 = ich.*kmod;
qch1 = qch.*kmod;
%channel estimation data generation ................
kndata = zeros(1,fftlen);
kndata0 = 2.*(rand(1,para)<0.5)-1;
kndata(1: para/2) = kndata0(1: para/2);
kndata((para/2)+1: para) = kndata0((para/2)+1: para);
ceich = kndata;
ceqch = zeros(1,para);
%data mapping .......................................
ich2 = [ceich.' ich1];
qch2 = [ceqch.' qch1];
%IDCT ...............................................
x = ich2 + qch2.*i;

y = idct(x);

CR = 4;

        clipping_threshold = (10^(CR/10))*sqrt(mean(ab
                        s(y).^2));
        tx_signal_Ang = angle(y);
        for ii = 1: length(y)
          if y(ii) = = y(ii);
              y(ii) = y(ii);
          elseif abs(y(ii)).^2> clipping_threshold
            y(ii) = clipping_threshold.*exp(sqrt(-1)*tx_
                    signal_Ang(ii));

        end
end

ich3 = real (y);
```

```
qch3 = imag (y);

%Guard interval insertion ...........................

[ich4,qch4] = giins(ich3,qch3,fftlen,gilen,nd+1);
fftlen2 = fftlen + gilen;

%Attenuation Calculation ...........................

spow = sum(ich4.^2+qch4.^2)/nd./para;
attn = 0.5*spow*sr/br*10.^(-ebno/10);
attn = sqrt (attn);
%fading ............................................
%******* Create Rayleigh fading channel object.*******
tstp = 1/sr/(fftlen+gilen);
itau = [0,2,3,4];
dlvll = [0,10,20,25];
n0 = [6,7,6,7];
th1 = [0,0,0,0];
itnd1 = [1000,2000,3000,4000];
now1 = 4;
itnd0 = nd*(fftlen+gilen)*20;
[ifade,qfade,ramp,rcos,rsin] = sefade(ich4,qch4,itau,d
lvll,th1,n0,itnd1,now1,length(ich4),tstp,fd,flat);
itnd1 = itnd1+itnd0;
ich4 = ifade;
qch4 = qfade;

%****************** AWGN addition ******************
%Receiver .........................................

%AWGN addition ....................................

  [ich5,qch5] = comb(ich4,qch4,attn);
  %perfect fading compensation ...............
................ ..........
ifade2 = 1./ramp.*(rcos(1,: ).*ich5+rsin(1,: ).*qch5);
qfade2 = 1./ramp.*(-rsin(1,: ).*ich5+rcos(1,: ).*qch5);
ich5 = ifade2;
qch5 = qfade2;

%Guard interval removal ...........................
[ich6,qch6] = girem (ich5,qch5,fftlen2,gilen,nd+1);
%DCT ..............................................
rx = ich6 + qch6.*j;
```

```
ry = dct(rx);
ich7 = real (ry);
qch7 = imag (ry);
%fading compensation by channel estimation symbol ....
ce = 1;
ice0 = ich2(:,ce);
qce0 = qch2(:,ce);
ice1 = ich7(:,ce);
qce1 = qch7(:,ce);
%calculate reverse rotation .........................
iv = real((1./(ice1.^2+qce1.^2)).*(ice0+i.*qce0).
*(ice1-i.*qce1));
qv = imag((1./(ice1.^2+qce1.^2)).*(ice0+i.*qce0).
*(ice1-i.*qce1));
%matrix for reverse rotation .........................
ieqv1 = [iv iv iv iv iv iv iv];
qeqv1 = [qv qv qv qv qv qv qv];
%reverse rotation ................ ........
icompen = real((ich7+i.*qch7).*(ieqv1+i.*qeqv1));
qcompen = imag((ich7+i.*qch7).*(ieqv1+i.*qeqv1));
ich7 = icompen;
qch7 = qcompen;
%channel estimation symbol removal ..................
knd = 1;                %number of known channel
                        estimation ofdm symbol
ich9 = ich7(:,knd+1: nd+1);
qch9 = qch7(:,knd+1: nd+1);

%Demodulation .......................................
ich10 = ich9./kmod;
qch10 = qch9./kmod;
outdemodata = qpskdemod (ich10,qch10,para,nd,m1);

function outdemodata = dwt_channel_estimation_no_
mapping_clipping(serdata,para,nd,m1,gilen,fftlen,sr,e
bno, br,fd,flat)

paradata = reshape(serdata,para,nd*m1);
%QPSK modulation .....................................
[ich,qch] = qpskmod(paradata,para,nd,m1);
% [ich0,qch0] = compoversamp(ich01,qch01,length
                (ich01),Ipoint);
kmod = 1/sqrt(2);
ich1 = ich.*kmod;
qch1 = qch.*kmod;
```

```
%channel estimation data generation ..................
kndata = zeros(1,fftlen);
kndata0 = 2.*(rand(1,para)<0.5)-1;
kndata(1: para/2) = kndata0(1: para/2);
kndata((para/2)+1: para) = kndata0((para/2)+1: para);
ceich = kndata;
ceqch = zeros(1,para);
%data mapping ........................................
ich2 = [ceich.' ich1];
qch2 = [ceqch.' qch1];


%IDWT ...............................................
x = ich2 + qch2.*j;
y = wavelet('D6',-1,x,'zpd'); % Invert 5 stages
%; % 2D wavelet transform
% R = wavelet('2D CDF 9/7',-2,Y); % Recover X from Y%
Forward transform with 5 stages
CR = 4;
        clipping_threshold = (10^(CR/10))*sqrt(mean(ab
                               s(y).^2));
        tx_signal_Ang = angle(y);
        for ii = 1: length(y)
              if y(ii) = = y(ii);
                    y(ii) = y(ii);
        elseif abs(y(ii)).^2> clipping_threshold
              y(ii) = clipping_threshold.*exp(sqrt
                    (-1)*tx_signal_Ang(ii));


        end
end

ich3 = real (y);
qch3 = imag (y);
%Guard interval insertion ...........................

[ich4,qch4] = giins(ich3,qch3,fftlen,gilen,nd+1);
fftlen2 = fftlen + gilen;

%Attenuation Calculation ............................

spow = sum(ich4.^2+qch4.^2)/nd./para;
attn = 0.5*spow*sr/br*10.^(-ebno/10);
attn = sqrt (attn);
%fading .............................................
%******* Create Rayleigh fading channel object.*******
```

```
tstp = 1/sr/(fftlen+gilen);
itau = [0,2,3,4];
dlvll = [0,10,20,25];
n0 = [6,7,6,7];
th1 = [0,0,0,0];
itnd1 = [1000,2000,3000,4000];
now1 = 4;
itnd0 = nd*(fftlen+gilen)*20;
[ifade,qfade,ramp,rcos,rsin] = sefade(ich4,qch4,itau,d
lvll,th1,n0,itnd1,now1,length(ich4),tstp,fd,flat);
itnd1 = itnd1+itnd0;
ich4 = ifade;
qch4 = qfade;


%*******************AWGN addition *******************
%Receiver ...........................................

%AWGN addition .......................................
  [ich5,qch5] = comb(ich4,qch4,attn);
  %perfect fading compensation .......................
  ifade2 = 1./ramp.*(rcos(1,: ).*ich5+rsin
          (1,: ).*qch5);
  qfade2 = 1./ramp.*(-rsin(1,: ).*ich5+rcos
          (1,:).*qch5);
  ich5 = ifade2;
  qch5 = qfade2;


%Guard interval removal ..............................
  [ich6,qch6] = girem (ich5,qch5,fftlen2,gilen,nd+1);
%DWT .................................................
rx = ich6 + qch6.*j;
ry = wavelet('D6',1,rx,'zpd');; % 2D wavelet
    transform
% R = wavelet('2D CDF 9/7',-2,Y); % Recover X from Y;
ich7 = real (ry);
qch7 = imag (ry);
%fading compensation by channel estimation symbol ....
ce = 1;
ice0 = ich2(:,ce);
qce0 = qch2(:,ce);
ice1 = ich7(:,ce);
qce1 = qch7(:,ce);
%calculate reverse rotation ..........................
iv = real((1./(ice1.^2+qce1.^2)).*(ice0+i.*qce0).
    *(ice1-i.*qce1));
```

```
qv = imag((1./(ice1.^2+qce1.^2)).*(ice0+i.*qce0).
    *(ice1-i.*qce1));
%matrix for reverse rotation .........................
ieqv1 = [iv iv iv iv iv iv iv];
qeqv1 = [qv qv qv qv qv qv qv];
%reverse rotation ................ ........
icompen = real((ich7+i.*qch7).*(ieqv1+i.*qeqv1));
qcompen = imag((ich7+i.*qch7).*(ieqv1+i.*qeqv1));
ich7 = icompen;
qch7 = qcompen;
%channel estimation symbol removal ..................
knd = 1;      %number of known channel estimation ofdm
symbol
ich9 = ich7(:,knd+1: nd+1);
qch9 = qch7(:,knd+1: nd+1);

%Demodulation ......................................
ich10 = ich9./kmod;
qch10 = qch9./kmod;
outdemodata = qpskdemod (ich10,qch10,para,nd,m1);

function outdemodata = fft_channel_offset_g(serdata,pa
ra,nd,m1,gilen,fftlen,sr,ebno, br,epsilon)

%serial to parallel conversion .....................
paradata = reshape(serdata,para,nd*m1);

%QPSK modulation .....................................
[ich,qch] = qpskmod(paradata,para,nd,m1);

% [ich0,qch0] = compoversamp(ich01,qch01,length
              (ich01),Ipoint);
kmod = 1/sqrt(2);
ich1 = ich.*kmod;
qch1 = qch.*kmod;

%IFFT .................................................
x = ich1 + qch1.*j;
y = ifft(x);
ich2 = real (y);
qch2 = imag (y);

%Guard interval insertion ..........................
[ich3,qch3] = giins1(ich2,qch2,fftlen,gilen,nd);
fftlen2 = fftlen + gilen;
```

```
%Attenuation Calculation ..............................
spow = sum(ich3.^2+qch3.^2)/nd./para;
attn = 0.5*spow*sr/br*10.^(-ebno/10);
attn = sqrt (attn);

%offset ...............................................
n = para/2;
offset = exp(j*2*pi*n*epsilon/para);
i_rx_signal = ich3.*offset;
q_rx_signal = qch3.*offset;
ich3a = i_rx_signal ;
qch3a = q_rx_signal ;

%Receiver ............................................
%AWGN addition .......................................
[ich4,qch4] = comb(ich3a,qch3a,attn);

%Guard interval removal ..............................
[ich5,qch5] = girem1 (ich4,qch4,fftlen2,gilen,nd);

%FFT .................................................
rx = ich5 + qch5.*j;
ry = fft(rx);
ich6 = real (ry);
qch6 = imag (ry);

%Demodulation ........................................
  ich7 = ich6./kmod;
  qch7 = qch6./kmod;
  outdemodata = qpskdemod (ich7,qch7,para,nd,m1);

function outdemodata = dct_channel_offset_g(serdata,pa
ra,nd,m1,gilen,fftlen,sr,ebno, br,epsilon)

%serial to parallel conversion .......................
paradata = reshape(serdata,para,nd*m1);

%QPSK modulation .....................................
[ich,qch] = qpskmod(paradata,para,nd,m1);

% [ich0,qch0] = compoversamp(ich01,qch01,length
                 (ich01),Ipoint);
kmod = 1/sqrt(2);
ich1 = ich.*kmod;
qch1 = qch.*kmod;
```

```
%IDCT ..............................................
x = ich1 + qch1.*j;
y = idct(x);
ich2 = real (y);
qch2 = imag (y);

%Guard interval insertion .........................
[ich3,qch3] = giins1(ich2,qch2,fftlen,gilen,nd);
fftlen2 = fftlen + gilen;

%Attenuation Calculation ...........................
spow = sum(ich3.^2+qch3.^2)/nd./para;
attn = 0.5*spow*sr/br*10.^(-ebno/10);
attn = sqrt (attn);

%offset ............................................
n = para/2;
offset = exp(j*2*pi*n*epsilon/para);
i_rx_signal = ich3.*offset;
q_rx_signal = qch3.*offset;
ich3a = i_rx_signal ;
qch3a = q_rx_signal ;

%Receiver ..........................................
%AWGN addition .....................................
  [ich4,qch4] = comb(ich3a,qch3a,attn);

%Guard interval removal ............................
[ich5,qch5] = girem1 (ich4,qch4,fftlen2,gilen,nd);

%DCT ...............................................
rx = ich5 + qch5.*j;
ry = dct(rx);
ich6 = real (ry);
qch6 = imag (ry);

%Demodulation ......................................
ich7 = ich6./kmod;
qch7 = qch6./kmod;
outdemodata = qpskdemod (ich7,qch7,para,nd,m1);
function outdemodata = dwt_channel_offset_g(serdata,pa
                       ra,nd,m1,gilen,dctlen,sr,ebno,
                       br,epsilon)

%serial to parallel conversion .....................
```

```
paradata = reshape(serdata,para,nd*m1);
%QPSK modulation ......................................
[ich,qch] = qpskmod(paradata,para,nd,m1);
% [ich0,qch0] = compoversamp(ich01,qch01,length
(ich01),Ipoint);
kmod = 1/sqrt(2);
ich1 = ich.*kmod;
qch1 = qch.*kmod;
%IDWT ................................................
x = ich1 + qch1.*j;
y = wavelet('D6',-1,x,'zpd'); % Invert 5 stages

% 2D wavelet transf
% R = wavelet('2D CDF 9/7',-2,Y); % Recover X from Y%
     Forward transform with 5 stages
ich2 = real (y);
qch2 = imag (y);

%Guard interval insertion ...........................
[ich3,qch3] = giins1(ich2,qch2,dctlen,gilen,nd);
dctlen2 = dctlen + gilen;

%Attenuation Calculation ............................
spow = sum(ich3.^2+qch3.^2)/nd./para;
attn = 0.5*spow*sr/br*10.^(-ebno/10);
attn = sqrt (attn);

%offset ..............................................
n = para/2;
offset = exp(j*2*pi*n*epsilon/para);
i_rx_signal = ich3.*offset;
q_rx_signal = qch3.*offset;
ich3a = i_rx_signal ;
qch3a = q_rx_signal ;

%Receiver ............................................
%AWGN addition .......................................
  [ich4,qch4] = comb(ich3a,qch3a,attn);
%Guard interval removal ..............................
  [ich5,qch5] = girem1 (ich4,qch4,dctlen2,gilen,nd);
%DWT .................................................
rx = ich5 + qch5.*j;
ry = wavelet('D6',1,rx,'zpd');; % 2D wavelet transform
% R = wavelet('2D CDF 9/7',-2,Y); % Recover X from Y;
ich6 = real (ry);
```

```
qch6 = imag (ry);

%Demodulation ........................................
ich7 = ich6./kmod;
qch7 = qch6./kmod;
outdemodata = qpskdemod (ich7,qch7,para,nd,m1);

function outdemodata = fft_channel_offset_zg(serdata,p
                       ara,nd,m1,gilen,fftlen,sr,ebno,
                       br,epsilon)

%serial to parallel conversion .......................
paradata = reshape(serdata,para,nd*m1);

%QPSK modulation .....................................
[ich,qch] = qpskmod(paradata,para,nd,m1);

% [ich0,qch0] = compoversamp(ich01,qch01,length
                (ich01),Ipoint);
kmod = 1/sqrt(2);
ich1 = ich.*kmod;
qch1 = qch.*kmod;

%IFFT ................................................
x = ich1 + qch1.*j;
y = ifft(x);
ich2 = real (y);
qch2 = imag (y);

%Guard interval insertion ............................
[ich3,qch3] = giins2(ich2,qch2,fftlen,gilen,nd);
fftlen2 = fftlen + gilen;

%Attenuation Calculation .............................
spow = sum(ich3.^2+qch3.^2)/nd./para;
attn = 0.5*spow*sr/br*10.^(-ebno/10);
attn = sqrt (attn);

%offset ..............................................
n = para/2;
offset = exp(j*2*pi*n*epsilon/para);
i_rx_signal = ich3.*offset;
q_rx_signal = qch3.*offset;
ich3a = i_rx_signal ;
qch3a = q_rx_signal ;
```

```
%Receiver ...............................................
%AWGN addition .........................................
[ich4,qch4] = comb(ich3a,qch3a,attn);

%Guard interval removal ............................
[ich5,qch5] = girem1 (ich4,qch4,fftlen2,gilen,nd);

%FFT ...................................................
rx = ich5 + qch5.*j;
ry = fft(rx);
ich6 = real (ry);
qch6 = imag (ry);

%Demodulation ........................................
  ich7 = ich6./kmod;
  qch7 = qch6./kmod;
  outdemodata = qpskdemod (ich7,qch7,para,nd,m1);

function outdemodata = dct_channel_offset_zg(serdata,p
                    ara,nd,m1,gilen,fftlen,sr,ebno,
                    br,epsilon)

%serial to parallel conversion ......................
paradata = reshape(serdata,para,nd*m1);

%QPSK modulation .....................................
[ich,qch] = qpskmod(paradata,para,nd,m1);

% [ich0,qch0] = compoversamp(ich01,qch01,length
               (ich01),Ipoint);
kmod = 1/sqrt(2);
ich1 = ich.*kmod;
qch1 = qch.*kmod;

%IDCT .................................................
x = ich1 + qch1.*j;
y = idct(x);
ich2 = real (y);
qch2 = imag (y);

%Guard interval insertion ...........................
[ich3,qch3] = giins2(ich2,qch2,fftlen,gilen,nd);
fftlen2 = fftlen + gilen;

%Attenuation Calculation .............................
```

```
spow = sum(ich3.^2+qch3.^2)/nd./para;
attn = 0.5*spow*sr/br*10.^(-ebno/10);
attn = sqrt (attn);

%offset ...........................................
n = para/2;
offset = exp(j*2*pi*n*epsilon/para);
i_rx_signal = ich3.*offset;
q_rx_signal = qch3.*offset;
ich3a = i_rx_signal ;
qch3a = q_rx_signal ;

%Receiver .........................................
%AWGN addition ....................................
  [ich4,qch4] = comb(ich3a,qch3a,attn);

%Guard interval removal ...........................
  [ich5,qch5] = girem1 (ich4,qch4,fftlen2,gilen,nd);

%DCT ..............................................
rx = ich5 + qch5.*j;
ry = dct(rx);
ich6 = real (ry);
qch6 = imag (ry);

%Demodulation .....................................
ich7 = ich6./kmod;
qch7 = qch6./kmod;
outdemodata = qpskdemod (ich7,qch7,para,nd,m1);

function outdemodata = dwt_channel_offset_zg(serdata,p
                       ara,nd,m1,gilen,dctlen,sr,ebno,
                       br,epsilon)

%serial to parallel conversion ......................
paradata = reshape(serdata,para,nd*m1);
%QPSK modulation ....................................
[ich,qch] = qpskmod(paradata,para,nd,m1);
% [ich0,qch0] = compoversamp(ich01,qch01,length
                (ich01),Ipoint);
kmod = 1/sqrt(2);
ich1 = ich.*kmod;
qch1 = qch.*kmod;
%IDWT ..............................................
x = ich1 + qch1.*j;
```

```
y = wavelet('D6',-1,x,'zpd'); % Invert 5 stages


% 2D wavelet transf
% R = wavelet('2D CDF 9/7',-2,Y); % Recover X from Y%
Forward transform with 5 stages
ich2 = real (y);
qch2 = imag (y);


%Guard interval insertion ..........................
[ich3,qch3] = giins2(ich2,qch2,dctlen,gilen,nd);
dctlen2 = dctlen + gilen;


%Attenuation Calculation ...........................
spow = sum(ich3.^2+qch3.^2)/nd./para;
attn = 0.5*spow*sr/br*10.^(-ebno/10);
attn = sqrt (attn);


%offset .............................................
n = para/2;
offset = exp(j*2*pi*n*epsilon/para);
i_rx_signal = ich3.*offset;
q_rx_signal = qch3.*offset;
ich3a = i_rx_signal ;
qch3a = q_rx_signal ;


%Receiver ..........................................
%AWGN addition .....................................
[ich4,qch4] = comb(ich3a,qch3a,attn);
%Guard interval removal ............................
[ich5,qch5] = girem1 (ich4,qch4,dctlen2,gilen,nd);
%DWT ...............................................
rx = ich5 + qch5.*j;
ry = wavelet('D6',1,rx,'zpd');; % 2D wavelet transform
% R = wavelet('2D CDF 9/7',-2,Y); % Recover X from Y;
ich6 = real (ry);
qch6 = imag (ry);


%Demodulation ......................................
ich7 = ich6./kmod;
qch7 = qch6./kmod;
outdemodata = qpskdemod (ich7,qch7,para,nd,m1);
function outdemodata = fft_channel_estimation_no_mappi
                       ng_g(serdata,para,nd,m1,gilen,f
                       ftlen,sr,ebno, br,fd,flat)
```

```
paradata = reshape(serdata,para,nd*m1);
%QPSK modulation ......................................
[ich,qch] = qpskmod(paradata,para,nd,m1);
% [ich0,qch0] = compoversamp(ich01,qch01,length(ich01)
,Ipoint);
kmod = 1/sqrt(2);
ich1 = ich.*kmod;
qch1 = qch.*kmod;
%channel estimation data generation .................
kndata = zeros(1,fftlen);
kndata0 = 2.*(rand(1,para)<0.5)-1;
kndata(1: para/2) = kndata0(1: para/2);
kndata((para/2)+1: para) = kndata0((para/2)+1: para);
ceich = kndata;
ceqch = zeros(1,para);
%data mapping .........................................
ich2 = [ceich.' ich1];
qch2 = [ceqch.' qch1];


%IFFT .................................................
x = ich2 + qch2.*i;
y = ifft(x);
ich3 = real (y);
qch3 = imag (y);


%Guard interval insertion .............................
[ich4,qch4] = giins1(ich3,qch3,fftlen,gilen,nd+1);
fftlen2 = fftlen + gilen;


%Attenuation Calculation ..............................
spow = sum(ich4.^2+qch4.^2)/nd./para;
attn = 0.5*spow*sr/br*10.^(-ebno/10);
attn = sqrt (attn);


%fading ...............................................
%******* Create Rayleigh fading channel object.*******
tstp = 1/sr/(fftlen+gilen);
itau = [0,2,3,4];
dlvl1 = [0,10,20,25];
n0 = [6,7,6,7];
th1 = [0,0,0,0];
itnd1 = [1000,2000,3000,4000];
now1 = 4;
itnd0 = nd*(fftlen+gilen)*20;
```

```
[ifade,qfade,ramp,rcos,rsin] = sefade(ich4,qch4,itau,d
     lvll,th1,n0,itnd1,now1,length(ich4),tstp,fd,flat);
itnd1 = itnd1+itnd0;
ich4 = ifade;
qch4 = qfade;


%******************* AWGN addition *******************
%Receiver .............................................
%AWGN addition ........................................
  [ich5,qch5] = comb(ich4,qch4,attn);
  %perfect fading compensation ..............
  ifade2 = 1./ramp.*(rcos(1,: ).*ich5+rsin
          (1,: ).*qch5);
  qfade2 = 1./ramp.*(-rsin(1,: ).*ich5+rcos
          (1,: ).*qch5);
  ich5 = ifade2;
  qch5 = qfade2;


%Guard interval removal ...............................
  [ich6,qch6] = girem1 (ich5,qch5,fftlen2,gilen,nd+1);
%FFT ..................................................
rx = ich6 + qch6.*j;
ry = fft(rx);
ich7 = real (ry);
qch7 = imag (ry);
%fading compensation by channel estimation symbol ....
ce = 1;
ice0 = ich2(:,ce);
qce0 = qch2(:,ce);
ice1 = ich7(:,ce);
qce1 = qch7(:,ce);
%calculate reverse rotation .........................
iv = real((1./(ice1.^2+qce1.^2)).*(ice0+i.*qce0).
     *(ice1-i.*qce1));
qv = imag((1./(ice1.^2+qce1.^2)).*(ice0+i.*qce0).
     *(ice1-i.*qce1));
%matrix for reverse rotation ......................
ieqv1 = [iv iv iv iv iv iv iv];
qeqv1 = [qv qv qv qv qv qv qv];
%reverse rotation ......................
icompen = real((ich7+i.*qch7).*(ieqv1+i.*qeqv1));
qcompen = imag((ich7+i.*qch7).*(ieqv1+i.*qeqv1));
ich7 = icompen;
qch7 = qcompen;
%channel estimation symbol removal ..................
```

```
knd = 1;   %number of known channel estimation ofdm
symbol
ich9 = ich7(:,knd+1: nd+1);
qch9 = qch7(:,knd+1: nd+1);

%Demodulation ........................................
ich10 = ich9./kmod;
qch10 = qch9./kmod;
outdemodata = qpskdemod (ich10,qch10,para,nd,m1);

function outdemodata = dct_channel_estimation_no_mappi
                       ng_g(serdata,para,nd,m1,gilen,f
                       ftlen,sr,ebno, br,fd,flat)

paradata = reshape(serdata,para,nd*m1);
%QPSK modulation ......................................
[ich,qch] = qpskmod(paradata,para,nd,m1);
% [ich0,qch0] = compoversamp(ich01,qch01,length
               (ich01),Ipoint);
kmod = 1/sqrt(2);
ich1 = ich.*kmod;
qch1 = qch.*kmod;
%channel estimation data generation ..................
kndata = zeros(1,fftlen);
kndata0 = 2.*(rand(1,para)<0.5)-1;
kndata(1: para/2) = kndata0(1: para/2);
kndata((para/2)+1: para) = kndata0((para/2)+1: para);
ceich = kndata;
ceqch = zeros(1,para);
%data mapping .........................................
ich2 = [ceich.' ich1];
qch2 = [ceqch.' qch1];

%IDCT .................................................
x = ich2 + qch2.*i;
y = idct(x);
ich3 = real (y);
qch3 = imag (y);

%Guard interval insertion .............................

[ich4,qch4] = giins1(ich3,qch3,fftlen,gilen,nd+1);
fftlen2 = fftlen + gilen;

%Attenuation Calculation ..............................
```

```
spow = sum(ich4.^2+qch4.^2)/nd./para;
attn = 0.5*spow*sr/br*10.^(-ebno/10);
attn = sqrt (attn);
%fading ..............................................
%******* Create Rayleigh fading channel object.*******
tstp = 1/sr/(fftlen+gilen);
itau = [0,2,3,4];
dlvll = [0,10,20,25];
n0 = [6,7,6,7];
th1 = [0,0,0,0];
itnd1 = [1000,2000,3000,4000];
now1 = 4;
itnd0 = nd*(fftlen+gilen)*20;
[ifade,qfade,ramp,rcos,rsin] = sefade(ich4,qch4,itau,d
lvll,th1,n0,itnd1,now1,length(ich4),tstp,fd,flat);
itnd1 = itnd1+itnd0;
ich4 = ifade;
qch4 = qfade;


%***************** AWGN addition *******************
%Receiver ...........................................

%AWGN addition .......................................
  [ich5,qch5] = comb(ich4,qch4,attn);
  %perfect fading compensation .......................
  ifade2 = 1./ramp.*(rcos(1,: ).*ich5+rsin
          (1,: ).*qch5);
  qfade2 = 1./ramp.*(-rsin(1,: ).*ich5+rcos
          (1,: ).*qch5);
  ich5 = ifade2;
  qch5 = qfade2;


%Guard interval removal ............................
  [ich6,qch6] = girem1 (ich5,qch5,fftlen2,gilen,nd+1);
%DCT ................................................
rx = ich6 + qch6.*j;
ry = dct(rx);
ich7 = real (ry);
qch7 = imag (ry);
%fading compensation by channel estimation symbol ....
ce = 1;
ice0 = ich2(:,ce);
qce0 = qch2(:,ce);
ice1 = ich7(:,ce);
qce1 = qch7(:,ce);
```

```
%calculate reverse rotation ...........................
iv = real((1./(ice1.^2+qce1.^2)).*(ice0+i.*qce0).
*(ice1-i.*qce1));
qv = imag((1./(ice1.^2+qce1.^2)).*(ice0+i.*qce0).
*(ice1-i.*qce1));
%matrix for reverse rotation ...........................
ieqv1 = [iv iv iv iv iv iv iv];
qeqv1 = [qv qv qv qv qv qv qv];
%reverse rotation .........................
icompen = real((ich7+i.*qch7).*(ieqv1+i.*qeqv1));
qcompen = imag((ich7+i.*qch7).*(ieqv1+i.*qeqv1));
ich7 = icompen;
qch7 = qcompen;
%channel estimation symbol removal ...................
knd = 1;      %number of known channel estimation ofdm
symbol
ich9 = ich7(:,knd+1: nd+1);
qch9 = qch7(:,knd+1: nd+1);

%Demodulation .........................................
ich10 = ich9./kmod;
qch10 = qch9./kmod;
outdemodata = qpskdemod (ich10,qch10,para,nd,m1);

function outdemodata = dwt_channel_estimation_no_mappi
                       ng_g(serdata,para,nd,m1,gilen,f
                       ftlen,sr,ebno, br,fd,flat)

paradata = reshape(serdata,para,nd*m1);
%QPSK modulation .......................................
[ich,qch] = qpskmod(paradata,para,nd,m1);
% [ich0,qch0] = compoversamp(ich01,qch01,length(ich01)
,Ipoint);
kmod = 1/sqrt(2);
ich1 = ich.*kmod;
qch1 = qch.*kmod;
%channel estimation data generation .................
kndata = zeros(1,fftlen);
kndata0 = 2.*(rand(1,para)<0.5)-1;
kndata(1: para/2) = kndata0(1: para/2);
kndata((para/2)+1: para) = kndata0((para/2)+1: para);
ceich = kndata;
ceqch = zeros(1,para);
%data mapping .........................................
ich2 = [ceich.' ich1];
```

```
qch2 = [ceqch.' qch1];

%IDWT ...............................................
x = ich2 + qch2.*j;
y = wavelet('D6',-1,x,'zpd'); % Invert 5 stages
%; % 2D wavelet transform
% R = wavelet('2D CDF 9/7',-2,Y); % Recover X from Y%
Forward transform with 5 stages

ich3 = real (y);
qch3 = imag (y);

%Guard interval insertion ...........................
[ich4,qch4] = giins1(ich3,qch3,fftlen,gilen,nd+1);
fftlen2 = fftlen + gilen;

%Attenuation Calculation .............................

spow = sum(ich4.^2+qch4.^2)/nd./para;
attn = 0.5*spow*sr/br*10.^(-ebno/10);
attn = sqrt (attn);
%fading ..............................................
%******* Create Rayleigh fading channel object.*******
tstp = 1/sr/(fftlen+gilen);
itau = [0,2,3,4];
dlvl1 = [0,10,20,25];
n0 = [6,7,6,7];
th1 = [0,0,0,0];
itnd1 = [1000,2000,3000,4000];
now1 = 4;
itnd0 = nd*(fftlen+gilen)*20;
[ifade,qfade,ramp,rcos,rsin] = sefade(ich4,qch4,itau,d
lvl1,th1,n0,itnd1,now1,length(ich4),tstp,fd,flat);
itnd1 = itnd1+itnd0;
ich4 = ifade;
qch4 = qfade;

%***************** AWGN addition ********************
%Receiver ............................................
%AWGN addition .......................................
  [ich5,qch5] = comb(ich4,qch4,attn);
  %perfect fading compensation .......................
  ifade2 = 1./ramp.*(rcos(1,: ).*ich5+rsin
          (1,: ).*qch5);
```

```
qfade2 = 1./ramp.*(-rsin(1,: ).*ich5+rcos
        (1,: ).*qch5);
ich5 = ifade2;
qch5 = qfade2;

%Guard interval removal ............................
[ich6,qch6] = girem1 (ich5,qch5,fftlen2,gilen,nd+1);
%DWT .................................................
rx = ich6 + qch6.*j;
ry = wavelet('D6',1,rx,'zpd');; % 2D wavelet
                                 transform
% R = wavelet('2D CDF 9/7',-2,Y); % Recover X from Y;
ich7 = real (ry);
qch7 = imag (ry);

%fading compensation by channel estimation symbol ....
ce = 1;
ice0 = ich2(:,ce);
qce0 = qch2(:,ce);
ice1 = ich7(:,ce);
qce1 = qch7(:,ce);
%calculate reverse rotation .........................
iv = real((1./(ice1.^2+qce1.^2)).*(ice0+i.*qce0).
*(ice1-i.*qce1));
qv = imag((1./(ice1.^2+qce1.^2)).*(ice0+i.*qce0).
*(ice1-i.*qce1));
%matrix for reverse rotation .........................
ieqv1 = [iv iv iv iv iv iv iv];
qeqv1 = [qv qv qv qv qv qv qv];
%reverse rotation ........................
icompen = real((ich7+i.*qch7).*(ieqv1+i.*qeqv1));
qcompen = imag((ich7+i.*qch7).*(ieqv1+i.*qeqv1));
ich7 = icompen;
qch7 = qcompen;
%channel estimation symbol removal ..................
knd = 1;     %number of known channel estimation ofdm
symbol
ich9 = ich7(:,knd+1: nd+1);
qch9 = qch7(:,knd+1: nd+1);

%Demodulation ........................................
ich10 = ich9./kmod;
qch10 = qch9./kmod;
outdemodata = qpskdemod (ich10,qch10,para,nd,m1);
```

```
function outdemodata = fft_channel_estimation_no_map-
                       ping_zg(serdata,para,nd,m1,gile
                       n,fftlen,sr,ebno, br,fd,flat)

paradata = reshape(serdata,para,nd*m1);
%QPSK modulation ......................................
[ich,qch] = qpskmod(paradata,para,nd,m1);
% [ich0,qch0] = compoversamp(ich01,qch01,length
(ich01),Ipoint);
kmod = 1/sqrt(2);
ich1 = ich.*kmod;
qch1 = qch.*kmod;
%channel estimation data generation ..................
kndata = zeros(1,fftlen);
kndata0 = 2.*(rand(1,para)<0.5)-1;
kndata(1: para/2) = kndata0(1: para/2);
kndata((para/2)+1: para) = kndata0((para/2)+1: para);
ceich = kndata;
ceqch = zeros(1,para);
%data mapping ......................................
ich2 = [ceich.' ich1];
qch2 = [ceqch.' qch1];

%IFFT ...............................................
x = ich2 + qch2.*i;
y = ifft(x);
ich3 = real (y);
qch3 = imag (y);

%Guard interval insertion ...........................
[ich4,qch4] = giins2(ich3,qch3,fftlen,gilen,nd+1);
fftlen2 = fftlen + gilen;

%Attenuation Calculation ............................

spow = sum(ich4.^2+qch4.^2)/nd./para;
attn = 0.5*spow*sr/br*10.^(-ebno/10);
attn = sqrt (attn);
%fading .............................................
%*******Create Rayleigh fading channel object.********
tstp = 1/sr/(fftlen+gilen);
itau = [0,2,3,4];
dlvll = [0,10,20,25];
n0 = [6,7,6,7];
th1 = [0,0,0,0];
```

```
itnd1 = [1000,2000,3000,4000];
now1 = 4;
itnd0 = nd*(fftlen+gilen)*20;
[ifade,qfade,ramp,rcos,rsin] = sefade(ich4,qch4,itau,d
lvll,th1,n0,itnd1,now1,length(ich4),tstp,fd,flat);
itnd1 = itnd1+itnd0;
ich4 = ifade;
qch4 = qfade;

%**************** AWGN addition ********************
%Receiver .........................................
%AWGN addition .....................................

  [ich5,qch5] = comb(ich4,qch4,attn);
  %perfect fading compensation ......................
  ifade2 = 1./ramp.*(rcos(1,: ).
          *ich5+rsin(1,: ).*qch5);
  qfade2 = 1./ramp.*(-rsin(1,: ).*ich5+rcos
                              (1,:).*qch5);
  ich5 = ifade2;
  qch5 = qfade2;

%Guard interval removal ............................
  [ich6,qch6] = girem1 (ich5,qch5,fftlen2,gilen,nd+1);
%FFT ...............................................
rx = ich6 + qch6.*j;
ry = fft(rx);
ich7 = real (ry);
qch7 = imag (ry);
%fading compensation by channel estimation symbol ....
ce = 1;
ice0 = ich2(:,ce);
qce0 = qch2(:,ce);
ice1 = ich7(:,ce);
qce1 = qch7(:,ce);
%calculate reverse rotation .........................
iv = real((1./(ice1.^2+qce1.^2)).*(ice0+i.*qce0).
      *(ice1-i.*qce1));
qv = imag((1./(ice1.^2+qce1.^2)).*(ice0+i.*qce0).
      *(ice1-i.*qce1));
%matrix for reverse rotation .......................
ieqv1 = [iv iv iv iv iv iv iv];
qeqv1 = [qv qv qv qv qv qv qv];
%reverse rotation ............... ........
icompen = real((ich7+i.*qch7).*(ieqv1+i.*qeqv1));
```

```
qcompen = imag((ich7+i.*qch7).*(ieqv1+i.*qeqv1));
ich7 = icompen;
qch7 = qcompen;
%channel estimation symbol removal ..................
knd = 1;      %number of known channel estimation ofdm
              symbol
ich9 = ich7(:,knd+1: nd+1);
qch9 = qch7(:,knd+1: nd+1);

%Demodulation ........................................
ich10 = ich9./kmod;
qch10 = qch9./kmod;
outdemodata = qpskdemod (ich10,qch10,para,nd,m1);

function outdemodata = dct_channel_estimation_no_map-
                       ping_zg(serdata,para,nd,m1,gile
                       n,fftlen,sr,ebno,  br,fd,flat)

paradata = reshape(serdata,para,nd*m1);
%QPSK modulation .....................................
[ich,qch] = qpskmod(paradata,para,nd,m1);
% [ich0,qch0] = compoversamp(ich01,qch01,length
               (ich01),Ipoint);
kmod = 1/sqrt(2);
ich1 = ich.*kmod;
qch1 = qch.*kmod;
%channel estimation data generation .................
kndata = zeros(1,fftlen);
kndata0 = 2.*(rand(1,para)<0.5)-1;
kndata(1: para/2) = kndata0(1: para/2);
kndata((para/2)+1: para) = kndata0((para/2)+1: para);
ceich = kndata;
ceqch = zeros(1,para);
%data mapping ........................................
ich2 = [ceich.' ich1];
qch2 = [ceqch.' qch1];

%IDCT ................................................
x = ich2 + qch2.*i;
y = idct(x);
ich3 = real (y);
qch3 = imag (y);
%Guard interval insertion ...........................
[ich4,qch4] = giins2(ich3,qch3,fftlen,gilen,nd+1);
fftlen2 = fftlen + gilen;
```

```
%Attenuation Calculation ...........................
spow = sum(ich4.^2+qch4.^2)/nd./para;
attn = 0.5*spow*sr/br*10.^(-ebno/10);
attn = sqrt (attn);
%fading ............................................
%******* Create Rayleigh fading channel object.*******
tstp = 1/sr/(fftlen+gilen);
itau = [0,2,3,4];
dlvll = [0,10,20,25];
n0 = [6,7,6,7];
th1 = [0,0,0,0];
itnd1 = [1000,2000,3000,4000];
now1 = 4;
itnd0 = nd*(fftlen+gilen)*20;
[ifade,qfade,ramp,rcos,rsin] = sefade(ich4,qch4,itau,d
lvll,th1,n0,itnd1,now1,length(ich4),tstp,fd,flat);
itnd1 = itnd1+itnd0;
ich4 = ifade;
qch4 = qfade;


%****************** AWGN addition ******************
%Receiver .........................................

%AWGN addition ....................................

  [ich5,qch5] = comb(ich4,qch4,attn);
  %perfect fading compensation ......................
  ifade2 = 1./ramp.*(rcos(1,: ).*ich5+rsin
                              (1,: ).*qch5);
  qfade2 = 1./ramp.*(-rsin(1,: ).*ich5+rcos
                              (1,: ).*qch5);
  ich5 = ifade2;
  qch5 = qfade2;

%Guard interval removal ............................
  [ich6,qch6] = girem1 (ich5,qch5,fftlen2,gilen,nd+1);
%DCT ...............................................
rx = ich6 + qch6.*j;
ry = dct(rx);
ich7 = real (ry);
qch7 = imag (ry);
%fading compensation by channel estimation symbol ....
ce = 1;
ice0 = ich2(:,ce);
qce0 = qch2(:,ce);
```

```
ice1 = ich7(:,ce);
qce1 = qch7(:,ce);
%calculate reverse rotation ........................
iv = real((1./(ice1.^2+qce1.^2)).*(ice0+i.*qce0).
*(ice1-i.*qce1));
qv = imag((1./(ice1.^2+qce1.^2)).*(ice0+i.*qce0).
*(ice1-i.*qce1));
%matrix for reverse rotation ........................
ieqv1 = [iv iv iv iv iv iv iv];
qeqv1 = [qv qv qv qv qv qv qv];
%reverse rotation ............... ........
icompen = real((ich7+i.*qch7).*(ieqv1+i.*qeqv1));
qcompen = imag((ich7+i.*qch7).*(ieqv1+i.*qeqv1));
ich7 = icompen;
qch7 = qcompen;
%channel estimation symbol removal ..................
knd = 1;   %number of known channel estimation ofdm
           symbol
ich9 = ich7(:,knd+1: nd+1);
qch9 = qch7(:,knd+1: nd+1);

%Demodulation .......................................
ich10 = ich9./kmod;
qch10 = qch9./kmod;
outdemodata = qpskdemod (ich10,qch10,para,nd,m1);

function outdemodata = dwt_channel_estimation_no_map-
                       ping_zg(serdata,para,nd,m1,gile
                       n,fftlen,sr,ebno, br,fd,flat)

paradata = reshape(serdata,para,nd*m1);
%QPSK modulation ....................................
[ich,qch] = qpskmod(paradata,para,nd,m1);
% [ich0,qch0] = compoversamp(ich01,qch01,length(ich01)
,Ipoint);
kmod = 1/sqrt(2);
ich1 = ich.*kmod;
qch1 = qch.*kmod;
%channel estimation data generation ................
kndata = zeros(1,fftlen);
kndata0 = 2.*(rand(1,para)<0.5)-1;
kndata(1: para/2) = kndata0(1: para/2);
kndata((para/2)+1: para) = kndata0((para/2)+1: para);
ceich = kndata;
ceqch = zeros(1,para);
```

```
%data mapping ........................................
ich2 = [ceich.' ich1];
qch2 = [ceqch.' qch1];


%IDWT ...............................................
x = ich2 + qch2.*j;
y = wavelet('D6',-1,x,'zpd'); % Invert 5 stages
%; % 2D wavelet transform
% R = wavelet('2D CDF 9/7',-2,Y); % Recover X from Y%
Forward transform with 5 stages

ich3 = real (y);
qch3 = imag (y);


%Guard interval insertion ...........................
[ich4,qch4] = giins2(ich3,qch3,fftlen,gilen,nd+1);
fftlen2 = fftlen + gilen;


%Attenuation Calculation ............................
spow = sum(ich4.^2+qch4.^2)/nd./para;
attn = 0.5*spow*sr/br*10.^(-ebno/10);
attn = sqrt (attn);
%fading .............................................
%****** Create Rayleigh fading channel object.********
tstp = 1/sr/(fftlen+gilen);
itau = [0,2,3,4];
dlvll = [0,10,20,25];
n0 = [6,7,6,7];
th1 = [0,0,0,0];
itnd1 = [1000,2000,3000,4000];
now1 = 4;
itnd0 = nd*(fftlen+gilen)*20;
[ifade,qfade,ramp,rcos,rsin] = sefade(ich4,qch4,itau,d
lvll,th1,n0,itnd1,now1,length(ich4),tstp,fd,flat);
itnd1 = itnd1+itnd0;
ich4 = ifade;
qch4 = qfade;


%****************** AWGN addition ******************
%Receiver ...........................................

%AWGN addition ......................................

  [ich5,qch5] = comb(ich4,qch4,attn);
  %perfect fading compensation .......................
```

```
  ifade2 = 1./ramp.*(rcos(1,: ).*ich5+rsin
         (1,: ).*qch5);
  qfade2 = 1./ramp.*(-rsin(1,: ).*ich5+rcos
         (1,: ).*qch5);
ich5 = ifade2;
qch5 = qfade2;

%Guard interval removal ..............................
  [ich6,qch6] = girem1 (ich5,qch5,fftlen2,gilen,nd+1);
%DWT ................................................
rx = ich6 + qch6.*j;
ry = wavelet('D6',1,rx,'zpd');; % 2D wavelet transform
% R = wavelet('2D CDF 9/7',-2,Y); % Recover X from Y;
ich7 = real (ry);
qch7 = imag (ry);

%fading compensation by channel estimation symbol ....
ce = 1;
ice0 = ich2(:,ce);
qce0 = qch2(:,ce);
ice1 = ich7(:,ce);
qce1 = qch7(:,ce);
%calculate reverse rotation .........................
iv = real((1./(ice1.^2+qce1.^2)).*(ice0+i.*qce0).
*(ice1-i.*qce1));
qv = imag((1./(ice1.^2+qce1.^2)).*(ice0+i.*qce0).
*(ice1-i.*qce1));
%matrix for reverse rotation .........................
ieqv1 = [iv iv iv iv iv iv iv];
qeqv1 = [qv qv qv qv qv qv qv];
%reverse rotation ............... ........
icompen = real((ich7+i.*qch7).*(ieqv1+i.*qeqv1));
qcompen = imag((ich7+i.*qch7).*(ieqv1+i.*qeqv1));
ich7 = icompen;
qch7 = qcompen;
%channel estimation symbol removal ..................
knd = 1; %number of known channel estimation ofdm
symbol
ich9 = ich7(:,knd+1: nd+1);
qch9 = qch7(:,knd+1: nd+1);

%Demodulation .......................................
ich10 = ich9./kmod;
qch10 = qch9./kmod;
outdemodata = qpskdemod (ich10,qch10,para,nd,m1);
```

```
function [iout,qout] = qpskmod(paradata,para,nd,m1)

%**************** variables **********************
% paradata : input data (para-by-nd matrix)
% iout : output Ich data
% qout : output Qch data
% para : Number of parallel channels
% nd : Number of data
% m1 : Number of modulation levels
%      (QPSK = 2 16QAM = 4)

%********************************************************

m2 = m1./2;

paradata2 = paradata.*2-1;
count2 = 0;
for jjj = 1: nd
      isi = zeros (para,1);
      isq = zeros (para,1);
      for ii = 1 : m2
            isi = isi + 2.^(m2 - ii).*paradata2
                  ((1: para),ii+count2) ;
            isq = isq + 2.^(m2 - ii).*paradata2
                  ((1: para),m2+ii+count2) ;
      end
            iout((1: para),jjj) = isi;
            qout((1: para),jjj) = isq;
            count2 = count2 + m1;
end
%******************** End of file ********************

function [demodata] = qpskdemod(idata,
                       qdata,para,nd,m1)

%**************** variables **********************

% idata : input Ich data
% qdata : input Qch data
% demodata : demodulated data
% para : Number of parallel channels
% nd : Number of data
% m1 : Number of modulation levels
%      (QPSK = 2 16QAM = 4)
%********************************************************
```

```
demodata = zeros (para,m1*nd);
demodata ((1: para), (1: m1 : m1*nd-1)) =
idata((1: para), (1: nd)) > = 0;
demodata ((1: para), (2: m1 : m1*nd)) =
qdata((1: para), (1: nd)) > = 0;


%****************** End of file *********************

function varargout = wavelet(WaveletName,Level,X,Ext,
                      Dim)
%WAVELET Discrete wavelet transform.
% Y = WAVELET(W,L,X) computes the L-stage discrete
  wavelet transform
% (DWT) of signal X using wavelet W. The length of X
  must be
% divisible by 2^L. For the inverse transform,
  WAVELET(W,-L,X)
% inverts L stages. Choices for W are
% 'Haar'                       Haar
% 'D1','D2','D3','D4','D5','D6' Daubechies'
% 'Sym1','Sym2','Sym3','Sym4','Sym5','Sym6' Symlets
% 'Coif1','Coif2'              Coiflets
% 'BCoif1'                     Coiflet-like [2]
% 'Spline Nr.Nd' (or 'bior Nr.Nd') for      Splines
% Nr = 0, Nd = 0,1,2,3,4,5,6,7, or 8
% Nr = 1, Nd = 0,1,3,5, or 7
% Nr = 2, Nd = 0,1,2,4,6, or 8
% Nr = 3, Nd = 0,1,3,5, or 7
% Nr = 4, Nd = 0,1,2,4,6, or 8
% Nr = 5, Nd = 0,1,3, or 5
% 'RSpline Nr.Nd' for the same Nr.Nd pairs Reverse
  splines
% 'S+P (2,2)','S+P (4,2)','S+P (6,2)',       S+P
  wavelets [3]
% 'S+P (4,4)','S+P (2+2,2)'
% 'TT'                         "Two-Ten" [5]
% 'LC 5/3','LC 2/6','LC 9/7-M','LC 2/10', Low
  Complexity [1]
% 'LC 5/11-C','LC 5/11-A','LC 6/14',
% 'LC 13/7-T','LC 13/7-C'
% 'Le Gall 5/3','CDF 9/7'      JPEG2000 [7]
% 'V9/3'                       Visual [8]
% 'Lazy'                       Lazy wavelet
% Case and spaces are ignored in wavelet names, for
  example, 'Sym4'
```

```
% may also be written as 'sym 4'. Some wavelets have
  multiple names,
% 'D1', 'Sym1', and 'Spline 1.1' are aliases of the
  Haar wavelet.
%
% WAVELET(W) displays information about wavelet W and
  plots the
% primal and dual scaling and wavelet functions.
%
% For 2D transforms, prefix W with '2D'. For example,
  '2D S+P (2,2)'
% specifies a 2D (tensor) transform with the S+P
  (2,2) wavelet.
% 2D transforms require that X is either MxN or MxNxP
  where M and N
% are divisible by 2^L.
%
% WAVELET(W,L,X,EXT) specifies boundary handling EXT.
  Choices are
% 'sym' Symmetric extension (same as 'wsws')
% 'asym' Antisymmetric extension, whole-point
  antisymmetry
% 'zpd' Zero-padding
% 'per' Periodic extension
% 'sp0' Constant extrapolation
%
% Various symmetric extensions are supported:
% 'wsws' Whole-point symmetry (WS) on both boundaries
% 'hshs' Half-point symmetry (HS) on both boundaries
% 'wshs' WS left boundary, HS right boundary
% 'hsws' HS left boundary, WS right boundary
%
% Antisymmetric boundary handling is used by default,
  EXT = 'asym'.
%
% WAVELET(...,DIM) operates along dimension DIM.
%
% [H1,G1,H2,G2] = WAVELET(W,'filters') returns the
  filters
% associated with wavelet transform W. Each filter is
  represented
% by a cell array where the first cell contains an
  array of
% coefficients and the second cell contains a scalar
  of the leading
```

```
% Z-power.
%
% [X,PHI1] = WAVELET(W,'phi1') returns an
  approximation of the
% scaling function associated with wavelet transform W.
% [X,PHI1] = WAVELET(W,'phi1',N) approximates the
  scaling function
% with resolution 2^-N. Similarly,
% [X,PSI1] = WAVELET(W,'psi1',...),
% [X,PHI2] = WAVELET(W,'phi2',...),
% and [X,PSI2] = WAVELET(W,'psi2',...) return
  approximations of the
% wavelet function, dual scaling function, and dual
  wavelet function.
%
% Wavelet transforms are implemented using the lifting
  scheme [4].
% For general background on wavelets, see for example
  [6].
%
%
% Examples:
% % Display information about the S+P (4,4) wavelet
% wavelet('S+P (4,4)');
%
% % Plot a wavelet decomposition
% t = linspace(0,1,256);
% X = exp(-t) + sqrt(t - 0.3).*(t > 0.3) - 0.2*
(t > 0.6);
% wavelet('RSpline 3.1',3,X); % Plot the decomposition
  of X
%
% % Sym4 with periodic boundaries
% Y = wavelet('Sym4',5,X,'per'); % Forward transform
  with 5 stages
% R = wavelet('Sym4',-5,Y,'per'); % Invert 5 stages
%
% % 2D transform on an image
% t = linspace(-1,1,128); [x,y] = meshgrid(t,t);
% X = ((x+1).*(x-1) - (y+1).*(y-1)) + real(sqrt(0.4 -
x.^2 - y.^2));
% Y = wavelet('2D CDF 9/7',2,X); % 2D wavelet transform
% R = wavelet('2D CDF 9/7',-2,Y); % Recover X from Y
% imagesc(abs(Y).^0.2); colormap(gray); axis image;
%
```

```
% % Plot the Daubechies 2 scaling function
% [x,phi] = wavelet('D2','phi');
% plot(x,phi);
%
% References:
% [1] M. Adams and F. Kossentini. "Reversible Integer-
    to-Integer Wavelet Transforms for Image Compression."
    IEEE Trans. on Image Proc., vol. 9, no. 6, Jun. 2000.
%
% [2] M. Antonini, M. Barlaud, P. Mathieu, and
    I. Daubechies. "Image Coding using Wavelet
    Transforms." IEEE Trans. Image Processing,vol. 1,
    pp. 205-220, 1992.
%
% [3] R. Calderbank, I. Daubechies, W. Sweldens, and
    Boon-Lock Yeo."Lossless Image Compression using
    Integer to Integer Wavelet Transforms." ICIP IEEE
    Press, vol. 1, pp. 596-599. 1997.
%
% [4] I. Daubechies and W. Sweldens. "Factoring Wavelet
    Transforms into Lifting Steps." 1996.
%
% [5] D. Le Gall and A. Tabatabai. "Subband Coding of
    Digital Images Using Symmetric Short Kernel Filters
    and Arithmetic Coding Techniques." ICASSP'88,
    pp.761-765, 1988.
%
% [6] S. Mallat. "A Wavelet Tour of Signal Processing."
    Academic Press, 1999.
%
% [7] M. Unser and T. Blu. "Mathematical Properties of
    the JPEG2000 Wavelet Filters." IEEE Trans. on Image
    Proc., vol. 12, no. 9, Sep. 2003.
%
% [8] Qinghai Wang and Yulong Mo. "Choice of Wavelet
    Base in JPEG2000." Computer Engineering, vol. 30,
    no. 23, Dec. 2004.
% Pascal Getreuer 2005-2006

if nargin < 1, error('Not enough input arguments.'); end
if ~ischar(WaveletName), error('Invalid wavelet
name.'); end

% Get a lifting scheme sequence for the specified
wavelet
```

```
Flag1D = isempty(findstr(lower(WaveletName),'2d'));
[Seq,ScaleS,ScaleD,Family] = getwavelet(WaveletName);

if isempty(Seq)
  error(['Unknown wavelet, ''',WaveletName,'''.']);
end
if nargin < 2, Level = ''; end
if ischar(Level)
   [h1,g1] = seq2hg(Seq,ScaleS,ScaleD,0);
   [h2,g2] = seq2hg(Seq,ScaleS,ScaleD,1);

   if strcmpi(Level,'filters')
      varargout = {h1,g1,h2,g2};
   else
   if nargin < 3, X = 6; end

   switch lower(Level)
   case {'phi1','phi'}
   [x1,phi] = cascade(h1,g1,pow2(-X));
   varargout = {x1,phi};
   case {'psi1','psi'}
   [x1,phi,x2,psi] = cascade(h1,g1,pow2(-X));
   varargout = {x2,psi};
   case 'phi2'
   [x1,phi] = cascade(h2,g2,pow2(-X));
   varargout = {x1,phi};
   case 'psi2'
   [x1,phi,x2,psi] = cascade(h2,g2,pow2(-X));
   varargout = {x2,psi};
   case ''
   fprintf('\n%s wavelet ''%s'' ',Family,WaveletName);
   if all(abs([norm(h1{1}),norm(h2{1})] - 1) < 1e-11)
   fprintf('(orthogonal)\n');
   else
   fprintf('(biorthogonal)\n');
   end

   fprintf('Vanishing moments: %d analysis,%d
   reconstruction\n',...
   numvanish(g1{1}),numvanish(g2{1}));
   fprintf('Filter lengths: %d/%d-tap\n',...
   length(h1{1}),length(g1{1}));
   fprintf('Implementation lifting steps: %d\n\n',...
   size(Seq,1)-all([Seq{1,: }] = = 0));
```

```
fprintf('h1(z) =%s\n',filterstr(h1,ScaleS));
fprintf('g1(z) =%s\n',filterstr(g1,ScaleD));
fprintf('h2(z) =%s\n',filterstr(h2,1/ScaleS));
fprintf('g2(z) =%s\n\n',filterstr(g2,1/ScaleD));

[x1,phi,x2,psi] = cascade(h1,g1,pow2(-X));
subplot(2,2,1);
plot(x1,phi,'b-');
if diff(x1([1,end])) > 0, xlim(x1([1,end])); end
title('\phi_1');
subplot(2,2,3);
plot(x2,psi,'b-');
if diff(x2([1,end])) > 0, xlim(x2([1,end])); end
title('\psi_1');
[x1,phi,x2,psi] = cascade(h2,g2,pow2(-X));
subplot(2,2,2);
plot(x1,phi,'b-');
if diff(x1([1,end])) > 0, xlim(x1([1,end])); end
title('\phi_2');
subplot(2,2,4);
plot(x2,psi,'b-');
if diff(x2([1,end])) > 0, xlim(x2([1,end])); end
title('\psi_2');
set(gcf,'NextPlot','replacechildren');
otherwise
error(['Invalid parameter, ''',Level,'''.']);
   end
end


return;
elseif nargin < 5
   % Use antisymmetric extension by default
   if nargin < 4
   if nargin < 3, error('Not enough input
   arguments.'); end
       Ext = 'asym';
   end

   Dim = min(find(size(X) ~ = 1));
   if isempty(Dim), Dim = 1; end
end

if any(size(Level) ~ = 1), error('Invalid
                   decomposition level.'); end
```

```
NumStages = size(Seq,1);
EvenStages = ~rem(NumStages,2);

if Flag1D % 1D Transfrom
  %%% Convert N-D array to a 2-D array with dimension
  Dim along the columns%%%
  XSize = size(X); % Save original dimensions
  N = XSize(Dim);
  M = prod(XSize)/N;
  Perm = [Dim: max(length(XSize),Dim),1: Dim-1];
  X = double(reshape(permute(X,Perm),N,M));
  if M == 1 & nargout == 0 & Level > 0
      % Create a figure of the wavelet decomposition
      set(gcf,'NextPlot','replace');
      subplot(Level+2,1,1);
      plot(X);
      title('Wavelet Decomposition');
      axis tight; axis off;

      X = feval(mfilename,WaveletName,Level,X,Ext,1);

      for i = 1: Level
      N2 = N;
      N = 0.5*N;
      subplot(Level+2,1,i+1);
      a = max(abs(X(N+1: N2)))*1.1;
      plot(N+1: N2,X(N+1: N2),'b-');
      ylabel(['d',sprintf('_%c',num2str(i))]);
      axis([N+1,N2,-a,a]);
      end

      subplot(Level+2,1,Level+2);
      plot(X(1: N),'-');
      xlabel('Coefficient Index');
      ylabel('s_1');
      axis tight;
      set(gcf,'NextPlot','replacechildren');
      varargout = {X};
      return;
  end

      if rem(N,pow2(abs(Level))), error('Signal length
      must be divisible by 2^L.'); end

      if N < pow2(abs(Level)), error('Signal length
      too small for transform level.'); end
```

```
if Level > = 0      % Forward transform
for i = 1: Level
      Xo = X(2: 2: N,: );
      Xe = X(1: 2: N,: ) +
      xfir(Seq{1,1},Seq{1,2},Xo,Ext);

      for k = 3: 2: NumStages
      Xo = Xo + xfir(Seq{k-1,1},Seq
      {k-1,2},Xe,Ext);
      Xe = Xe + xfir(Seq{k,1},Seq{k,2},Xo,Ext);
      end

if EvenStages
Xo = Xo + xfir(Seq{NumStages,1},
      Seq{NumStages,2},Xe,Ext);
end

X(1: N,: ) = [Xe*ScaleS; Xo*ScaleD];
N = 0.5*N;
end
else   % Inverse transform
N = N * pow2(Level);

for i = 1: -Level
      N2 = 2*N;
      Xe = X(1: N,: )/ScaleS;
      Xo = X(N+1: N2,: )/ScaleD;
      if EvenStages

      Xo = Xo - xfir(Seq{NumStages,1},Seq{NumSta
      ges,2},Xe,Ext);
      end

      for k = NumStages - EvenStages: -2: 3
      Xe = Xe - xfir(Seq{k,1},Seq{k,2},Xo,Ext);
      Xo = Xo - xfir(Seq{k-1,1},Seq{k-
            1,2},Xe,Ext);
      end
      X([1: 2: N2,2: 2: N2],: ) = [Xe - xfir
      (Seq{1,1},Seq{1,2},Xo,Ext); Xo];
      N = N2;
      end
end

X = ipermute(reshape(X,XSize(Perm)),Perm);
      % Restore original array dimensions
```

```
else % 2D Transfrom
N = size(X);
if length(N) > 3 |
any(rem(N([1,2]),pow2(abs(Level))))
error('Input size must be either MxN or MxNxP
where M and N are divisible by 2^L.');
end

if Level > = 0 % 2D Forward transform
     for i = 1: Level
          Xo = X(2: 2: N(1),1: N(2),: );
          Xe = X(1: 2: N(1),1: N(2),: ) +
          xfir(Seq{1,1},Seq{1,2},Xo,Ext);

for k = 3: 2: NumStages
     Xo = Xo + xfir(Seq{k-1,1},Seq{k-1,2},
     Xe,Ext);
     Xe = Xe + xfir(Seq{k,1},Seq{k,2},Xo,Ext);
end

if EvenStages
     Xo = Xo + xfir(Seq{NumStages,1},Seq{NumSta
     ges,2},Xe,Ext);
end

X(1: N(1),1: N(2),: ) = [Xe*ScaleS; Xo*ScaleD];

Xo = permute(X(1: N(1),2: 2: N(2),: ),[2,1,3]);
Xe = permute(X(1: N(1),1: 2: N(2),: ),[2,1,3])...
+ xfir(Seq{1,1},Seq{1,2},Xo,Ext);
     for k = 3: 2: NumStages
     Xo = Xo + xfir(Seq{k-1,1},Seq{k-1,2},
     Xe,Ext);
     Xe = Xe + xfir(Seq{k,1},Seq{k,2},Xo,Ext);
end
if EvenStages
     Xo = Xo + xfir(Seq{NumStages,1},Seq{NumSta
     ges,2},Xe,Ext);
end

     X(1: N(1),1: N(2),: ) =
     [permute(Xe,[2,1,3])*ScaleS,...
     permute(Xo,[2,1,3])*ScaleD];
N = 0.5*N;
end
```

```
else   % 2D Inverse transform
       N = N*pow2(Level);

       for i = 1: -Level
              N2 = 2*N;
              Xe = permute(X(1: N2(1),1: N(2),:
              ),[2,1,3])/ScaleS;
              Xo = permute(X(1: N2(1),N(2)+1: N2(2),:
              ),[2,1,3])/ScaleD;

       if EvenStages
       Xo = Xo - xfir(Seq{NumStages,1},Seq{NumStages,2}
       ,Xe,Ext);
       end

       for k = NumStages - EvenStages: -2: 3
       Xe = Xe - xfir(Seq{k,1},Seq{k,2},Xo,Ext);
       Xo = Xo - xfir(Seq{k-1,1},Seq{k-1,2},Xe,Ext);
       end
       X(1: N2(1),[1: 2: N2(2),2: 2: N2(2)],: ) =...
       [permute(Xe - xfir(Seq{1,1},Seq{1,2},Xo,
       Ext),[2,1,3]),...
       permute(Xo,[2,1,3])];
       Xe = X(1: N(1),1: N2(2),: )/ScaleS;
       Xo = X(N(1)+1: N2(1),1: N2(2),: )/ScaleD;

       if EvenStages
       Xo = Xo - xfir(Seq{NumStages,1},Seq{NumStages,2}
       ,Xe,Ext);
       end
              for k = NumStages - EvenStages: -2: 3
              Xe = Xe - xfir(Seq{k,1},Seq{k,2},Xo,Ext);
              Xo = Xo - xfir(Seq{k-1,1},Seq{k-
              1,2},Xe,Ext);
              end
              X([1: 2: N2(1),2: 2: N2(1)],1: N2(2),: )
              =...
              [Xe - xfir(Seq{1,1},Seq{1,2},Xo,Ext); Xo];
              N = N2;
              end
       end
end

varargout{1} = X;
return;
```

```
function [Seq,ScaleS,ScaleD,Family] =
getwavelet(WaveletName)
%GETWAVELET Get wavelet lifting scheme sequence.
% Pascal Getreuer 2005-2006


WaveletName = strrep(WaveletName,'bior','spline');
ScaleS = 1/sqrt(2);
ScaleD = 1/sqrt(2);
Family = 'Spline';


switch strrep(strrep(lower(WaveletName),'2d',''),' ','')
case {'haar','d1','db1','sym1','spline1.1','rspline1.1'}
      Seq = {1,0; -0.5,0};
      ScaleD = -sqrt(2);
      Family = 'Haar';
case {'d2','db2','sym2'}
      Seq = {sqrt(3) ,0; [-sqrt(3),2-sqrt(3)]/4,0;
      -1,1};
      ScaleS = (sqrt(3)-1)/sqrt(2);
      ScaleD = (sqrt(3)+1)/sqrt(2);
      Family = 'Daubechies';
case {'d3','db3','sym3'}
      Seq = {2.4254972439123361,0; [-0.352387657680182
      3,0.0793394561587384] ,0;
      [0.5614149091879961,-2.8953474543648969],2;
      -0.0197505292372931,-2};
      ScaleS = 0.4318799914853075;
      ScaleD = 2.3154580432421348;
      Family = 'Daubechies';
case {'d4','db4'}
      Seq = {0.3222758879971411,-1;[0.3001422587485443
      ,1.1171236051605939],1;
      [-0.1176480867984784,0.0188083527262439],-1;
      [-0.6364282711906594,-2.1318167127552199],1;
      [0.0247912381571950,-
      0.1400392377326117,0.4690834789110281],2};
      ScaleS = 1.3621667200737697;
      ScaleD = 0.7341245276832514;
      Family = 'Daubechies';
case {'d5','db5'}
      Seq = {0.2651451428113514,-1;
      [-0.2477292913288009,-0.9940591341382633],1;
      [-0.2132742982207803,0.5341246460905558],1;
      [0.7168557197126235,-0.2247352231444452],-1;
```

```
        [-0.0121321866213973,
        0.0775533344610336],3;0.035764924629411,-3};
        ScaleS = 1.3101844387211246;
        ScaleD = 0.7632513182465389;
        Family = 'Daubechies';
case {'d6','db6'}
        Seq = {4.4344683000391223,0;[-0.214593449940913,
        0.0633131925095066],0;
        [4.4931131753641633,-9.970015617571832],2;
        [-0.0574139367993266,0.0236634936395882],-2;
        [0.6787843541162683,-2.3564970162896977],4;
        [-0.0071835631074942,0.0009911655293238],-4;
        -0.0941066741175849,5};
        ScaleS = 0.3203624223883869;
        ScaleD = 3.1214647228121661;
        Family = 'Daubechies';
case 'sym4'
        Seq = {-0.3911469419700402,0; [0.339243991864945
        1,0.1243902829333865],0;
        [-0.1620314520393038,1.4195148522334731],1;
        -[0.1459830772565225,
        0.4312834159749964],1;1.049255198049293,-1};
        ScaleS = 0.6366587855802818;
        ScaleD = 1.5707000714496564;
        Family = 'Symlet';
case 'sym5'
        Seq = {0.9259329171294208,0; -[0.131923027028234
        1,0.4985231842281166],1;
        [1.452118924420613,0.4293261204657586],0;
        [-0.2804023843755281,0.0948300395515551],0;
        -[0.7680659387165244,1.9589167118877153],1;
        0.1726400850543451,0};
        ScaleS = 0.4914339446751972;
        ScaleD = 2.0348614718930915;
        Family = 'Symlet';
case 'sym6'
        Seq = {-0.2266091476053614,0;
[0.2155407618197651,-1.2670686037583443],0;
        [-4.2551584226048398,0.5047757263881194],2;
        [0.2331599353469357 =,0.0447459687134724],-2;
        [6.6244572505007815,-18.389000853969371],4;
        [-0.0567684937266291,0.1443950619899142],-4;
        -5.5119344180654508,5};
        ScaleS = -0.5985483742581210;
        ScaleD = -1.6707087396895259;
```

```
        Family = 'Symlet';
case 'coif1'
        Seq = {-4.6457513110481772,0;[0.205718913884,
        0.1171567416519999],0;
        [0.6076252184992341,-7.468626966435207],2;
        -0.0728756555332089,-2};
        ScaleS = -0.5818609561112537;
        ScaleD = -1.7186236496830642;
        Family = 'Coiflet';
case 'coif2'
        Seq = {-2.5303036209828274,0;
        [0.3418203790296641,-0.2401406244344829],0;
        [15.268378737252995,3.1631993897610227],2;
        [-0.0646171619180252,0.005717132970962],-2;
        [13.59117256930759,-63.95104824798802],4;
        [-0.0018667030862775,0.0005087264425263],-4;
        -3.7930423341992774,5};
        ScaleS = 0.1076673102965570;
        ScaleD = 9.2878701738310099;
        Family = 'Coiflet';
case 'bcoif1'
        Seq = {0,0; -[1,1]/5,1;[5,5]/14,0;
        -[21,21]/100,1};
        ScaleS = sqrt(2)*7/10;
        ScaleD = sqrt(2)*5/7;
        Family = 'Nearly orthonormal Coiflet-like';
case {'lazy','spline0.0','rspline0.0','d0'}
        Seq = {0,0};
        ScaleS = 1;
        ScaleD = 1;
        Family = 'Lazy';
case {'spline0.1','rspline0.1'}
        Seq = {1,-1};
        ScaleD = 1;
case {'spline0.2','rspline0.2'}
        Seq = {[1,1]/2,0};
        ScaleD = 1;
case {'spline0.3','rspline0.3'}
        Seq = {[-1,6,3]/8,1};
        ScaleD = 1;
case {'spline0.4','rspline0.4'}
        Seq = {[-1,9,9,-1]/16,1};
        ScaleD = 1;
case {'spline0.5','rspline0.5'}
        Seq = {[3,-20,90,60,-5]/128,2};
```

```
        ScaleD = 1;
case {'spline0.6','rspline0.6'}
        Seq = {[3,-25,150,150,-25,3]/256,2};
        ScaleD = 1;
case {'spline0.7','rspline0.7'}
        Seq = {[-5,42,-175,700,525,-70,7]/1024,3};
        ScaleD = 1;
case {'spline0.8','rspline0.8'}
        Seq = {[-5,49,-245,1225,1225,-245,49,-
        5]/2048,3};
        ScaleD = 1;
case {'spline1.0','rspline1.0'}
        Seq = {0,0; -1,0};
        ScaleS = sqrt(2);
        ScaleD = -1/sqrt(2);
case {'spline1.3','rspline1.3'}
        Seq = {0,0; -1,0; [-1,8,1]/16,1};
        ScaleS = sqrt(2);
        ScaleD = -1/sqrt(2);
case {'spline1.5','rspline1.5'}
        Seq = {0,0; -1,0; [3,-22,128,22,-3]/256,2};
        ScaleS = sqrt(2);
        ScaleD = -1/sqrt(2);
case {'spline1.7','rspline1.7'}
        Seq = {0,0; -1,0;
        [-5,44,-201,1024,201,-44,5]/2048,3};
        ScaleS = sqrt(2);
        ScaleD = -1/sqrt(2);
case {'spline2.0','rspline2.0'}
        Seq = {0,0; -[1,1]/2,1};
        ScaleS = sqrt(2);
        ScaleD = 1;
case {'spline2.1','rspline2.1'}
        Seq = {0,0; -[1,1]/2,1;0.5,0};
        ScaleS = sqrt(2);
case {'spline2.2','rspline2.2','cdf5/3','legall5/3','s
+p(2,2)','lc5/3'}
        Seq = {0,0; -[1,1]/2,1;[1,1]/4,0};
        ScaleS = sqrt(2);
case {'spline2.4','rspline2.4'}
        Seq = {0,0; -[1,1]/2,1;[-3,19,19,-3]/64,1};
        ScaleS = sqrt(2);
case {'spline2.6','rspline2.6'}
        Seq = {0,0; -[1,1]/2,1;[5,-39,162,162,-
                39,5]/512,2};
```

```
              ScaleS = sqrt(2);
case {'spline2.8','rspline2.8'}
              Seq = {0,0; -[1,1]/2,1; [-35,335,-1563,
                     5359,5359,-1563,335,-35]/16384,3};
              ScaleS = sqrt(2);
case {'spline3.0','rspline3.0'}
              Seq = {-1/3,-1; -[3,9]/8,1};
              ScaleS = 3/sqrt(2);
              ScaleD = 2/3;
case {'spline3.1','rspline3.1'}
              Seq = {-1/3,-1; -[3,9]/8,1;4/9,0};
              ScaleS = 3/sqrt(2);
              ScaleD = -2/3;
case {'spline3.3','rspline3.3'}
              Seq = {-1/3,-1; -[3,9]/8,1; [-3,16,3]/36,1};
              ScaleS = 3/sqrt(2);
              ScaleD = -2/3;
case {'spline3.5','rspline3.5'}
              Seq = {-1/3,-1; -[3,9]/8,1; [5,-34,128,34,-5]
                     /288,2};
              ScaleS = 3/sqrt(2);
              ScaleD = -2/3;
case {'spline3.7','rspline3.7'}
              Seq = {-1/3,-1; -[3,9]/8,1; [-35,300,-1263,
                     4096,1263,-300,35]/9216,3};
              ScaleS = 3/sqrt(2);
              ScaleD = -2/3;
case {'spline4.0','rspline4.0'}
              Seq = {-[1,1]/4,0; -[1,1],1};
              ScaleS = 4/sqrt(2);
              ScaleD = 1/sqrt(2);
              ScaleS = 1; ScaleD = 1;
case {'spline4.1','rspline4.1'}
              Seq = {-[1,1]/4,0; -[1,1],1;6/16,0};
              ScaleS = 4/sqrt(2);
              ScaleD = 1/2;
case {'spline4.2','rspline4.2'}
              Seq = {-[1,1]/4,0; -[1,1],1; [3,3]/16,0};
              ScaleS = 4/sqrt(2);
              ScaleD = 1/2;
case {'spline4.4','rspline4.4'}
              Seq = {-[1,1]/4,0; -[1,1],1; [-5,29,29,-5]
                     /128,1};
              ScaleS = 4/sqrt(2);
              ScaleD = 1/2;
```

```
case {'spline4.6','rspline4.6'}
      Seq = {-[1,1]/4,0; -[1,1],1;[35,-265,998,998,-
           265,35]/4096,2};
      ScaleS = 4/sqrt(2);
      ScaleD = 1/2;
case {'spline4.8','rspline4.8'}
      Seq = {-[1,1]/4,0; -[1,1],1;[-63,595,-2687,
           8299,8299,-2687,595,-63]/32768,3};
      ScaleS = 4/sqrt(2);
      ScaleD = 1/2;
case {'spline5.0','rspline5.0'}
      Seq = {0,0; -1/5,0; -[5,15]/24,0; -[9,15]/10,1};
      ScaleS = 3*sqrt(2);
      ScaleD = sqrt(2)/6;
case {'spline5.1','rspline5.1'}
      Seq = {0,0; -1/5,0; -[5,15]/24,0;
           -[9,15]/10,1;1/3,0};
      ScaleS = 3*sqrt(2);
      ScaleD = sqrt(2)/6;
case {'spline5.3','rspline5.3'}
      Seq = {0,0; -1/5,0; -[5,15]/24,0;
           -[9,15]/10,1;[-5,24,5]/72,1};
      ScaleS = 3*sqrt(2);
      ScaleD = sqrt(2)/6;
case {'spline5.5','rspline5.5'}
      Seq = {0,0; -1/5,0; -[5,15]/24,0;
           -[9,15]/10,1;[35,-230,768,230,-35]
           /2304,2};
      ScaleS = 3*sqrt(2);
      ScaleD = sqrt(2)/6;
case {'cdf9/7'}
      Seq = {0,0; [1,1]*-1.5861343420693648,1;[1,1]
           *-0.0529801185718856,0;
      [1,1]*0.8829110755411875,1;[1,1]*0.4435068520511
      142,0};
      ScaleS = 1.1496043988602418;
      ScaleD = 1/ScaleS;
      Family = 'Cohen-Daubechies-Feauveau';
case 'v9/3'
      Seq = {0,0; [-1,-1]/2,1;[1,19,19,1]/80,1};
      ScaleS = sqrt(2);
      Family = 'HSV design';
case {'s+p(4,2)','lc9/7-m'}
      Seq = {0,0; [1,-9,-9,1]/16,2;[1,1]/4,0};
      ScaleS = sqrt(2);
```

```
            Family = 'S+P';
case 's+p(6,2)'
            Seq = {0,0; [-3,25,-150,-150,25,-3]
                    /256,3;[1,1]/4,0};
            ScaleS = sqrt(2);
            Family = 'S+P';
case {'s+p(4,4)','lc13/7-t'}
            Seq = {0,0; [1,-9,-9,1]/16,2;[-1,9,9,-1]/32,1};
            ScaleS = sqrt(2);
            Family = 'S+P';
case {'s+p(2+2,2)','lc5/11-c'}
            Seq = {0,0; [-1,-1]/2,1;[1,1]/4,0; -[-1,1,1,-1]
                    /16,2};
            ScaleS = sqrt(2);
            Family = 'S+P';
case 'tt'
            Seq = {1,0; [3,-22,-128,22,-3]/256,2};
            ScaleD = sqrt(2);
            Family = 'Le Gall-Tabatabai polynomial';
case 'lc2/6'
            Seq = {0,0; -1,0; 1/2,0; [-1,0,1]/4,1};
            ScaleS = sqrt(2);
            ScaleD = -1/sqrt(2);
            Family = 'Reverse spline';
case 'lc2/10'
            Seq = {0,0; -1,0; 1/2,0; [3,-22,0,22,-3]/64,2};
            ScaleS = sqrt(2);
            ScaleD = -1/sqrt(2);
            Family = 'Reverse spline';
case 'lc5/11-a'
            Seq = {0,0; -[1,1]/2,1;[1,1]/4,0; [1,-1,-1,1]
                    /32,2};
            ScaleS = sqrt(2);
            ScaleD = -1/sqrt(2);
            Family = 'Low complexity';
case 'lc6/14'
            Seq = {0,0; -1,0; [-1,8,1]/16,1;[1,-6,0,6,-1]
                    /16,2};
            ScaleS = sqrt(2);
            ScaleD = -1/sqrt(2);
            Family = 'Low complexity';
case 'lc13/7-c'
            Seq = {0,0; [1,-9,-9,1]/16,2;[-1,5,5,-1]/16,1};
            ScaleS = sqrt(2);
            ScaleD = -1/sqrt(2);
```

```
        Family = 'Low complexity';
        otherwise
        Seq = {};
return;
end


if ~isempty(findstr(lower(WaveletName),'rspline'))
[Seq,ScaleS,ScaleD] = seqdual(Seq,ScaleS,ScaleD);
Family = 'Reverse spline';
end


return;

function [Seq,ScaleS,ScaleD] =
seqdual(Seq,ScaleS,ScaleD)
% Dual of a lifting sequence

L = size(Seq,1);

for k = 1: L
        % f'(z) = -f(z^-1)
        Seq{k,2} = -(Seq{k,2} - length(Seq{k,1}) + 1);
        Seq{k,1} = -fliplr(Seq{k,1});
end

if all(Seq{1,1} = = 0)
        Seq = reshape({Seq{2: end,: }},L-1,2);
else
        [Seq{1: L+1,: }] = deal(0,Seq{1: L,1},0,Seq{1:
        L,2});
end

ScaleS = 1/ScaleS;
ScaleD = 1/ScaleD;
return;

function [h,g] = seq2hg(Seq,ScaleS,ScaleD,Dual)
% Find wavelet filters from lifting sequence
if Dual, [Seq,ScaleS,ScaleD] =
seqdual(Seq,ScaleS,ScaleD); end
if rem(size(Seq,1),2), [Seq{size(Seq,1)+1,: }] =
deal(0,0); end

h = {1,0};
g = {1,1};
```

```
for k = 1: 2: size(Seq,1)
      h = lp_lift(h,g,{Seq{k,: }});
      g = lp_lift(g,h,{Seq{k+1,: }});
end

h = {ScaleS*h{1},h{2}};
g = {ScaleD*g{1},g{2}};

if Dual
      h{2} = -(h{2} - length(h{1}) + 1);
      h{1} = fliplr(h{1});

      g{2} = -(g{2} - length(g{1}) + 1);
      g{1} = fliplr(g{1});
end

return;

function a = lp_lift(a,b,c)
% a(z) = a(z) + b(z) c(z^2)

d = zeros(1,length(c{1})*2-1);
d(1: 2: end) = c{1};
d = conv(b{1},d);
z = b{2}+c{2}*2;
zmax = max(a{2},z);
f = [zeros(1,zmax-
a{2}),a{1},zeros(1,a{2} - length(a{1}) - z +
length(d))];
i = zmax-z + (1: length(d));
f(i) = f(i) + d;

if all(abs(f) < 1e-12)
      a = {0,0};
else
      i = find(abs(f)/max(abs(f)) > 1e-10);
i1 = min(i);
a = {f(i1: max(i)),zmax-i1+1};
end
return;

function X = xfir(B,Z,X,Ext)
%XFIR Noncausal FIR filtering with boundary handling.
% Y = XFIR(B,Z,X,EXT) filters X with FIR filter B with
leading
```

```
% delay -Z along the columns of X. EXT specifies the
boundary
% handling. Special handling is done for one and two-
tap filters.

% Pascal Getreuer 2005-2006

N = size(X);

% Special handling for short filters
if length(B) = = 1 & Z = = 0
      if B = = 0
            X = zeros(size(X));
      elseif B ~ = 1
            X = B*X;
      end
      return;
end

% Compute the number of samples to add to each end of
the signal
pl = max(length(B)-1-Z,0); % Padding on the left end
pr = max(Z,0);             % Padding on the right end

switch lower(Ext)
case {'sym','wsws'} % Symmetric extension, WSWS
if all([pl,pr] < N(1))
            X = filter(B,1,X([pl+1: -1: 2,1:
                N(1),N(1)-1: -1: N(1)-pr],:,: ),[],1);
            X = X(Z+pl+1: Z+pl+N(1),:,: );
            return;
      else
            i = [1: N(1),N(1)-1: -1: 2];
            Ns = 2*N(1) - 2 + (N(1) = = 1);
            i = i([rem(pl*(Ns-1): pl*Ns-1,Ns)+1,1:
                N(1),rem(N(1): N(1)+pr-1,Ns)+1]);
      end
case {'symh','hshs'} % Symmetric extension, HSHS
      if all([pl,pr] < N(1))
            i = [pl: -1: 1,1: N(1),N(1): -1:
                N(1)-pr+1];
      else
            i = [1: N(1),N(1): -1: 1];
            Ns = 2*N(1);
```

```
                    i = i([rem(pl*(Ns-1): pl*Ns-1,Ns)+1,1:
                        N(1),rem(N(1): N(1)+pr-1,Ns)+1]);
        end
case 'wshs'  % Symmetric extension, WSHS
if all([pl,pr] < N(1))
                    i = [pl+1: -1: 2,1: N(1),N(1): -1:
                        N(1)-pr+1];
        else
                    i = [1: N(1),N(1): -1: 2];
                    Ns = 2*N(1) - 1;
                    i = i([rem(pl*(Ns-1): pl*Ns-1,Ns)+1,1:
                        N(1),rem(N(1): N(1)+pr-1,Ns)+1]);
        end
case 'hsws'  % Symmetric extension, HSWS
        if all([pl,pr] < N(1))
                    i = [pl: -1: 1,1: N(1),N(1)-1: -1:
                    N(1)-pr];
            else
                    i = [1: N(1),N(1)-1: -1: 1];
                    Ns = 2*N(1) - 1;
                    i = i([rem(pl*(Ns-1): pl*Ns-1,Ns)+1,1:
                    N(1),rem(N(1): N(1)+pr-1,Ns)+1]);
            end
case 'zpd'
        Ml = N; Ml(1) = pl;
        Mr = N; Mr(1) = pr;
        X = filter(B,1,[zeros(Ml);X;zeros(Mr)],[],1);
        X = X(Z+pl+1: Z+pl+N(1),:,: );
        return;
case 'per'   % Periodic
        i = [rem(pl*(N(1)-1): pl*N(1)-1,N(1))+1,1:
            N(1),rem(0: pr-1,N(1))+1];
case 'sp0'   % Constant extrapolation
        i = [ones(1,pl),1: N(1),N(1)+zeros(1,pr)];
case 'asym'  % Asymmetric extension
        i1 = [ones(1,pl),1: N(1),N(1)+zeros(1,pr)];

        if all([pl,pr] < N(1))
                    i2 = [pl+1: -1: 2,1: N(1),N(1)-1: -1:
                        N(1)-pr];
            else
                    i2 = [1: N(1),N(1)-1: -1: 2];
                    Ns = 2*N(1) - 2 + (N(1) = = 1);
                    i2 = i2([rem(pl*(Ns-1): pl*Ns-1,Ns)+1,1:
                        N(1),rem(N(1): N(1)+pr-1,Ns)+1]);
```

```
      end

      X = filter(B,1,2*X(i1,:,: ) - X(i2,:,: ),[],1);
      X = X(Z+pl+1: Z+pl+N(1),:,: );
      return;
otherwise
      error(['Unknown boundary handling,
      ''',Ext,'''.']);
end

X = filter(B,1,X(i,:,: ),[],1);
X = X(Z+pl+1: Z+pl+N(1),:,: );
return;

function [x1,phi,x2,psi] = cascade(h,g,dx)
% Wavelet cascade algorithm

c = h{1}*2/sum(h{1});
x = 0: dx: length(c) - 1;
x1 = x - h{2};
phi0 = 1 - abs(linspace(-1,1,length(x))).';

ii = []; jj = []; s = [];

for k = 1: length(c)
      xk = 2*x - (k-1);
      i = find(xk > = 0 & xk < = length(c) - 1);
      ii = [ii,i];
      jj = [jj,floor(xk(i)/dx)+1];
      s = [s,c(k)+zeros(size(i))];
end

% Construct a sparse linear operator that iterates the
dilation equation

Dilation = sparse(ii,jj,s,length(x),length(x));

for N = 1: 30
phi = Dilation*phi0;
if norm(phi - phi0,inf) < 1e-5, break; end
phi0 = phi;
end

if norm(phi) = = 0
      phi = ones(size(phi))*sqrt(2); % Special case
            for Haar scaling function
```

```
else
        phi = phi/(norm(phi)*sqrt(dx)); % Rescale result
end

if nargout > 2
        phi2 = phi(1: 2: end); % phi2 is approximately
               phi(2x)
        if length(c) = = 2
                L = length(phi2);
        else
                L = ceil(0.5/dx);
        end

        % Construct psi from translates of phi2
        c = g{1};
        psi = zeros(length(phi2)+L*(length(c)-1),1);
        x2 = (0: length(psi)-1)*dx - g{2} - 0*h{2}/2;

        for k = 1: length(c)
        i = (1: length(phi2)) + L*(k-1);
                psi(i) = psi(i) + c(k)*phi2;
        end
end
return;

function s = filterstr(a,K)
% Convert a filter to a string
[n,d] = rat(K/sqrt(2));
if d < 50
        a{1} = a{1}/sqrt(2); % Scale filter by sqrt(2)
        s = '(';
else
        s = '';
end

Scale = [pow2(1: 15),10,20,160,280,inf];

for i = 1: length(Scale)
        if norm(round(a{1}*Scale(i))/Scale(i) -
        a{1},inf) < 1e-9
                a{1} = a{1}*Scale(i); % Scale filter by a
                power of 2 or 160
                s = '(';
                break;
        end
end
```

```
z = a{2};
LineOff = 0;

for k = 1: length(a{1})
        v = a{1}(k);
        if v ~ = 0 % Only display nonzero coefficients
                if k > 1
                        s2 = [' ',char(44-sign(v)),' '];
                v = abs(v);
        else
                s2 = '';
        end
                s2 = sprintf('%s%g',s2,v);
        if z = = 1
                s2 = sprintf('%s z',s2);
                        elseif z ~ = 0
                s2 = sprintf('%s z^%d',s2,z);
        end
        if length(s) + length(s2) > 72 + LineOff % Wrap
        long lines
                s2 = [char(10),'    ',s2];
                LineOff = length(s);
        end
        s = [s,s2];
        end
z = z - 1;
end

if s(1) = = '('
        s = [s,')'];
        if d < 50, s = [s,' sqrt(2)']; end

        if i < length(Scale)
                s = sprintf('%s/%d',s,Scale(i));
        end
end

return;

function N = numvanish(g)
% Determine the number of vanishing moments from
highpass filter g(z)

for N = 0: length(g)-1 % Count the number of roots at
z = 1
```

```
        [g,r] = deconv(g,[1,-1]);
        if norm(r,inf) > 1e-7, break; end
end
return;


function [iout,qout] = delay(idata,qdata,nsamp,idel)
iout = zeros(1,nsamp);
qout = zeros(1,nsamp);
if idel ~ = 0
        iout(1: idel) = zeros(1,idel);
        qout(1: idel) = zeros(1,idel);
end
iout(idel+1: nsamp) = idata(1: nsamp-idel);
qout(idel+1: nsamp) = qdata(1: nsamp-idel);


function [iout,qout] = comb(idata,qdata,attn)
iout = randn(1,length(idata)).*attn;
qout = randn(1,length(qdata)).*attn;
iout = iout+idata(1: length(idata));
qout = qout+qdata(1: length(qdata));


function [iout,qout] = giins(idata,qdata,fftlen,gilen
                        ,nd);
idata1 = reshape(idata,fftlen,nd);
qdata1 = reshape(qdata,fftlen,nd);
idata2 = [idata1(fftlen-gilen+1: fftlen,: );idata1];
qdata2 = [qdata1(fftlen-gilen+1: fftlen,: );qdata1];
iout = reshape(idata2,1,(fftlen+gilen)*nd);
qout = reshape(qdata2,1,(fftlen+gilen)*nd);


function [iout,qout] = girem(idata,qdata,fftlen2,gilen
                        ,nd);
idata2 = reshape(idata,fftlen2,nd);
qdata2 = reshape(qdata,fftlen2,nd);
iout = idata2(gilen+1: fftlen2,: );
qout = qdata2(gilen+1: fftlen2,: );


function [iout,qout] = giins1(idata,qdata,fftlen,
                        gilen,nd);
idata1 = reshape(idata,fftlen,nd);
qdata1 = reshape(qdata,fftlen,nd);
gg = gilen/2;
idata2 = [zeros(gg,nd);idata1(fftlen-gg+1: fftlen,: )
        ;idata1];
```

```
qdata2 = [zeros(gg,nd);qdata1(fftlen-gg+1:
        fftlen,: );qdata1];
iout = reshape(idata2,1,(fftlen+gilen)*nd);
qout = reshape(qdata2,1,(fftlen+gilen)*nd);

function[iout,qout,ramp,rcos,rsin] = sefade(idata,qdat
a,itau,dlvl,th,n0,itn,n1,nsamp,tstp,fd,flat)
iout = zeros(1,nsamp);;
qout = zeros(1,nsamp);;
total_attn = sum(10.^(-1 .0. *dlvl./10.0));
for k = 1: n1
        atts = 10.^(-0.05.*dlvl(k));
        theta = th(k).*pi./180.0;
        [itmp,qtmp] = delay(idata,qdata,nsamp,itau(k));
        [itmp3,qtmp3,ramp,rcos,rsin] = fade(itmp,qtmp,ns
        amp,tstp,fd,n0(k),itn(k),flat);
        iout = iout+atts.*itmp3./sqrt(total_attn);
        qout = qout+atts.*qtmp3./sqrt(total_attn);
end

function [iout,qout] = girem1(idata,qdata,fftlen2,gile
                        n,nd);
idata2 = reshape(idata,fftlen2,nd);
qdata2 = reshape(qdata,fftlen2,nd);
iout = idata2(gilen+1: fftlen2,: );
qout = qdata2(gilen+1: fftlen2,: );

function [iout,qout] = giins1(idata,qdata,fftlen,gilen
                        ,nd);
idata1 = reshape(idata,fftlen,nd);
qdata1 = reshape(qdata,fftlen,nd);
idata2 = [zeros(gilen,nd);idata1];
qdata2 = [zeros(gilen,nd);qdata1];
iout = reshape(idata2,1,(fftlen+gilen)*nd);
qout = reshape(qdata2,1,(fftlen+gilen)*nd);
```

INDEX